Salt Warriors

WINNER OF THE 2007

ROBERT A. CALVERT BOOK PRIZE

Number Eleven:

CANSECO-KECK HISTORY SERIES

Jerry Thompson, General Editor

Salt

Warriors

INSURGENCY ON THE RIO GRANDE

PAUL COOL

Texas A&M
University Press
College Station

First edition

This paper meets the requirements of ANSI/NISO
Z39.48-1992 (Permanence of Paper).
Binding materials have been chosen for durability.
⊗ ♻

Library of Congress Cataloging-in-Publication Data
Cool, Paul, 1950–
Salt warriors : insurgency on the Rio Grande / Paul Cool.—1st ed.
p. cm.—(Canseco-Keck history series ; no. 11)
"Winner of the 2007 Robert A. Calvert Book Prize."
Includes bibliographical references and index.
ISBN-13: 978-1-60344-016-5 (cloth : alk. paper)
ISBN-10: 1-60344-016-X (cloth : alk. paper)
1. El Paso (Tex.)—History—19th century. 2. Rio Grande Region—History—19th century. 3. Mexican Americans—Texas—El Paso—Politics and government—19th century. 4. Salt industry and trade—Political aspects—Texas—El Paso—History—19th century. 5. Insurgency—Texas—El Paso—Politics and government—19th century. 6. Government, Resistance to—Texas—El Paso—Politics and government—19th century. 7. El Paso (Tex.)—Politics and government—19th century. 8. Rio Grande Region—Politics and government—19th century. 9. El Paso (Tex.)—Ethnic relations—History—19th century. 10. Rio Grande Region—Ethnic relations—History—19th century. I. Title.
F394.E4C66 2008
976.4'96061—dc22
2007037944

FRONTISPIECE

Assault on El Molino. *Courtesy of the artist, Bob Boze Bell.*

For Karen, who made it possible

CONTENTS

ILLUSTRATIONS

MAPS

PREFACE

Though the Salt War was, at bottom, a struggle over conflicting economic and political rights, it was permeated from first to last by strong feelings of racial and ethnic bias, even hatred. Anglos regularly used epithets and slurs to describe Mexicans and Mexican Americans. Anti-Anglo terms such as "gringo" surface only occasionally because we do not have the insurgent party's firsthand accounts. In a work which relies heavily upon contemporary sources, it is impossible to avoid the frequent appearance of repugnant terms and ideas. When such language is outside quotations, I have attempted to ensure the reader understands the thoughts emanate from the subjects.

The terminology used to describe people and places is also confusing. In accounts contemporary with the Salt War, "Mexican" might mean either a citizen of Mexico or a U.S. citizen of Mexican ancestry; "El Paso" generally meant the city in Mexico now called Juarez. Contemporary documents regarding El Paso, Texas, referred equally to El Paso or Franklin. For consistency, I use "Franklin," since the previously incorporated city of El Paso had no functioning government in 1877–78. The glossary below is intended to keep the reader straight regarding the author's use of identifiers.

American: in contemporary documents, generally refers to Anglos.
Anglo: a white, non-Mexicano citizen of the United States.
El Paso: Generally El Paso County. The modern city of El Paso was born of several small American communities, the most prominent of which was Franklin.
Fanega: An measurement of volume common in Spanish America, equal to two bushels, more or less.
Franklin: The American settlement that formed the core of modern El Paso.
Mexican: a citizen of Mexico.
Mexicano: a person of Mexican descent and citizen of either nation.
Paseño: a Mexicano who lives in the valley of the Rio Grande/Río Bravo, on either side of the river at or near the Pass of the North.
Paso del Norte: literally, "Pass of the North." The geographic pass between the mountains is generally referred to simply as "the Pass." Paso del Norte is generally used to identify the Mexican city opposite El Paso, Texas, known since 1887 as Ciudad Juarez.

Pelado: derogatory term of inexact meaning, applied toward the poorer classes.

Salinero: "salt gatherer," a Paseño who trekked to the Guadalupe Lakes for salt.

Tejano: a Texan of Mexican ethnicity.

Finally, frontier spelling was frequently a matter of personal choice. To avoid the too-frequent use of "[*sic*]," I have usually corrected misspellings, including names. So as not to be dogmatic, I occasionally break my rules when I think the original deserves to be retained. Similarly, punctuation has been modernized in the interest of readability. I have not "fixed" punctuation if doing so would obscure the meaning. I have not otherwise changed the transcription of primary source documents.

ACKNOWLEDGMENTS

Many have encouraged and supported *Salt Warriors* since its inception. I am grateful for their guidance and interventions.

My friend Peter Brand introduced me to the Salt War and served as a sounding board for many of my theories. For that gift I will always be thankful. In *And Die in the West,* Paula Mitchell Marks pointed the way to narrating how complex personal, political, economic, and social forces could lead to lawlessness and violence along the U.S.-Mexico border. In telling of a very different borderlands conflict, I have tried to emulate her model. Tombstone scholars Jeff Morey, Gary L. Roberts, and Casey Tefertiller taught me how to sift through tall tales, regurgitated lies, and half-baked truths.

In honing my historiography of the Salt War, Bluce A. Glasrud bettered my understanding of the sources. Rick Hendricks and Rick Miller were especially generous with their knowledge and the fruits of their research. Bob Alexander, Mark Boardman, Manuel Callahan, Steve Gatto, Allen Hatley, Steven Lubet, Frederick Nolan, Rubén Osorio, Robert Palmquist, Chuck Parsons, Jean Smith, Robert M. Utley, and Harold J. Weiss also shared their knowledge, took time to discuss the people and events of West Texas and New Mexico, and corrected errors. Sam and Ben Sanchez of the San Elizario Genealogy and Historical Society and archeologist Harry W. "Skip" Clark helped me to better understand San Elizario as it existed in 1877.

Donaly Brice (Texas State Library and Archives Commission), Claudia Rivers and Anne Allis (University of Texas at El Paso) went "above and beyond" to uncover fresh Salt War evidence. Evan Hocker and Steven Williams (Center for American History, University of Texas at Austin), Dolores B. Oliveras and Thomas Shelton (Institute of Texas Cultures, University of Texas at San Antonio), Christina Stokpa (Texas Ranger Hall of Fame and Museum), Lynn Russell (El Paso County Historical Society), Nan J. Card and John Ransom (Rutherford B. Hayes Presidential Center), Susan Berry (Silver City Museum), Max Van Balgooy (Cooper Regional History Museum), Jim Bradshaw (Nita Stewart Haley Memorial Library and J. Evetts Haley History Center), Robert N. Jones (Texas General Land Office), and others helped me wade through their archival collections. Many professionals and students at the National Archives and various archival libraries responded to my numerous requests for help.

Leon Metz, Ida Saunders, and Barbara Dent helped open a long-shut pioneer trunk for me. Ida also scrolled through many a microfilm roll to find elusive newspaper articles. Nicholas Houser generously shared his cabinets of research into El Paso's Salt War era. Both Ida and Nick and their spouses welcomed me into their homes, helping underwrite my research.

Antoinette Duran, Alex Loya, Gus J. Lujan, Conrado Montes, Pat Mullin, Ray Sanchez, Jennifer Schoolcraft, Olga Trujillo y Hernandez, and Larry Zickefoose shared their families' memories of Salt War El Paso.

Manuella Cárdenas-Oellermann and Consuelo Gilbert translated Spanish language documents. Mark Dworkin courageously waded through the first draft, making a thousand welcome suggestions.

Robert J. McCubbin, Clint Porter of the El Paso County Sheriff's Department, and Ray Sanchez of "Los Desperados" contributed photographs to this book. Larry Zickefoose provided the "bad hat day" photo of his Texas Ranger ancestors. Photographs of the insurgents, including their leaders, are unfortunately lacking. The boldness, determination, and other martial qualities of the Paseño people are instead dramatically captured in original paintings by Bob Boze Bell.

Allen Barra guided me into the arcane world of publishing, made valuable suggestions on manuscript revision, and pointed me to the right publisher.

Arnoldo DeLeón, whose body of work continues to shape my understanding of the Tejano experience, provided insight, encouragement, and public support that made this book happen.

Salt Warriors was sharpened and shepherded by my wise and enthusiastic editor, Mary Lenn Dixon, who made safe the full story of community, corruption, and chaos. The book and I also owe much to the staff and anonymous readers of Texas A&M University Press.

Through their joint award of the 2007 Robert A. Calvert Book Prize, the university press and history department of Texas A&M have helped this book reach out to its intended audience.

I learned about the heroism of soldiers and civilians caught in battle zones from my parents. Colonel Paul E. Cool, a much decorated World War II pilot, provided a standard to measure the duty, honor, and bravery expected of soldiers in war. Anne Roche Cool survived bullets in the streets of Civil War Dublin, Nazi bombs dropped upon London, and the sudden American withdrawal from Nanjing, China. She taught me that "non-combatants" are warriors no less than those bearing arms.

My wife Karen gave me love, patience, and license to take all those trips to Texas and the Rio Grande.

THE "SALT WARRIORS"

The following list of key individuals is arranged alphabetically. For Paseños, the name is followed by date of birth. For immigrants to the area, the name is followed by place of birth and year of arrival at the Pass of the North.

CIPRIANO ALDERETE (Paseño, born at the Pass of the North in 1804 or 1807?), friend of Louis Cardis, insurgent sympathizer

DESIDARIO APODACA (Paseño, born 1832), farmer, insurgent leader

JOHN ATKINSON (New York, arrived at the Pass of the North in 1862), state police lieutenant, school commissioner, merchant

FRANCISCO "CHICO" BARELA (Paseño, born 1827?), farmer, Texas Ranger, insurgent leader

ALLEN BLACKER (Ohio, arrived 1869), Democrat, district judge

THOMAS BLAIR (Scotland, arrived 1877), Captain, 15th Infantry

ANTONIO SEVERO BORRAJO (Spain, arrived 1851), parish priest of San Elizario and later Guadalupe, Mexico

PETER BOURGADE (France, arrived 1876), Parish priest of San Elizario

LOUIS CARDIS (Italy, arrived 1864), popular political leader, state legislative representative

PRICE COOPER (New York, arrived 1843?), stagecoach driver, freighter, deputy sheriff, Texas Ranger

CHARLES ELLIS (Vermont, arrived 1862), sheriff, school commissioner, merchant

JOHN C. FORD (Louisiana, arrived 1877?), deputy sheriff, Texas Ranger, salt lakes agent

ALBERT JENNINGS FOUNTAIN (New York, arrived 1862), Radical Republican state senator, newspaper editor

GREGORIO N. GARCIA I (Paseño, born 1819), Indian fighter, Texas Ranger captain, community leader

GREGORIO N. GARCIA II (Paseño, born 1855), son of Captain Garcia, Texas Ranger

MIGUEL GARCIA (Paseño, born 1844), Texas Ranger

VIDAL GARCIA (Paseño, born 1835), farmer, insurgent sympathizer

LINO GRANILLO (Paseño, born 1815?), farmer, insurgent leader

JAMES PRICE HAGUE (Missouri, arrived 1871), Cardis ally, insurgent sympathizer

EDWARD HATCH (Maine, arrived 1877),Colonel, 9th Cavalry

CHARLES HOWARD (Virginia, arrived 1872), Democrat, district judge, salt lakes agent

JOHN B. JONES (Texas, arrived 1877), Major, Frontier Battalion (Texas Rangers)

CHARLES KERBER (Germany, arrived 1869), Union officer, school teacher, sheriff

JOHN KINNEY (New York, arrived 1877), "rustler boss," Silver City volunteer

MARK LUDWICK (Missouri, arrived 1878), Texas Ranger

JUAN MAURO LUJAN (Paseño, born circa 1839), farmer, community leader, insurgent sympathizer

H.O. "ROCKY" MATHEWS (Ireland, arrived 1877?), Texas Ranger

JIM MCDANIELS (Tennessee, arrived 1875), rustler, Texas Ranger

PABLO MEJÍA (Paseño, born 1832 or 1842), state policeman, Texas Ranger, deputy sheriff

WILLIAM WALLACE MILLS (Indiana, arrived 1858), Conservative Republican, U.S. customs officer

TELESFORO MONTES (Paseño, born 1820), school law resister, Texas Ranger lieutenant

AGATÓN PORRAS (Paseño, born 1840), laborer, insurgent leader

LOUIS H. RUCKER (Illinois, arrived 1877), First Lieutenant, 9th Cavalry

SISTO SALCIDO (Paseño, born 1837 to 1843), farmer, Texas Ranger, insurgent leader

JOHN B. TAYS (Nova Scotia, arrived 1872), Texas Ranger lieutenant

SALOMÉ TELLES (Paseña, born 1835), resister and witness

THOMAS ZICKEFOOSE (West Virginia, arrived 1877?), Texas Ranger

GEORGE B. ZIMPELMAN (Germany, arrived 1876), Travis County sheriff, salt lakes claimant

Salt Warriors

Guadalupe Salt Lakes, West Texas. From the author's collection

INTRODUCTION

In the mid-nineteenth century, the east-west route along the base of Texas' Guadalupe Mountains brought American explorers to glistening white salt lakes. Not everyone who approached comprehended the view. To Boundary Commissioner John Russell Bartlett, "The plain . . . was interspersed with what looked like silvery and tranquil lakes, glittering in the sun, seeming, as it were, to tempt the weary traveler to their brink. Our young men cried out, 'Water!' delighted with the idea of again enjoying this luxury after a long day's ride. But the whole turned out a delusion; what appeared to be the glassy surface of a lake or pond, being nothing but the saline incrustations of a dried up lake."[1]

One U.S. Army officer saw something more. "It is quite impossible," wrote Captain John Pope, "to estimate the extent or value of these dry lakes; but the quantity of salt is so enormous, and so easily obtained, that, with cheap communications with a market, they would supply the United States at a cost far below the present market prices." More than two decades later, in June 1877, one party of entrepreneurs following Pope's advice saw in those shimmering beds the promise of monopoly wealth. The profiteers would have done well to heed Texas frontiersman John S. "Rip" Ford's warning to the thirsty wayfarer: "Should he follow the direction of the sparkling guide, his disappointment will be complete and vexatious."[2]

For Charles H. Howard, who followed "the sparkling guide" in June 1877, the result was not merely disappointment, but catastrophe, for himself, for his hated rival, and for the community of Paseños, the Mexican Americans living at *El Paso del Norte,* the Pass of the North. That catastrophe was the El Paso Salt War, sometimes described as a bloody riot by a howling mob but in reality a complex political, social, and military struggle. Newspaper editors from New York to San Francisco argued at the time over whether the conflict was merely a local dispute or *casus belli* for a second Mexican-American War. Though today the Salt War is little remembered, the episode has relevancy to a nation grappling with racial and ethnic tensions, religious intolerance and fanaticism, entrepreneurial greed, a porous border, and the role of the military in times of civil disorder.

Paseños, who had owned the Guadalupe Salt Lakes in common since the days of Spanish rule, used their precious yield for curing and cooking food, for bartering, and for replenishing what evaporation took from their bodies and those of their animals. To entrepreneurs, salt was a cash crop, not just for the mundane purposes of the local people, but for the lucrative business of silver mining. Silver refining required salt—tons of salt. If one could own the lakes and pay a nominal amount for the recovery of that salt and its transportation to the mines, one might make a tidy profit.

The war had its origins in a vicious struggle among Anglo politicians and speculators eager to seize ownership of the lakes from the people. For years, no one succeeded, but in 1877, Austin banker George B. Zimpelman staked a legal claim, and his son-in-law Charles Howard demanded that Paseños begin paying Zimpelman for *his* salt. The result was the climax of the Salt War, a popular insurgency that disrupted government in El Paso County for twelve weeks. Before the year was over, hundreds armed with guns and machetes rushed to the area from all points of the compass, turning the peaceful valley into a battle zone. It was an unusual war. Soldiers avoided battle, leaving the violence to angry farmers, untested Texas Rangers, religious festival-goers, and a mercenary sheriff's posse out for a bloody lark. Scores were killed or wounded, and others were raped or robbed, prompting hundreds of American citizens to flee to Mexico.

Contrary to previous accounts, the war did not end when the U.S. Army arrived to restore "order." It simply moved into a new phase. Lawlessness prevailed for another year until Paseño voters and Anglo lawmen defeated an alliance between the local sheriff and a criminal gang, a precursor to more infamous such combinations in the New Mexico and Arizona territories. Only then was the battered county able to focus its attention on the future.

The "theater of war" was *la Isla,* a sometimes island located a few miles southeast of El Paso formed by one of the wandering Rio Grande's periodic changes of course. An accident of history placed *la Isla,* site of the Mexican farming communities of San Elizario, Socorro and Ysleta, on the American side of the international border upon the conclusion of the Mexican-American War. To newcomers from other states, Paseño farmers did not look, dress, eat, think, or worship God like proper Americans. While a few leading "Mexicans" with light features, old "Spanish blood," or money were accepted as intelligent men, the vast majority were frequently referred to and usually treated as fit only for their lowly stations. Charles Howard was not alone in regarding Paseños as a "deluded and credulous people." Consequently, their resistance to injustice was considered mindless, either as the work of a lying Italian "vampire" who duped them or the wild vengeance of "an ignorant and hot-blooded race."[3]

The insurgency of 1877 was not the action of a mindless rabble, but the product of a deliberate, community-based decision squarely in the tradition of the American nation's original fight for self-government. Like Thomas Jefferson, these descendants of Spaniards and Indians firmly believed that governments derived "their just powers from the consent of the governed." Like that Virginian, they also agreed that occasional rebellion was a necessary thing. Generations of Indian warfare had given Paseños rudimentary military know-how and the courage to use it. Like the New Englanders of 1775, they were "minute men" ready to stand in defense of their rights to property, religion, and self-rule against corrupt officials and businessmen supported by a distant and unsympathetic government. This too was an "American" trait the founding fathers would have recognized.[4]

These American insurgents understood their rights under the Constitution, a compact created in part to "provide for the common defense [and] promote the general welfare." But nothing had prepared them for the small army of lawyers and capitalists who swooped down upon the Pass after the American Civil War to exercise another American right, what historian Walter A. McDougall calls the "full freedom to hustle." If salt was the excuse for war, the underlying reason was this struggle between the rights of the community and those of hustling individualists.[5]

The few score Anglos of El Paso saw opportunity at this place where the mountains parted to form a verdant valley with natural corridors for people and commerce. The first Americans had arrived well before the Civil War, integrating easily into the Paseño culture. Union soldiers followed in 1862 and, liking what they saw, took their discharges here. They and others who followed were men on the make, aiming to grab whatever profit could be obtained from the land and its people. Northern carpetbaggers, Texas capitalists, and European immigrants waged an all-out, unremitting struggle with one another over power, patronage, business, and real estate. They stuffed ballot boxes, forged documents, and stole from the public on a grand scale. They committed acts of malfeasance, intimidation, libel, assault, conspiracy to murder, and assassination. They corrupted the county and themselves to control an impoverished El Paso that promised to be the next El Dorado.

Jefferson once wrote, "The smaller the society, the bitterer the dissensions." Never was that observation better exemplified than among the politicians of post-war El Paso. Every political alliance represented a potential personal hatred. Enemies lost all sense of proportion. When honor was at stake, a rival's defeat was seldom enough. His personal ruin or death became the preferred solution. This was the fate of Louis Cardis and Charles Howard, two strong-willed and ill-matched allies whose eventual collision triggered war in El Paso.[6]

The Howard-Cardis mutual hatred society and the willingness of Paseño farmers to fight injustice exemplify the combustible mixture of personal and historical forces that brought on the Salt War and determined its course. The conflict occurred at the confluence of several violent streams of American and Mexican life: the bitter politics of Reconstruction; the postwar Texan proclivity for answering grievances with the gun; persistent Mexican political instability; widespread outlawry; and Indian warfare on both sides of the river. This brutal world intersected with the dawning era of unbridled capitalism we call the Gilded Age. Attitudes of ethnic and racial superiority and hostility displayed by many (though not all) Anglos at El Paso, Texas, also directly led to the conflict, as did the geographic isolation of the Pass from the governments and military forces of two nations, and the U.S. government's newfound reluctance to respond militarily to civil disturbance.[7]

And so El Paso was a simmering cauldron in the 1870s, filled with the ingredients for conflict from the forces of history, economics, politics, sociology, and geography. But individuals made the Salt War happen. And for all its seemingly irresoluble issues, the conflict would not have followed its sanguinary course had certain individuals displayed less foolishness, more integrity, a greater sense of public duty, or a better appreciation of consequences. Instead, the manipulations of Cardis, the self-centeredness of Howard, the intolerance of Padre Borrajo, the racism of Sheriff Kerber, and the dishonor of Captain Blair helped turn animosity into tragedy. The unwillingness of men like Francisco Barela, Sisto Salcido, and Lino Granillo to submit meekly to injustice also played its part in events, as such resistance has throughout history.

It has been written that the conflict's only result was the permanent reestablishment of Fort Bliss. This assessment ignores the casualties and the hundreds exiled, many permanently. For families living in or on the brink of poverty, these personal losses, compounded by the failure of untended crops, was disastrous. In the long term, the distrust and marginalization of Paseño citizens by Anglos was deepened. When the railroad came in 1881, Paseño towns were left behind. If there was any positive result to the sacrifices made in 1877, it came generations later, when the people of *la Isla* began looking with pride to their ancestors' fight for justice as a foundation on which to build a future.

Salt Warriors is the most extensive study of the El Paso Salt War since a military commission took testimony in 1878. It is an attempt to provide a balanced view of the causes, course, and results of the conflict and to correct, through available evidence, the many falsehoods and innocent errors about *all* the participants repeated since the events took place.

As the title suggests, the book focuses on the populations and individuals who brought about the conflict and suffered through it. Considerable attention is devoted to the Paseño's culture, economy, and strong traditions of self-government and military self-defense so that the reader can better understand their desire to keep control of their salt and their destiny. El Paso's dozen years of corrupt, violent, and highly personal Anglo politics leading to the war is also discussed at length. A culture of threatened, attempted, and successful "politics by assassination," as much as the Anglo hustling to seize ownership of the Guadalupe Lakes salt, persuaded the Paseño population in 1877 to defend their rights through extra-legal action.

The lives of key and representative citizen-soldiers, hustling individualists, rustler-lawmen, priests, and other "salt warriors" before and after the struggle are revealed to help explain for the first time their decisions and actions during the moments of crisis. Inevitably, Anglo thoughts and deeds receive greater attention, greater detail, and less filtering. This is understandable, since Anglos controlled those agents of history: government and the press. The insurgents' story is far less documented. Their cause is unfortunately seen through the eyes of others, and their words, where we have them, too often come from those who did not comprehend or had reason to mislead. Nevertheless, the chards of evidence, pieced together, replace the false, competing images of howling mobs and noble victims with Paseños who were knowing, willing, and courageous "agents of their own fate."[8]

While the book focuses on Anglo and Mexicano participation in the Salt War, other populations also played roles in the drama. The Mescalero Apache were instrumental in the Paseños' retention of a warrior culture, while Tigua Indians living on *la Isla* supplied some of the insurgency's soldiers. African Americans provided the military force that ended the insurgency's hopes for success. Contemporary records regarding the activity and influence of El Paso's women, though sparse, bear witness to the travails and courage of Paseño and Anglo women alike. The sources are silent regarding participation by Paseño *soldaderas*, but their presence in the Salt War cannot be discounted.[9]

The significant gaps in the discovered written record and the author's point of view ensure that this work is far from the last word on the Salt War. But it should raise new questions and spur new research into a tragic, captivating, and timely American story, one that begins in the Chihuahuan Desert of West Texas.

PART ONE
Community

Carreta in front of Our Lady of Guadalupe Mission at Paso del Norte, 1880.
Courtesy of UTSA's Institute of Texas Cultures, No. 71-353, Courtesy of Calleros Estate

PROLOGUE SALT ROADS

The train of salt-laden ox carts (*carretas*) made its way along the final miles back to San Elizario, moving only as fast as the slowest animal. Perhaps eighty men drove, rode, and walked alongside the sixteen carts, pulled by sixty yoke of oxen and loaded down with the weight of eight hundred bushels of salt.[1]

It was mid-December 1877, and the *salineros,* salt gatherers, were nearing the end of their return trip. They were armed. One report said they carried "new breech-loading carbines," but it is more likely they possessed a variety of weapons, including modern Winchesters, ancient muskets, pistols of every design and reliability, and even bows and arrows. It was standard for salt trains to travel in large, well-protected groups. The Guadalupe Salt Lakes lay at the edge of the Mescalero Apaches' mountain fastness, and these Indians had raided all the way to San Elizario on many occasions over the years. Paseños had made such trips to these dry beds for at least fifteen years according to some accounts, and for generations according to others. The round trip, 180 or more miles from the Rio Grande to the lakes and back again, was in its thirteenth day. It was a slow journey, for the men "made frequent stops to feed the oxen and to repair their creaking wagons. . . . Every night they drew the wagons in a circle and slept within to protect themselves."[2]

The cargo was halite, rock salt, used to preserve food, cure hides, and replenish bodies. It was also a "commodity crop." As one historian put it, "On the western frontier the mineral most sought after was not gold but salt." Some salt was sold to the U.S. Army. Most was hauled 225 more miles to Ciudad Chihuahua. There *salineros* sold or bartered the salt for necessities, including "clothing, blankets, hardware, pottery [and] chocolate," for direct use or resale in merchandise stores at the Pass.[3]

Mining entrepreneurs were also vitally interested in salt. Most Chihuahuan silver was refined by the "patio process," an amalgamation of ore with mercury, copper sulfide, and salt. A sixteen-cart salt train, like the one rolling to San Elizario, was capable of carrying enough salt to refine 220 tons of ore down to more than 220 kilograms of silver, worth as much as eighty-eight hundred dollars. The *salineros'* total payment by middlemen might have been as much as five hundred dollars, worth nearly seventy thousand dollars in 2005 dollars, enough to make the difficult trek worth their while.[4]

Drought struck the Southwest in 1877. In far-off Tucson, the wheat crop was failing. In New Mexico Territory, little rain had fallen on the Organ Mountains; permanent springs in Soledad *Cañon* had gone dry. Corn crops in Lincoln County were half the normal yield. The Rio Grande was low at Las Cruces and Mesilla, and little water flowed through the irrigation ditches, called *acequias,* to the cornfields, orchards, and vineyards. By the end of September, which should have marked the end of the rainy season, the river was already dry at El Paso. Crops were failing in the unworked fields. The salt these Paseños hauled spelled an answer to hunger.[5]

As they approached their homes near the Rio Grande, the men had reason to be on their guard. They were breaking the law by "stealing" Charles Howard's salt. A month before, a company of Texas Rangers had been assembled in El Paso County to prevent them from doing just that. Howard himself had only arrived in the county in 1872, ready to begin his political career. He never took the time to master the language of his constituents nor understand their thinking. But in June 1877 he took the time to "locate" the salt lakes and, as a right afforded private citizens by the Texas constitution, lay claim to them on behalf of his father-in-law, an Austin banker. It was not a law the men with the ox carts recognized.

Under the laws of Spain, the king had title to all mines and minerals, including salt, not expressly granted. Title to land carried no mineral rights absent such an express grant. By royal decree of King Philip III, made on September 3, 1611, the inhabitants of Spain's New World colonies were granted "the right to gather salt free of charge or tax." Salt deposits existed for the benefit of the community. Mexico continued this policy, as did the Republic of Texas, and it remained in force after Texas became a State. This was the only law the *salineros* recognized.[6]

One of the most ardent voices opposing Howard's pronouncement of his legal right to be paid for their salt belonged to the leader of this train, sixty-two-year-old Lino (Leon) Granillo. Ten weeks earlier, he had helped organize a popular revolt to restore the rule of Paseño law, overturning Howard's claims. The Paseños had taken the Anglo prisoner, forced him to sign away his claim, and compelled him also to leave the county. Howard, a stubborn man, had not gotten the message, but Granillo was his equal in intransigence. For ten weeks, he had represented San Elizario in the *juntas* that governed the people, while Howard's puppet judges and the county's "greaser"-hating sheriff were officials without authority, at least none they dared enforce.[7]

Now, as Granillo led the *salineros* back home, word came that the people of San Elizario and other towns had risen up again. Howard had returned to

El Paso, intending to use those new Texas Rangers to enforce his property rights. Ironically, Howard and Granillo wanted the same thing, an end to threats and posturing and a resolution of the questions, "Whose salt was it?" Because neither man would back down, another question remained, "Who would be driven forever from El Paso or die fighting for salt?"

The Guadalupe Salt Lakes draw their name from the Guadalupe Mountains, a forbidding escarpment several miles to the east and running parallel to the alkali flat. The mountains are the exposed remains of an ancient limestone reef, whose highest peaks rise five thousand feet above the thirty-six-hundred-foot elevation where *salineros* once labored. The lakes' glistening appearance is the result of eleven thousand years of evaporation, as intense desert heat carries water-borne concentrated gypsum and brine to the surface, creating the bright white playas that stretch for one hundred miles. While the mineral deposits run deep in places, evaporated sodium chloride forms only a "thin veneer . . . several centimeters thick." Some lakes were endowed with particularly pure salt, while some appeared to regenerate an inexhaustible supply.[8]

The *salineros* knew nothing about the capillary action that created and regenerated the salt. They did know that the vast expanse of white crystals stretching before them was legally theirs. They also knew that what seemed "so easily obtained" by the exploring Captain Pope came at the price of work, discomfort, and danger. The poorest among them were indifferently clothed in white cotton shirt and trousers. They wore sombreros or high, flat-crowned, wide brimmed leather hats and may have worn shoes, sandals, or no footwear at all. The weather had been cold since they left—overnight temperatures below freezing were recorded at Mesilla at this time—so they also brought their serapes or rough wool coats for the 180-mile roundtrip trek. Those who were better off rode horses and, like vaqueros, generally wore clothing suitable for riding: low, flat-crowned leather hats, a *chaqueta* (short working jacket), canvas trousers with a leather seat, a colorful sash serving as a belt, leather chaps, and a vaquero's good high-heeled, embossed leather boots.[9]

Some men watched for Indians, others collected water, tended the oxen, or repaired the carts. The rest gathered salt. Like their ancestors, most probably used wooden shovels, though the American merchants had introduced steel tools to those who could afford them. There was much work to do, shoveling the salt, ladling it into canvas or cotton bags, and piling those bags high on the carts. They worked hard, and with dispatch, for it did not pay to linger in Apache country.[10]

Their *carretas* were two-wheeled, spokeless vehicles constructed "from incipiency to climax with [cotton] wood, iron being shunned with a scrupulous and studious regard." Copious lengths of rawhide lashed the cart together. A *caretta* in motion gave out a distinctive screech. "The wheels are never greased, and as they are driven along they make an unearthly sound, which echoes through the mountains far and near, being a respectable tenor for a double-bass horse-fiddle."[11]

The carts were ox-drawn, probably with three or four yoke (six to eight oxen) apiece, while extra animals accompanied the train to replace any ox that failed. Oxen were slow, traveling just two miles an hour. But they were steady animals, more reliable than mules and less apt to stray. Though they tired more easily than mules, and could make fewer miles per day, they could bear far heavier loads. This trait was especially valued when the train reached the sandy soil surrounding *la Isla*. The oxen of this sixteen-cart train hauled an average of fifty bushels (twenty-eight hundred pounds) of salt per cart.[12]

With bags of salt stacked high in their sturdy *carretas,* the salt gatherers yoked their oxen and turned them homeward. Depending on the expedition's origin, *salineros* would use one of several roads that had brought them from the Rio Grande. Paseños from the towns lowest down the river, Guadalupe and San Ignacio, generally came by way of Fort Quitman. They crossed the Rio Grande from Mexico at a point below San Elizario, headed down river, then north to the salt basin before reaching Fort Quitman. The so-called Fort Quitman route was popular with Mexican citizens seeking to avoid American customs inspectors.[13]

Salt gatherers from Paso del Norte and other towns close to the Pass used a much older route. This road, probably the original one to the lakes, may have been used since the seventeenth century. This road first ran from Paso del Norte to the Hueco Tanks twenty-four miles away, before continuing another twenty-five miles northeasterly to the seeping springs of Ojo de los Alamos, and then to the tanks of the Las Cornudas Mountains, another nine miles distant, just across the line in New Mexico. The "tanks" in the Hueco and Cornudas mountains included immense basins filled with thousands of gallons of water, many in caves sheltered from the sun. The road along this stretch was firm. Excellent grazing for horses, mules, and oxen was also available along the way. The prevalence of nutritious grama-grass, "which exists in the most profuse abundance over the entire surface of these table-lands," meant that the Paseños did not have to transport grain or fodder either coming or going.[14]

From the Cornudas, the road dipped back into Texas until it reached the

Ojo del Cuerpo, or Crow Spring. The water was sulphurous but drinkable. Grazing was not good here, but Crow Spring was at the northern end of Crow Flats, also called the Salt Basin, the place of the Guadalupe Salt Lakes.[15]

An alternate route served the villagers of San Elizario, Socorro, and Ysleta. This spur pointed northeast from *la Isla* on its way to the main route just east of the Hueco Mountains. At ninety miles, the journey was shorter than alternative paths, but this route lacked water from the Rio Grande to the springs of Ojo de los Alamos. Care was taken by travelers to avoid the direct sun and to pre-position water at the half-way point for the return journey. This salt road, constructed in 1863, was the route taken by Granillo's train.[16]

One other *salinero* road existed prior to 1863, a 120-mile highway running directly north from Paso del Norte to the San Andres Salt Lakes, located east of the San Andres Mountains of New Mexico, and northeast of White Sands. Texas speculators, relying on the fact that the San Andres deposits lay inside the Rio Grande boundary originally claimed by the republic of Texas, attempted to secure ownership. In 1852, the "rights" passed to El Paso merchant James Wiley Magoffin. Two years later, backed by a cannon aimed by a sheriff's posse far from its jurisdiction, Magoffin drove off a large *salinero* party, killing or seriously wounding eight. The Texans herded the oxen back to El Paso, leaving both the loaded carts and over 120 New Mexicans, some wounded, stranded in the desert.[17]

The episode created a firestorm. Magoffin was indicted in Dona Ana County on a charge of assault with intent to kill. Pressured, he surrendered the animals and made full restitution to the *salineros*. The territorial legislature declared, "That all the salt lakes in this Territory . . . is, and shall be free to all citizens, and each one shall have power to collect salt on any occasion free from molestation or disturbance." Interference with this right was a felony.[18]

A point of contention in testimony taken after the Salt War was whether the Paseños' use of the Guadalupe Lake deposits was only a recent phenomenon. Testimony by Father Ramon Ortiz of Guadalupe Church in Paso del Norte appeared to support that view, though his chronology defies understanding.[19]

Magoffin Salt War historian J. J. Bowden suggests that the increased demands for salt generated by the Civil War prompted Paseños to look once more to the closer Guadalupe Salt Lakes as a source for the precious mineral. This view is in accord with the Salt War testimony of various Anglos that Paseños ignored this source until 1862. Union veteran Caleb Miller made the preposterous claim that he had personally discovered the lakes in 1862.

"[The] people seemed to be surprised at it, and had not known of the existence of the lakes before. . . . After that date there was a road made to the salt lakes, which was called the Guadalupe Lakes, and people began to go there and get salt."[20]

Albert Jennings Fountain, a Union army sergeant, first visited the lakes on patrol in late 1862. He later claimed that "one of the guides, an intelligent man, it may have been Captain Garcia . . . had a long conversation with me in regard to the salt question." Garcia (Fountain's identification is probably correct) allegedly explained that Paseños had drawn their salt supply from the Tularosa [i.e., San Andres] salt lakes until the "Magoffin war" of 1854. The Guadalupe Lakes provided higher quality salt, and the distance from the Rio Grande was much shorter. For this reason, said Garcia, "the people on the river contemplated opening a wagon road to these lakes and obtaining their salt from them." Fountain observed "no wagon tracks visible about the lakes, and no wagon-road in that vicinity that I could discover, as we were looking for Indian sign I scrutinized the country closely; had there been any road or wagon tracks about there I could not have failed to discover them."[21]

Several facts regarding Fountain's statement deserve examination. First, the Union column, which traversed the lakes via the old Butterfield Trail at the north end of the lakes, did not use or cross the salt road leading from San Elizario to the south end of the lake beds. Second, it is not certain that Sergeant Fountain's patrols saw the lower road either. Third, Garcia's alleged statement to Fountain clearly indicates his countrymen were already aware of the route's existence and efficacy for commercial use, especially with improvements. Fourth, Fountain had already been rejected by Paseños over the issue of lake ownership several years before making this claim. His Salt War newspaper reporting and editorials evidence no sympathy for the insurgents, who argued ancient rights to these beds. Fifth, Fountain's claim does not address the possibility that the residents of *la Isla* used the roundabout but relatively water-rich road past the Hueco and Cornudas tanks.

The Anglo argument is that Paseños had relied upon the Salina de San Andres because the journey to the Guadalupe beds, although shorter, "was much more hazardous, as the marauding Apaches infested the entire area between the Hueco Tanks and the Guadalupe Mountains."[22] However, an item appearing in the *New York Daily Times* in February 1857 indicates that the citizen-soldiers of San Elizario exerted their power in the shadow of the Guadalupe escarpment even before the direct road from *la Isla* to the lakes was opened: "About the middle of last month, a party of Mexican citizens, of San Elizario, Texas, proceeded to the Guadalupe mountains, about 180 miles east of that place [actually 90], and on their return reported to Captain Walker that they had surprised a small encampment of Indians, supposed to have been Mescaleros, taking their scalps (of women, it is believed), and capturing or destroying all the arms and other property found in the camp."[23]

While the *Times* clearly described a retaliatory mission, not a mining venture, the episode reveals that Paseños were willing to hazard a journey into Mescalero territory. *Salinero* trains were much less mobile than militia columns, but they generally included sufficient armed men to protect themselves. Given the nearness of the Guadalupe lakes, the quality of the salt there, and the Paseños' ability to operate within the Apache domain, it is unlikely that *salineros* ever completely abandoned these deposits.

The leader of that 1857 punitive expedition might have been one of the town's most notable warriors, Don Gregorio Nacianceno Garcia, who also served as a justice of the peace and as county commissioner. He was wealthy by Paseño standards, with assets amounting to fifty-five hundred dollars in real and personal property in 1860. His elegant home bordering the town plaza reflected his status as one of San Elizario's first citizens.[24]

Garcia's position also derived from a reputation as the town's foremost defender against the ongoing Apache threat. He likely learned the soldier's craft from Capt. Jose Ignacio Ronquillo, commandant of the presidio after 1830 and the foremost Paseño soldier of the Mexican era. That officer ranged over large tracts of land in pursuit of Comanche, Kiowa, and Apache war parties. He scored several notable victories, earning promotion and a sizeable land grant in what is now west Texas. Garcia, drafted for military service in his fifteenth year, learned from Ronquillo how to track and fight Indians.[25]

Garcia served in Mexico's army against the Americans, probably in the debacle at Brazitos on Christmas Day 1846. In the 1850s, he solidified his reputation as an Indian fighter. Union Brig. Gen. James Carleton's choice of Garcia to command twenty Paseño scouts who would guide a column of cavalry deep into Mescalero territory in late 1862 bears witness to his local renown. Led by Garcia's scouts, Captain Nathaniel J. Pishon's column of blue-coated troopers scouted "no less than nine different ranchos" that Garcia knew the Indians to use, but their quarry had long since departed. It was on a patrol to the lakes that an inquisitive young sergeant, Albert Jennings Fountain, first spoke to scout leader Garcia about the salt deposits.[26]

Without doubt, the presence of Union forces convinced Paseños to undertake the subscription and construction of an improved direct wagon road to the lakes. Carleton's offensive against the Mescaleros temporarily reduced the threat of depredations, and the federal army provided the *salineros* with a significant, reliable customer. The deed was accomplished no later than 1863. From that time forward, the people of *la Isla* used this direct route. There was no water along the way, except which Paseños deposited on the trek outward. The Apaches ratcheted up their raiding in the summer of 1865, but the Guadalupe Salt Lakes were now, from *every* account, the Paseños' chief source of salt.

The building of the San Elizario salt road in the midst of civil war initially had only a local impact. With the war's end and the promise of new railways constructed across Texas, the Guadalupe Salt Lakes were bound to come to wider attention. The Confederate defeat brought to El Paso a small army of discharged Union veterans intent on making money through aggressive entrepreneurial capitalism. Those who settled in El Paso could see the gold glittering in the *salineros'* salt.

CHAPTER 1

"This country is destined to become of importance."

Paleo Indians first arrived in what is now El Paso County perhaps ten thousand years ago. Spaniards began colonizing the area in 1598. Americans who stayed and took local wives began arriving in the 1820s. By the 1870s, perhaps fourteen thousand people lived in the various communities at or near the Pass of the North, including nearly four thousand on the U.S. side of the Rio Grande. The largest town, and the center of Mexican government authority, was Paso del Norte (modern Ciudad Juarez), located on the south side of the river at the point where the pass was narrowest. Just across the Rio Grande from the city's plaza was the burg of El Paso (also called Franklin; population 764 in 1870), dominated by Anglos but largely Mexican in population. The citizens of Ysleta (800), Socorro (635), and San Elizario (1,120), the next towns downriver on the Texas side, were nearly all Mexican in ethnicity, language, and culture.[1]

Many early immigrants left to posterity their impressions of the Pass of the North. John Russell Bartlett, the U.S. government's boundary commissioner following the Mexican-American War, described Paseño homes at the dawn of American influence as one story, built with adobe (mud bricks) "baked in the sun. This material, with slight repairs, will endure for centuries. . . ." Not every adobe structure was so impervious to the elements. Maj. William H. Emory, the U.S. commissioner charged with surveying the border, reported that heavy rains in 1855 left most homes "damaged and rendered leaky." As to interiors, Major Emory observed that rooms were "never more than twenty or thirty feet wide, the span of a stick of timber, without the aid of king-posts. The houses of the better classes are large, and built in the form of a hollow square. The walls are from two to three foot in thickness, and have but few openings. When plastered and whitewashed, they look very neat and make very comfortable dwellings. All the floors are laid with mud, concrete or brick. Such a thing as a wooden floor is unknown in the country." Bartlett asserted that, "Moorish capitals and ornaments are still visible both in the fine dwelling and the humble cottage in northern Mexico."[2]

Bartlett was struck by the absence of window glass. "The ordinary dwellings of the poorer class have no windows. The larger ones are entered by a large gateway, and have a few barred openings on the street. The other three sides present externally an unbroken and prison-like appearance. To all other parts of the house the light is admitted through windows or doors opening on an inner area [i.e., a patio]. . . . [The] occupants make themselves sufficiently warm by covering the opening with muslin or white cotton." He noted that Paseños used fire little, except for cooking, and not even to warm their homes on cold winter days. Fuel was provided by the abundant mesquite.[3]

To protect their livestock from theft, villagers penned in the cattle at night. Often, their corrals were actually the patios enclosed within the larger adobe houses. Though nearly impregnable to attack, stealthy Indians sometimes crept over corral walls at night and sawed through the adobe with rawhide ropes. Settlers learned to counter by mixing sharp pieces of cut glass and bones into their walls.[4]

In 1877, a patio home in San Elizario served as a fortress for Texas Rangers against repeated assaults by Paseño insurgents. It no longer exists, but one such home and the remains of another still stand in San Elizario. The Gandara house, positioned directly behind the present church, was built around 1840. The patio contained two courtyards in the central core, a *placita* (or small square) in front, and a corral in the rear. The high walls gave way in front to a gate wide enough for wagons to pass. The complex served as a Butterfield Stage station in the late 1850s. The stately Casa Ronquillo (the home was purchased by Captain Ronquillo in 1850) originally consisted of "twelve or more rooms in three wings with an interior courtyard surrounded by a portal." (Only one wing of five rooms remains.) The exterior walls were two-and-one-half to three feet thick. Twelve solid wood frames created doorways to the patio area. A portion of collapsed roof reveals the construction. Pine beams called *vigas* were set into the adobe walls some ten feet off the floor (fifteen feet up before twentieth-century renovations). Thick adobe completed the ceiling.[5]

The people living in these dwellings, both before the Mexican-American War and at the time of the Salt War, were largely farm laborers, with a sprinkling of tradesmen and some servants. There also existed a small "middling class" of citizens, an "identifiable crust of society" considered by Anglos as a "cut above" the poor. They had land, perhaps a mercantile business or legal or military office, domestic servants or other employees, some formal education, and a strong sense of "civic mindedness."[6]

Regardless of their place in society, the Paseños were virtually all Roman Catholics. Their priests wielded an influence over parishioners that went beyond the care of souls. The most important cleric in the valley was Father Ramón Ortiz, an educated landowner, businessman, and for nearly sixty years the pastor of Our Lady of Guadalupe Mission at Paso del Norte and the surrounding missions. Though consistent and unflinching in opposition to any weakening of traditional Roman Catholic precepts and practices to accommodate newcomers in a changing world, he counseled goodwill toward Americans after Mexico's defeat in 1848.[7]

Bartlett described the day-to-day wear of the typical Paseño: "The upper class dress as we do. Among the inferior classes, the men wear a short jacket with large white cotton drawers, over which are drawn pantaloons, open at the outer side, from the hip down. Along this are rows of gilt buttons and other ornaments. Around the waist a red silk sash is generally worn. The whole is covered with a serape or blanket in cold weather. All the women wear the *reboso*—a scarf thrown over the head and around the shoulders: it is made of silk or cotton, and costs from one to thirty dollars. The most respectable ladies generally appear in the street in black, but at evening parties the richest and most gaudy articles are worn." Of Paseño women, another early settler recalled, "Their chief attire was red calico skirts and heavy mantilla shawls."[8]

Every resident of the valley was dependent upon the Rio Grande: the "American Nile," in the words of one nineteenth-century publicist; "the longest and poorest watered river of the North American continent," according to an immigrant. Before overuse and concrete banks forever altered its character, the river was erratic in its course and capricious in its reliability. In most years, melting snows in the mountains of Colorado and New Mexico turned the river into a turbulent, muddy mess at the pass in May or June. At this point, "the river plays some fantastic pranks, occasionally overflowing its banks from foot-hill to foot-hill." Around 1830 to 1831, the wandering river cut a second and ultimately deeper channel south of the old one. The land in between, an island twenty miles long and two to four miles wide and encompassing Ysleta, Socorro, and San Elizario, was simply called *la Isla*. The river banks were lined with cottonwood trees, some of immense size. The timber was not used in building homes, except for the roof-beams. It was reserved instead for farming tools, the *carretas,* and other necessary objects.[9]

The Rio Grande was generally fordable except when flooding. Mountain runoff raised the river level beginning in early May, keeping it high in ordinary years until mid-August, after which the river became a trickle—or a dry bed—while water was rerouted for irrigation. Water from the mountains was supplemented only slightly by rain, which generally measured less than eight inches annually, virtually all of it falling from July to September. Drought and deluge were commonplace. The dry years of 1849 to 1851, for example, preceded several years of heavy rain. Major Emory would always remember his first experience with a desert monsoon. "[A]wakened by a roaring noise," he stepped out of his tent to find himself "knee-deep in water" that soon rose "up to the chin." Emory barely escaped up a slope with his precious surveying equipment.[10]

Farmers were dependent on the construction and maintenance of a dam and network of irrigation ditches, called *acequias*. These canals diverted anywhere from half to all of the river's water. The main dam was constructed two miles north of Paso del Norte. Water collected at this point was fed into the Acequia Madre, a canal fifteen feet wide and twelve miles long, and from that channel to smaller *acequias* and thence to the fields. As late as 1877, the year of the Salt War, the dam was built "in the most primitive fashion being constructed of loose rock, straw, brush, willow baskets filled with stones and held together by logs." In time of drought the porous wall failed to hold the water for the *acequias*. The precious liquid simply seeped through without being used. At other times, one observer noted, "the entire population is in a continual state of worry and work to keep the dam in repair, and it happens now and then that some portion gives way and the entire harvest is endangered."[11]

According to Frank Edwards, a U.S. Army private during the Mexican-American War, "Each farmer has a day allowed him to use the water . . . , but cannot touch it at other times without the permission of the special alcalde of sakos [i.e., *acequias*]. . . . This officer has the powers of a judge in all things relating to his department. Every person is required to keep his sako in repair; and should any damage occur to his neighbor's property by inattention, the delinquent has to make good the damage."[12]

Wherever the river flowed, it deposited a fertile layer of slimy mud. Although the soil was usually no more than four or five inches thick, atop a bed of sand, the yearly renewal of nutrients enabled the farmers to grow crops year-round, year after year. Captain Pope observed that the crude implements fashioned by the Paseño farmers were "conclusively shown" to be the best. The iron plough, Pope noted, dug too deeply into the sand, mixing it in with the rich sediments, diluting fertility. He observed that traditional methods would result in a yield of "thirty to forty bushels of corn to the acre."[13]

In 1852, J. D. B. De Bow, a leading antebellum proponent of southern industrial and commercial development, called El Paso "one continuous orchard and vineyard." William M. Pierson, the American vice consul in Paso del Norte, agreed. The Mexicans, he observed in 1872, had cultivated a "solid mass" of land ten miles long and six to ten miles wide. In comparison, the land under cultivation on the U.S. side was quite limited, perhaps five square miles each around Ysleta, Socorro, and San Elizario, with less around Franklin. In general, said Pierson, family farms were small, from ten to fifty acres in size. Beyond these farms, lay "a barren waste for many miles, where a few jack rabbits and horned frogs struggle for existence."[14]

In the valley, Pierson reported, "The cereals attain great perfection and most especially all varieties of wheat. . . . Vegetables of all kinds known to the temperate zones develop finely, and attain a thrift and size rarely known in other portions of the United States. The famous El Paso Onion often reaches three to four pounds in weight, and is far more mild and palatable than the common onion of the North, while its yield to the acre is enormous." The vice consul extolled the "apple, pear, peach, quince, plum, cherry, raspberry, and strawberry" as equaling the best found anywhere in the States.[15]

Every newcomer had his own favorite food and beverage. James G. Bell, a cattleman driving beeves to market in California in 1854, remembered "onions as large as my fist . . . and Muscal liquor [with] a taste between whiskey and brandy, and considerable intoxicating power . . . Had 4 different dishes of onion, today for dinner; raw onion with vinegar, salt & pepper; fried onions; onions, meat and egg in one dish; and beef soup seasoned with onions. . . . Onions! Onions!! Onions!!!"[16]

Another newcomer left a different picture of the Rio Grande diet. In a letter to his family in Germany, young Ernst Kohlberg observed that meat was eaten at all three meals, washed down with coffee in the morning and tea in the evening. He made particular mention of the "chile or Spanish pepper, the principle crop here. It is eaten green when roasted, though the Mexicans even eat it raw, or it is eaten cooked with cheese when it is ripe, and as red pepper (*chile Coronado*) in many different ways particularly with enchiladas which are always served at balls and considered a good remedy for a hangover. . . . At first chile and everything connected with it was a hellish kind of food for me, but one becomes accustomed to anything and now I can almost swallow it like a Mexican, and I miss it if it is not served."[17]

According to the hyperbolic De Bow, "The most important productions of the valley is [*sic*] the grape, from which annually is manufactured not less than two hundred thousand gallons of perhaps the richest and best wine in the world. . . . I doubt not that they are far superior to the best wines ever

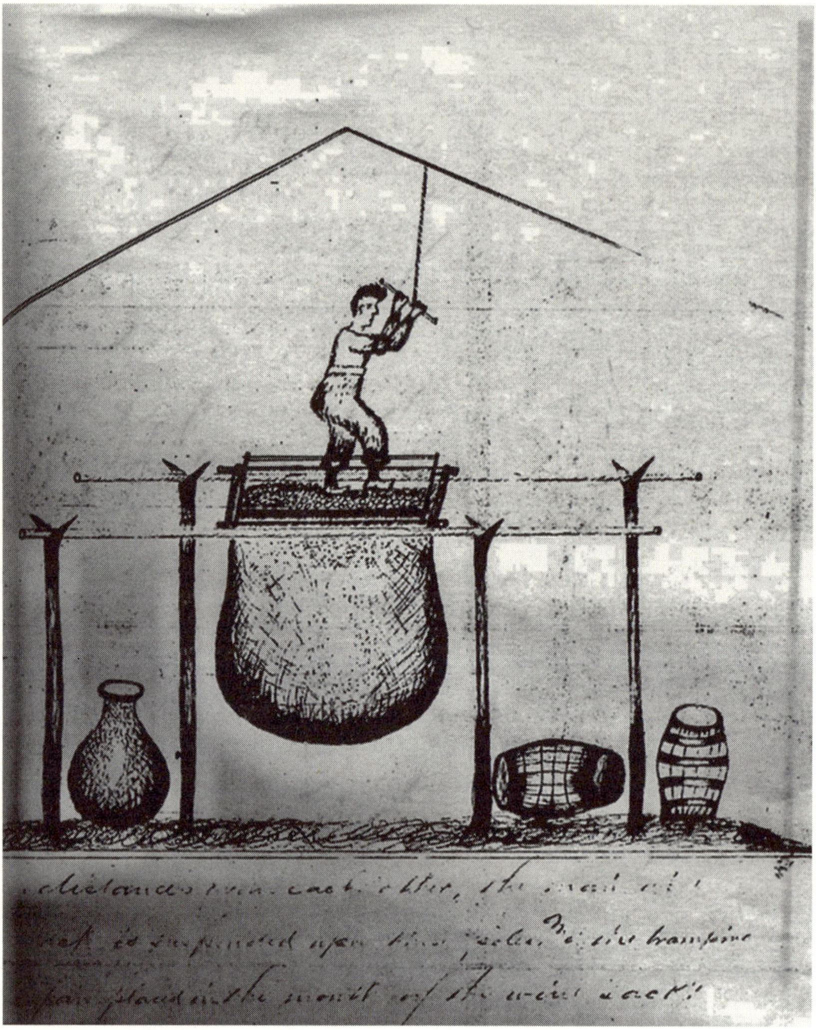

Vice Consul William E. Pierson's drawing of a "stalwart Mexican" pressing grapes in a Paseño wine house. National Archives

produced in the valley of the Rhine, or on the sunny hills of France." At two dollars per gallon, wine provided Paso del Norte's principle source of revenue. The vineyards, producing wine of "the first rank of North American specimens," were of special interest to Pierson. In one report, he drew the image of a "stalwart Mexican" pressing grapes with his "remarkably brawny feet."[18]

Produce not eaten by local residents traveled along trade routes stretching to Santa Fe in the north and Chihuahua in the south. Whereas crops provided a reliable livelihood in most years, raising livestock was more problematic. At first glance, the land seemed more than suitable. According to Pope, who explored the lands above the valley floor in 1854, the "profuse abundance" of grama-grass "present[s] advantages for raising and maintaining stock as remarkable as they are limitless." Unfortunately, the West Texas uplands lacked the wells, windmills, and dams needed to make ranching profitable.[19]

Mescalero Apaches were a second hindrance to pasturing cattle beyond the Rio Grande Valley. This tract of the Chihuahuan Desert stretching north and east from the valley to the Sacramento and Guadalupe mountains belonged to the Apaches. Indians raided repeatedly into the valley, sometimes into the very heart of Paseño towns. In some years, the losses in lives and animals were staggering.[20]

Paseños learned to live with Apache depredations, but a new enemy altered their way of life forever. In December 1846, Col. Alexander Doniphan and eight hundred Missouri volunteers seized Paso del Norte. One company occupied San Elizario, where they "committed many outrages and violences," such as destruction of the town's archives, including records relating to Paseños' historic rights to the Guadalupe Salt Lakes. The U.S. soldiers camped inside the ruins of the crumbling presidio. Doniphan's column marched south to Chihuahua in February, leaving behind one company at Paso del Norte and two more at San Elizario. This continuing U.S. military presence signaled an even more dramatic change in the valley. By the terms of the Treaty of Guadalupe Hidalgo (1848), the Rio Grande became an international boundary. Where the river created multiple channels, the deepest would separate the nations. Authorities in the United States determined that the deepest channels included the one placing Ysleta, Socorro, and San Elizario on the U.S. side.[21]

On January 24, 1848, gold was discovered in California, prompting thousands of fortune hunters to traverse the southern route to the gold fields. After navigating six hundred miles of the worst terrain West Texas had to offer, some four thousand survivors descended like locusts on the valley. Some paid for provisions, equipment, and animals, thus driving up prices; others did not, looting anything farmers failed to hide away. For a time, Socorro was beset by "a set of ruffians, who . . . had become so formidable, that the life of no one was considered safe beyond the walls of his own house." The town rapidly depopulated until the worst criminals were brought to justice. El Paso's wide-open reputation continued to draw "gamblers, murderers and thieves" for many years.[22]

A more permanent impact was made by U.S. surveyors who affixed the new land's boundaries and uncovered their assets and potential for development. Texas, eager to claim all territory as far as Santa Fe, dispatched its own expedition under John S. "Rip" Ford. The federal government organized a score of expeditions throughout the new territory in the years before the Civil War. Bartlett, Emory, Pope, and others all wrote positively about the opportunities to be made once the roads to El Paso were laid, forts built, water sources developed, likely railways identified, and Anglo Americans moved in. If most Anglos passed through, enough remained to begin inexorable change. Among these were the officers and men of the U.S. Army. The first infantry companies arrived in September 1849. Ultimately, the "Post opposite El Paso" was rechristened Fort Bliss.[23]

U.S. businessmen began the real transformation. Among the first were Simeon Hart, Benjamin Franklin Coons, James Wiley Magoffin, and Hugh Stephenson. Hart, Magoffin, and Stephenson all married the daughters of upper-class Mexicano families and learned the ways of their adoptive relations. Together, these traders established several U.S. settlements at the Pass of the North. One of these was Franklin, named for Coons's mercantile store and located in what is now central El Paso. Magoffinsville, located between Franklin and Ysleta, "became the headquarters of the largest, most important commercial operation in the American southwest [specializing] in European cotton goods and fabrics." This string of trading centers, rented army posts, and family residences evolved into the U.S. city of El Paso. Despite the merchants' best efforts, trade at the Pass of the North remained a precarious business prior to the war. The high cost of fitting out a train, the great risk of loss through Indian depredations or natural catastrophe, and Mexican protectionist policies all made profits problematic. In 1856, trade came "to a near halt, triggering an economic crisis in Paso del Norte." Better times came and went, but it would take the railroad to more dramatically transform commerce. Until then, the Anglo community remained small in numbers, if significant in impact.[24]

The first stagecoach arrived in 1853, bringing scheduled mail and passenger service from San Antonio aboard George H. Giddings's San Antonio–San Diego line. In September 1858, the Butterfield Overland Mail line inaugurated service between St. Louis and San Francisco via stations at San Elizario and Franklin.[25]

Families were needed to complete the "Americanization" of El Paso. The first pioneers, overwhelmingly male, found local brides; but homes, apparel, cuisine, and other facets of daily life began to change under the influence of women from the States. As late as 1860, only seven Anglo females lived in Franklin (the foremost Anglo town in the valley). According to one account, the first was Bavarian immigrant Emily Rohmann. Her husband, Anton Bernard Rohmann, a successful merchant, had moved his business from Galena, Illinois, to Santa Fe and then, in 1853, to El Paso. When Emily Rohmann stepped off the stagecoach at San Elizario, "with its mud houses, its thinly scattered cottonwood trees, its Mexicans, strange to her and jabbering in a language with which she was unfamiliar, she sat down and cried, asking her husband why he had sent for her to come to so Godforsaken a country."[26]

More challenges awaited Emily Rohmann. Tallow candles provided the only lighting other than sunlight through the glassless window openings. The Rohmanns installed the first wood floor and the first carpeted floor in Franklin. Water was hauled in barrels from a nearby spring. Food was cooked in "huge Mexican pots" hung in the fireplace over burning mesquite roots or *tornillo* wood. Daughter Amelia recalled in 1916 that her mother was "the first person to buy a cooking stove in the village," purchased from an officer at Fort Bliss. There was plenty of fruit and, because Emily brought seeds with her, their garden "proved to be a . . . pleasing change from the regular peas and beans."

Unable to speak Spanish and schooled at home, the Rohmann children—Amelia, her older brother William, and Alice Samuels, their motherless cousin—were bored in their new surroundings. Alice recalled, "Thus cut off from our friends and to be placed where there was no outside companionship to be had, no diversions, no schools, no culture or any of the things that make up for the necessaries of life, it is easily imagined the heart aches and loneliness we underwent." If Alice's memory served correctly, no other American women settled in the valley for several years. When at last they arrived, "they appeared to be almost angels to us."[27]

The Rohmanns were not the only newcomers unimpressed by San Elizario. Bartlett declared the church and presidio in "a ruined state" in August 1852. Two years later, *El Sabio Sembrador*, Frederick Augustus Percy's handwritten newspaper, remarked, "In San Elizario the glory of other days has moulded, decayed and fallen to the ground, a crumbling portion of the outward wall is all that remains to designate the spot of Christian rendezvous a century ago."[28]

The crumbling town actually underwent renewal in the 1850s. The people constructed a new church in 1853. Sometime in the late 1850s, Gregorio N. Garcia built a new home, constructed with the usual adobe brick walls and cottonwood rafters. The home was well situated, facing the town plaza to the north. The principle feature of the front façade was a long covered gallery, or portal, giving the home its name, Los Portales.[29]

Other signs of revival were visible. San Elizario was selected as the first county seat. While Anglos held the county's chief legal offices in the antebellum years, including the offices of district and county judge, district attorney, and sheriff, Paseños in the "middling class" dominated the ranks of justices of the peace and county constables. Among the notables "who displayed vigorous leadership and deep concern for the welfare of the community and its people" were Garcia and Jose Ignacio Ronquillo, retired from military service following Mexico's defeat. Both served as justices of the peace. Luis and Petra Lujan Aranda, both of notable families, took advantage of the increased commerce with San Antonio to send their ten-year-old son, Maximo, there to Saint Mary's College in 1855. Enterprising and educated young Paseños like Maximo learned retailing and real estate while clerking for Anglo merchants.[30]

Dislocation was a factor, of course, as Paseños adapted to the new controlling elite, with its foreign ideas about everything; but the few Anglos who moved here in the 1850s did their share of assimilating. It was a slow transformation, not without problems, but not beyond the capacity of people to accept. All that changed beginning in 1861. An unremitting whirlwind of civil war, overheated Reconstruction politics, and Gilded Age capitalism wrenched the sleepy towns of *la Isla* off their sluggish path toward progress, hurtling them into a world without tradition, without rules, with few restraints on human behavior. It was a world made for Americans of the most ambitious sort, and the most ruthless.

CHAPTER 2

"The inhabitants rallied their fighting men."

In *The Magnificent Seven,* John Sturges's translation of Akiro Kurasawa's *Seven Samurai* to the Western film genre, peaceful Mexican farmers rise up to save their village from a small army of bandits. Roused from instinctive and comfortable docility, they hire professional gunmen from north of the border who teach them which end of a gun to shoot and how to find their courage. Unlike these movie farmers, Paseños were anything but docile. Whereas Sturges's cinematic village had but one old man who knew how to fight, the real villages of San Elizario, Socorro, and Ysleta counted hundreds. It is the old man of *The Magnificent Seven,* not his frightened neighbors, who reflect the reality of the Paseño towns.

What Mexican revolutionary historian Rubén Osorio writes about the State of Chihuahua applies specifically to the Pass of the North: "The peasant class had come into being under very special conditions during the colonial period, as settler-soldiers, or literally as an armed peasantry living in presidios constructed by the Spanish government against the 'barbarous Indians.'" Whereas Ysleta and Socorro were organized around missions, San Elizario began in 1789 as such a presidio, or fort, the town growing up around it. Presidio soldiers campaigned against Indians and regularly escorted caravans to the provincial capital in Santa Fe. Its troops also contributed to the wider defense of the Mexican Republic. After the garrison was permanently reassigned to Santa Fe in 1843, San Elizario became a farming center. Its residents tore down the old presidio's adobe walls and buildings to construct their own homes and pens. Because Apache depredations continued, however, these farmers, ex-soldiers, and descendants of settler-soldiers became citizen warriors, frequently defending their homes against Apaches or taking the fight to their enemy's lair.[1]

The relationship between settlers along the Rio Grande and their nomadic neighbors ran in cycles. The Spanish and Mexican regimes alternated between waging war with the Apaches and settling them in "peace establishments" collocated with military posts, including San Elizario's presidio. Whenever government rations stopped, the Apaches resumed raiding, at times sweeping the valley clean of livestock.[2]

In the years between America's Mexican and civil wars, Mescaleros continued to raid commerce in and out of El Paso County. According to W. W. Mills, in 1858, "The Americans and Mexicans were secure only near the military posts, or villages, or large settlements, and when they traveled from place to place, they traveled in companies strong enough for defense, or at night and by stealth, trusting to Providence, or luck, each according to his faith." The garrison at Fort Bliss seldom included sufficient mounted troops to effectively counter the Indians. To bolster them, Paseños willingly joined the soldiers as guides and auxiliaries.

One such expedition took place in early 1859. When word reached Fort Bliss that Apaches had stolen cattle, horses, and mules from San Elizario, 1st Lt. Henry M. Lazelle and thirty mounted rifleman pursued. They first stopped at San Elizario, where the officer added several citizens to augment and guide his command. Among these was Mexican Army veteran Gregorio Garcia. The chase stretched for 160 winding miles to New Mexico's Sacramento Mountains. On February 8, Lazelle's thirsty men passed through towering cliffs into rock-strewn and cactus-choked Dog Cañon. Here thirty Apaches greeted the soldiers under a white flag. After a fruitless discussion, the fighting began. The soldiers were soon flanked, prompting their withdrawal. In a fighting retreat to the canyon's mouth, the pursuit party lost three men killed and six wounded, including Lazelle, shot twice through the lungs. The army's official report of the action failed to mention the Paseño role, but a later account indicates, "[Gregorio] Garcia, with some citizens of San Elizario, prevented the massacre of the United States troops in Dog Cañon by his skill and courage." The survivors returned without the stolen livestock. The defeat illustrated both the difficulty of waging war on Apaches and the martial qualities of Paseño citizen soldiers.[3]

In July 1861, the American Civil War came to El Paso. Confederate troops arriving that summer and autumn, coming off a six-hundred-mile desert march, stole what they needed, forcibly appropriating crops, animals, and other possessions from local citizens. While Brig. Gen. Henry Hopkins Sibley swiftly forwarded his Texans to New Mexico as they appeared, their strung-out march prolonged the Paseños' agony.[4]

For his part, Sibley found the Pass of the North "perfectly overrun with desperadoes, gamblers, &c. . . . who incited citizens of the valley to rebel against the proper authorities." A prime target of Confederate wrath was William Wallace Mills, a 25-year-old native of Indiana. W. W., as he was known, had arrived in El Paso in 1858, on the heels of his older brother Anson. When Texas seceded, Anson went east to become a brigadier general, while William watched and reported on rebel activity from Paso del Norte. Simeon Hart, acting for the Confederacy, had Mills seized and dragged back to Texas. Mills's captivity was temporary but a humiliation he remembered to repay.[5]

Sibley's would-be invasion of Colorado was stopped in its tracks when his supply train was captured at Glorieta Pass. The Confederates were forced into a headlong retreat, and soon their "provisions, forage and ammunition [were] completely exhausted." By May 3 the tattered and starving rebel army was back at Fort Bliss, with Col. William Steele's rear guard at Fort Fillmore (near Mesilla). "Of sugar and coffee we have scarcely a pound," Sibley wrote, "all other rations are alike deficient; whilst of salt meat we have not an ounce. Our ammunition may be said to be exhausted. . . . Clothing completely exhausted, with no means of renewing the supply."[6]

The Confederates transferred their misery to the people of El Paso County. A Mexican merchant reported that the Confederates were "committing outrages upon the inhabitants they meet upon the highway. They are almost on the point of starvation. . . . Sibley has pressed [i.e., seized] all the spare provisions on the American side of the river, and has given orders to take all the necessary transportation and provisions from the villages below. . . . there is no supply, not even of live stock, and there will be a scanty wheat crop to say the best of it. It is believed there will be famine among the people."[7]

The Mexicanos of both Dona Ana and El Paso counties began to push the ravenous Texans out. The first Paseño uprising may have occurred in Ysleta. There a Confederate requisition for beeves was not fully met by the people. When the Texans refused to pay, "the inhabitants rallied their fighting men, from 2 to 300 strong, pursued our men, killed some 15 or 20, mostly stragglers, and captured about 15. [Reinforcements] came to the rescue, killed 30 Mexicans and liberated the prisoners." On June 15, a company of Confederates "had quite a lively fight," in Socorro, killing twenty Paseños, wounding many more, destroying the church, and tearing up the town.[8]

The Confederate retreat was hastened along by Brig. Gen. James Henry Carleton's "California Column" closing in from the west. From early March to early August 1862, this force of some 2,350 Volunteer infantry and cavalry marched from Yuma to the Rio Grande Valley, twice skirmishing with Confederates and fighting one tough battle with Apaches. But it was the intense heat, scarcity, poor quality of water and grass, and abundant alkali dust along the route, that made their march epic. As Carleton's advance elements approached the Rio Grande, Colonel Steele's rear guard seized supplies and transportation around Mesilla, touching off renewed "armed resistance" to his foraging parties. A "desperate fight" took place on July 2 between the Confederates, who lost seven men, and *Nuevo Mexicanos,* who lost about forty.[9]

When the Californians were halted by the Rio Grande's floodwaters and tired mounts, the pause gave Steele time to hurry away from the Pass of the North. Nevertheless, Paseños pressed their attacks. On July 3, a Texan detail met with resistance while attempting to confiscate animals at Socorro. Undaunted by their lopsided defeat less than three weeks earlier, some fifty Paseños charged the detail, killing one soldier and taking nine prisoners. On July 12, Steele, facing mutiny, made ready "to start for San Antonio with very limited means of transportation, and insufficient supply of breadstuff and beef . . . , and with troops in many instances almost naked." When a Union cavalry company at last reached El Paso, the Texans were gone.[10]

Upon reaching Las Cruces on August 10, 1862, Carleton sent cavalry into Franklin and pushed out patrols to Forts Quitman and Davis. He also issued General Orders No. 15, declaring that the people "may now rest assured that the era of anarchy and misrule—when there was no protection to life or property . . . has passed away." Many Paseños were at first suspicious of the newcomers. "They had been taught by the Texans," Carleton reported, "that we were coming among them as marauders and as robbers. . . . They found that we treated them kindly and paid them a fair price for all the supplies we required to find, [and] as they came under the old flag once more, that they could now have protection and will be treated justly."[11]

Not every Paseño was so paid or treated. A large number of beef cattle belonging to Garcia were confiscated and consumed by a Union column. If George Wythe Baylor was correct, Garcia's herd was singled out for seizure because that Paseño had "served under Col. John R. Baylor in the campaign against the Federal troops at Fort Filmore [*sic*]." But now the Yankees needed scouts familiar with the ground and under the right leader. This military necessity would direct Carleton's attention to the question of Garcia's confiscated beeves.[12]

Commissioned in 1839, Carleton spent his pre–Civil War career with dragoons in the west, acquiring an understanding of guerilla warfare matched by few peers. He believed in conducting war harshly. Responding to a rumor that Colonel Baylor was in San Antonio raising six thousand men "to retake and hold New Mexico at all hazards," Carleton instructed Col. Joseph R. West to prepare for their return. In addition to stockpiling grain at Mesilla and driving all livestock to safety, West was to urge Paseños to "wage a partisan warfare upon the invader." West was to "animate" the people "to attack the enemy from every cover; to shoot down his teams, to stampede his stock when grazing, to destroy the bridges over the *acequias;* to hover by night around his camps; to set fire to the grass and all kinds of fodder which his animals might otherwise get; to shoot down his men by night and then before day to scatter singly in all directions, and then to repeat this over and over again until he feels the just vengeance of a people who never did him harm, but whom he wantonly and ruthlessly insulted, oppressed, and robbed." Carleton intended to arm the Paseños only at the last moment. Fortunately, the rebels never returned and destructive guerrilla warfare was never launched. But Carleton had accurately gauged the martial capabilities of these farmers.[13]

Meanwhile, Indians throughout New Mexico had taken advantage of the disintegration in U.S. military power in 1861. Virtually unopposed, they killed many settlers and forced others to abandon homes. (It was probably to help keep the Apaches at bay that prompted Garcia to join Baylor's Confederate force.) In early September 1862, raiders penetrated to deadly effect as far as San Elizario. In reply, Carleton planned a multipronged offensive to "punish and control the Mescaleros." While Col. Christopher "Kit" Carson was to attack them from Fort Stanton, Colonel West was to dispatch two columns moving east through Dog Cañon and northeast from Franklin, respectively. The Franklin command, under the orders of Capt. Nathaniel J. Pishon, included two companies and twenty Paseño scouts.[14]

Carleton's choice to lead these scouts was Garcia. In case the Union army's earlier confiscation of Garcia's herd might color the *don's* thinking, the general instructed West, "The sixty-six head of beeves, and one hundred head of sheep and goats belonging to the United States, now in the hands of Lieut. Coleman, you can use to replace those received of Mr. Gregorio Garcia at San Elizario by Lieut. Haden, which have been consumed as fresh meat." With this resolved, the general's adjutant wrote to Garcia, "[T]he General Comdg. trusts that you will use every exertion to advance the interests of the Government."[15]

Garcia needed no urging to hunt down Apaches. Awakened by noises one night in early November 1862, he rushed outside to discover that Apache raiders had stolen into the heart of San Elizario and released from their corrals both his cattle and those of a neighbor. Spotting Garcia, the Indians left the cattle and rode out on three of his horses, including a two-year-old black Kentucky horse worth two hundred dollars. Soldiers pursued, but the horses were never recovered.[16]

Captain Pishon's column, with Garcia scouts in front, moved out on November 15, 1862. After extending their search into the mountains of southern New Mexico, the formation returned to El Paso in late December without finding and engaging their enemy. However bloodless the patrol, it brought to the lakes soldiers with differing views on fulfilling the economic promise of the Guadalupe Lakes. For Paseños such as Garcia, the lakes were a community treasure chest needing only a community subscription road to unlock their full potential. For the bright, energetic, and ambitious young sergeant, it was a place where a man might make a personal fortune. Other men of equal ambition and greater greed soon had similar reactions, ultimately igniting a community's passion for justice and inciting the citizen-soldiers of San Elizario, Socorro, and Ysleta to take up arms in a citizen's rebellion without equal in post–Civil War American history.[17]

PART TWO
Corruption

U.S. Customs Collector William Wallace Mills. Robert G. McCubbin Collection

CHAPTER 3

"Salt lakes on the brain"

When the Confederate army abandoned El Paso, W. W. Mills, the leading prewar Unionist, returned as U.S. customs collector for New Mexico and Trans-Pecos Texas. By controlling the Customs House, Mills was in a position to grant or withhold favors to local merchants. With the army's support and El Paso's commerce and capital subject to his pressure, he dominated local politics, rewarded friends, and repaid enemies. Mills's allies, known as the Custom House ring, were placed in key government offices. Prominent among these were California Column veterans James A. Zabriskie, Albert H. French, and A. J. Fountain, respectively El Paso's new district attorney, county judge, and county surveyor. Another Californian, John J. Atkinson, became a customs inspector at San Elizario, while a Paseño ally, Juan Mauro Lujan of San Elizario, was elected county clerk. In December 1865, Mills and his friends "indulged in a dizzy round of judicial skullduggery," using the New Mexico courts to sequester Confederate properties in El Paso. The ring snapped up publicly auctioned properties at ridiculous prices and quickly sold them for tidy sums. Mills also profited from a fifty thousand dollar judgment against Simeon Hart, whom he blamed for his temporary incarceration in 1861. Mills also used his position to secure army supply contracts, including one for salt from the Guadalupe Lake deposits.[1]

A. J. Fountain was born simply Albert Jennings on Staten Island in October 1838. During a tour of Europe with four other students in 1857 or 1858, the young men deserted their tutor and boarded ships bound for Capetown, Calcutta, and eventually Canton. After the U.S. consul general rescued the five from a Chinese jail (they had boarded an opium cargo ship), they sailed to San Francisco. Assuming his mother's surname, Fountain prospected, freighted, clerked, and worked as a newspaper reporter. Fountain enlisted in the Union army in August 1861. He marched with the First California Infantry during the hot, choking passage across the Sonoran Desert, earning his First Sergeant's stripes. He distinguished himself in the desperate battle against Cochise for control of the springs at Apache Pass. When the column reached the Rio Grande, Fountain was initially posted to Fort Fillmore, outside Mesilla.

The young sergeant fell in love and on October 27 married Mariana Pérez. In May 1863 he became Lieutenant Fountain.[2]

Like hundreds of Californians, Fountain remained in New Mexico when his regiment mustered out in 1864. Before long he was back in uniform as captain of New Mexico Volunteers, commanding a company raised along the Rio Grande Valley. He survived his closest call of the war in June 1865. While tracking a large party of Navajo with a single scout, he was ambushed in a narrow lava-walled pass. Trapped beneath the weight of his dead horse, he was struck by a bullet and two arrows. Fountain spent a tense night holding off his attackers. Luckily, the scout escaped to bring a rescue column the next morning. Fountain's wounds were treated at Fort Bliss, though a fever nearly killed him in the heat of summer.[3]

It was many months before Fountain fully recovered. Nearly destitute, he allied himself with Mills. The ring leader secured several posts for his protégé, including district surveyor and chief assistant to the customs collector. Mills also saw to it that the penniless lawyer was retained to investigate the titles to the businesses and properties owned by the targeted secessionists. As trustee of the Butterfield Overland Mail buildings, Fountain moved his family onto the El Paso property and arranged quarters for the families of Gaylord Clarke, Mills's college chum, and James Price Hague.[4]

One of Fountain's first jobs as district surveyor was to "locate" the Guadalupe Salt Lakes for Texas land baron Samuel Maverick. Under Texas law, grant land certificates were not issued for any specific tract. It was the responsibility of the certificate holder to find available land from the public domain, have the desired vacant land surveyed, and locate the certificate located upon the surveyed land. The field notes from the survey would be filed with the Texas General Land Office. After any other conditions of the grant were met, the grantee, or his assignee, could apply for a patent (original title) on the land from the land office.[5]

In the fall of 1866, Maverick filed an application for survey. Fountain initially refused the job, "upon the ground that the saline deposits were not subject to location under the constitution and laws of Texas." He was then informed that a provision of the 1866 Constitution (Article VII, sec. 39) stated, "That the State of Texas hereby releases to the owner of the soil all mines and mineral substances." This clause overturned the Paseños' communal mining property rights grandfathered safely through two centuries and six flags.

Maverick's salt scheme was a failure. The certificate missed several hundred acres and Fountain's deputy bungled the survey, locating the certificate as much as twelve miles from the salt beds. According to Fountain, the people "ceased taking salt for a time, but recommenced it when it became generally known that Maverick's survey did not take in all of the lakes."[6]

The Custom House Ring now stepped in. In 1868, Mills, Fountain, French, Lujan, Clarke, Benjamin F. Williams, saloon owner Benjamin Dowell, and new District Surveyor Charles H. Comley formed a company to secure ownership of the lakes. They became known as the *Salt Ring*. At first all went well. Mills obtained the required certificate from Dowell, Comley surveyed the lakes, and the certificate was filed. When word got around that the ring was attempting to corral the salt, their scheme "aroused considerable bad feeling among the native population." The adverse reaction was so strong that the partners began quarrelling among themselves over whether to pursue the matter. Fountain claimed that from this point he "had nothing more to do with it." When it was discovered that the certificate was defective, "the whole matter fell through." The people of *la Isla* remained in possession of their salt.[7]

The quarrel irrevocably split the ring. Mills retained the support of French, Zabriskie, and Clarke. Fountain became the leader of what was known as *the Anti-Salt Ring*. With him went Williams and a new ally, fellow New Yorker William Pierson. The split reflected the widening rift between the Conservative and Radical wings of the Republican Party in Texas, which differed on economic issues and the treatment of ex-Confederates and their former slaves. The party's divisions became irreconcilable during the state's Constitutional Convention, which met at Austin from June 1868 to February 1869. Mills, El Paso's representative, supported A. J. Hamilton, leader of the Conservatives, or moderates, against the Radicals under Edmund J. Davis, president of the Convention. The bond between Mills and Hamilton was reinforced in early 1869, when Mills wed Hamilton's daughter, Mary.[8]

During Mills's prolonged absence from El Paso, Fountain assumed leadership of the local Republican Party. He jumped aboard the Radical bandwagon, putting himself in opposition to his mentor. The rupture between the Radical and Conservative factions led to separate nominating conventions for the upcoming fall elections. The Conservative nominees included Hamilton for governor and Mills for legislative representative from El Paso. The Radicals nominated Davis for governor and Fountain for state senator from El Paso. Mills sacked Fountain as deputy collector, and Fountain orchestrated Mills' removal from his customs post. Mills's replacement was Dr. Dwight C. Marsh, a thirty-five-year-old physician who had come to Texas with the federal occupying force.[9]

During his senate campaign, Fountain reminded the overwhelmingly

State Senator Albert Jennings Fountain. Robert G. McCubbin Collection

Paseño electorate that their free access to Guadalupe salt was still in jeopardy from Mills's "Salt Ring." He promised to protect their traditional community rights. With "get-out-the-vote" help from Father Antonio Borrajo, the parish priest of San Elizario and Socorro, Fountain won his election to the Texas state senate. Mills was defeated in his own election for legislative representative, though the vote totals were much disputed. Mills's supporter Judge French charged that the Fountain forces had "opened the box and scratched our one hundred and forty-three votes for themselves. F's name represents yours on the scratched tickets." Mills now dedicated himself to ruining Fountain.[10]

Fountain quickly assumed key leadership roles in the state senate. He chaired two committees of interest to his West Texas constituents, including Indian Affairs and Frontier Protection. In addition, the young legislator was elected senate majority leader. Shortly after the legislative session got underway, the senator submitted a bill transferring the state's title to the Guadalupe Salt Lakes to the County of El Paso "for the use of her citizens forever." The bill should have passed, but a petition from the citizens of El Paso came out against Fountain's bill! The source of Fountain's grief was the alliance of Louis Cardis and Padre Borrajo.

Antonio Severo Borrajo, born around 1820 in the mountainous District of Orense, in Spain's ancient northwest province of Galicia, grew up in a region marked by the strongest devotion to Catholic Christianity. Orense itself, the site of a fourteenth-century Gothic cathedral, was not far from Santiago de Compostela, believed to be the resting place of the remains of St. James the Apostle and thus the terminus of one of the great Christian pilgrimage routes. Borrajo's birth occurred only a few years after Napoleon's armies were driven out of Spain. It is more than likely that Antonio's family lost members or friends in the struggle and that he listened to his elders tell stories of French cruelty and godlessness and of Spanish courage against the occupiers. Borrajo's adult life suggests that the time and place of his youth had a profound effect on him.[11]

Borrajo received an excellent education, earning a doctorate in theology. His letters to religious superiors are forceful, well reasoned (from his point of view), and models of clarity, the latter quality stemming from his complete confidence in the truth of his own views. Ordained in 1849, the Spaniard was recruited to serve the Church in Texas. By 1851, he was at the Pass of the North, where he worked as an assistant to the parish priest at Ysleta and Socorro, then still within the bishopric of Durango, Mexico. Borrajo was shuttled about the diocese before being permanently assigned to the Pass in 1863. It was at this time that Father Ramón Ortiz, parish priest at Paso del Norte, appointed Borrajo pastor at San Elizario and Socorro.[12]

Borrajo was unflinching in opposition to any weakening of traditional Roman Catholic precepts and practices by other local clerics seeking to accommodate to a changing world. An "uncompromising dogmatist," he was determined to thwart encroaching American influence and power at every turn. When Mormons began proselytizing among the Paseños, Borrajo identified their emissaries as a plague "that destroyed the very soul." The Spaniard's single mindedness ultimately contributed to one of the bloodiest civil disturbances in the long, difficult history of the border.[13]

More than a half century after the Salt War, Father Borrajo was described to historian C. L. Sonnichsen by several Paseños who remembered him, though they seem to recollect a man in extreme old age, rather than in his vigorous fifties: "In person he was a tall, slender old man with bent shoulders, long gray hair, and black blazing eyes set in a thin white face. Temperamentally he resembled a volcano—was always sure he was right, and was always determined to have his own way."[14]

If Borrajo had a weakness, it was horses. He reportedly owned a set of "well matched and high-spirited geldings," which pulled his buggy across his parishes at a good speed. Such horses did not come cheap, but they may have been a gift from wealthy benefactors. If Borrajo had two weaknesses, the second, according to his Anglo enemies, was money. Shortly after the 1869 election, the senator-elect was allegedly approached by Padre Borrajo with a very curious suggestion. The rector of San Elizario asked him what he proposed to do about the salt lakes. Fountain replied that he felt bound to fulfill his campaign promise to secure the lakes for his constituents. Borrajo then suggested that "it would be better if the lakes fell into private hands," specifically his own and Fountain's. The senator-elect, said Borrajo, should locate the lakes in his own name, and impose a small fee for the salt, which the priest would advise his parishioners to pay. The two would split the proceeds, and each would "derive a snug revenue." Fountain rejected the scheme out of hand. Enraged, Borrajo threatened Fountain, who ordered him out of his house. Later, Borrajo approached Fountain through an intermediary, Judge French. The second proposal was similar to the first, only this time a third partner would share in the proceeds. Fountain rejected the new proposal.[15]

The third man in Borrajo's equation was the priest's ally, Louis Cardis. He was described by a contemporary as a "tall, sinewy Italian, hair black and tinged with gray." Hailing from Piedmont, Italy, forty-four-year-old Cardis was said to have served as a captain in Giuseppe Garibaldi's army before immigrating to America in 1854. Ten years later, he made his way to El Paso, where, in December 1865, he took advantage of the carpetbag scheme to buy and resell for profit ex-Confederates's confiscated property. Cardis also made a living as a retail liquor merchant, apparently dealing from Simeon Hart's old home.[16]

Like other El Paso politicians, Cardis was prone to violence, usually with a gun rather than his fists. On at least three occasions he was charged in the El Paso courts with assault with intent to murder. The first of these occurred in 1869, when he was alleged to have shot a man in the left side and used the pistol to beat him on the face and head. The surviving records are incomplete, but it is doubtful that Cardis was convicted on anything more than a simple assault, if that. Assaults were common in El Paso, convictions were rare, and the penalty could be as little as five dollars.[17]

Since politics was the key to advancement in El Paso, in 1869 he became a naturalized citizen of the United States. His command of Spanish gave him a powerful advantage as the would-be leader of El Paso's voters. By every account, Cardis gained the complete trust of the ordinary farmers and laborers along the Rio Grande Valley. It was said that the people had "unbounded confidence" in him. To those who despised him, he was a shrewd manipulator, who deceived his constituents as if they were simple children. Now, Cardis and Borrajo determined to capsize any bill Senator Fountain might submit to reserve the lakes for the people.[18]

According to Fountain, he "called a meeting of the people at San Elizario, and offered to present them with a certificate for 320 acres of land, conditioned that they would pay the expense of having it located on the unlocated portion of the Guadalupe salt lakes . . . , and that the location should be for the use and benefit of the citizens of the county of El Paso." The people would select a board to hold the patent in trust for the citizenry. "Borrajo and Cardis exerted their influence successfully to have this proposition rejected." Undeterred, Fountain then met with leading citizens of San Elizario to discuss an alternative formulation. Fountain, whose command of Spanish enabled him to speak directly to the assembly, read aloud his proposed resolution. It provided for a transfer of title to the county "for the use of her citizens forever, all right, title, and interest . . . in and to the unlocated portion of the Guadaloupe salt lakes." The assembled group expressed their approval.[19]

Satisfied that this support would successfully counter the influence of Cardis and Borrajo, Fountain introduced his joint resolution in the senate. A few days later, Senator Marmion Henry Bowers from Travis County handed Fountain a petition signed by Cardis and over four hundred other citizens of

El Paso County, protesting the passage of any bill or resolution affecting the salt lakes. Fountain withdraw his resolution.[20]

According to Fountain's account, he acted in the interests of the Paseños who, like naïve children, blithely follow trusted leaders into acting against their interests. Another reading of his proposal is that transfer of title to the county government would have allowed Anglos controlling that government to dispose of the lakes as they wished. Fountain would soon pull a legislative maneuver to dispossess the Tigua Indians of their communally owned land in Ysleta. In the opinion of San Elizario historian Rick Hendricks, "Borrajo and the other Hispanic community leaders were clearly suspicious of Fountain and anything that smacked of a title to commonly held or used land."[21]

As senate majority leader, Fountain was more successful in securing the passage of much of Governor Davis's legislative priorities, including three measures designed to address the pandemic violence that plagued Reconstruction Texas: the reestablishment of a ranger force and a state militia, and the creation of a state police. El Paso County was authorized by the governor to raise a Frontier Force (ranger) company of sixty men and was allotted a state police detachment.[22]

A legislative recess in the second half of 1870 permitted the senator to do some political fence mending back home. He doled out patronage jobs to former political enemies like Gaylord Clarke, appointed district judge, and Albert French, who became lieutenant of state police.[23]

French "desired to effect a general reconciliation, to adjust and settle all past differences, and reorganize the Republican Party in the county, under [Fountain's] leadership." French advised Fountain that Cardis also desired to reconcile. The three men met to determine the division of spoils. Cardis announced that he also represented Father Borrajo, who would ratify any decision agreed to by Cardis. Fountain asked how Cardis and Borrajo "had succeeded in inducing the people to sign the 'protest' against their own interest." Cardis, Fountain claimed, replied that he and the priest had warned the people that Fountain's purpose was to "trick" the people out of ownership of the lakes.[24]

Cardis then offered his plan for party reorganization: Fountain would engineer the removal of D. C. Marsh as customs collector and his replacement by French. Cardis would be made mounted inspector of customs. In exchange, Fountain would lead a united party. Father Borrajo's interests remained. The priest "insisted upon having the entire control of the public schools including the selection of teachers." And there was still the matter of the salt lakes. Cardis said that Borrajo had "salt lakes on the brain." The priest allegedly wanted Fountain to acquire the lakes in his own name, charge a fee for salt taken, and divide the proceeds with Borrajo. The priest would advise the people to submit to any fee Fountain charged, confident that his parishioners would do so.[25]

Why, Fountain asked, didn't Cardis or Borrajo locate the lakes in their own name? No, such an action "would completely destroy their influence with the people." What security would Cardis and Borrajo have that Fountain would actually split the proceeds? Cardis admonished him, "We know that you would not do anything of the kind, because it would cost you your life, and you know it." Cardis emphasized that "Borrajo was determined to have the lakes, and that no other person should enjoy their possession adversely to him." Fountain would not agree to any of these propositions, and the meeting broke up. "Cardis, Borrajo, and their faction at once commenced a bitter warfare upon me, which lasted as long as I remained in the county."[26]

This account is based solely on the one-sided memory of A. J. Fountain. Previous and subsequent events lend some credence to the story. Fountain himself does not appear the saint, but rather a pragmatic politician, refraining from robbing his constituents on a grand scale, sticking instead to divvying up government spoils in the sanctioned fashion. The most astonishing allegation is that of Father Borrajo's greed. El Paso's Anglos generally considered Borrajo a malevolent character, but some simply detested the Catholic religion. The question is whether they demonized the priest for his activities in matters most Americans considered strictly secular.

Fountain, who had married into a *Nuevo Mexicano* family, and who mixed easily with this population, cannot be accused of deep-rooted racial prejudice. He did passionately oppose the idea of priests teaching in the public schools, and in this regard he may have shared the Anglos's special antipathy felt toward Padre Borrajo. But if the senator falsely accused the priest of avaricious motives, then he may have lied about Borrajo's actions, or Cardis's, or his own. Possibly it was Cardis who lied about Borrajo's personal designs.[27]

The answer may lie in a communication dated June 3, 1866, from the *Jefe Politico* of Paso del Norte to Sebastián Lerdo de Tejada, then minister of foreign relations in Benito Juarez's administration. "The foreigner Antonio Borrajo [i.e., a Spaniard by birth], so-called Catholic priest, came to our village from the other side of the Río Grande, in United States territory." During a mass celebrated in Paso del Norte's Guadalupe Mission, Borrajo attacked the Mexican government for authorizing religious tolerance. The police apprehended Borrajo the same day, along with a signed copy of his sermon. This document, reported the *jefe politico,* "shows the undisguised ignorance of

Borrajo, not only of civic laws but also of religious principles. What one can discover in his sermon—amongst the falsehoods and ignorance—is the resentment that the foreigner harbors against government laws that give liberty to all to celebrate religious activities." Borrajo's true motivation for this assault on religious freedom, the *jefe* charged, was the "less than evangelical" greed exhibited by some priests. "[The] foreigner Borrajo" was only attempting to avoid "the danger of administering the sacraments free of charge, like Jesus and the Apostles always administered them." Allegedly, the priest demanded payment in advance of administering any sacrament.[28]

Borrajo himself attacked *norte Americanos* and Europeans who "only come to live among the Mexicans for the love of gold and riches. They only believe in the Almighty Dollar." But here, in the words of a Mexican government official, we find a second accusation of Borrajo's "greed." There is no question of anti-Mexicanism here. It must be noted that the *jefe* was an agent of Benito Juarez's government, which in 1866 was still at war with the French occupiers and thus greatly in need of the people's loyalty. The accusation of Borrajo's greed may have been accurate, but it was also a common charge laid against Mexico's clerics. The Catholic Church had, for centuries, been wealthier than the Mexican government. The Church's wealth allowed the institution to serve the masses as banker and distributor of largesse. Juarez wanted to break the Church's stranglehold through the provisions of the Constitution of 1857 disestablishing and disendowing that institution. As for Borrajo's motives, he was unyielding in his opposition to civil "intrusion" into any traditionally religious sphere. Other than an alleged fondness for high-stepping horses, there is no evidence at all that he preferred to live high. The priest's alleged deep interest in the salt lake revenue may have stemmed only from a desire to control funds he could dispense in a way that would restore and cement the traditional authority of the Church.[29]

Whatever his motives, Borrajo's sermon at Guadalupe Mission opposed the authority of the Juarez government. As a "foreigner" at a time when the Mexican government was fighting for its life against foreign occupation, the priest was an easy target. President Juarez ordered him expelled from Mexico. Juarez eventually lifted the penalty, and in 1869, Borrajo was permitted back in Mexico. The padre was not through causing trouble in the valley, however. His religious zealotry would wreak havoc on El Paso County for years to come.[30]

CHAPTER 4

"What chance is there for reform?"

The fatal shootings of two politicians and the attempted murder of a third marked the opening salvos in El Paso's era of politics by assassination. The bloody morning served notice that the rewards of political success needed to be weighed carefully against the risks. The trouble began with Fountain's failure to find a patronage job for Frank Williams, a staunch supporter in the senator's feud with Mills. Left out of the spoils, Williams took to publicly and drunkenly berating Fountain and Judge Gaylord Clarke. The lid blew off when Williams and Clarke exchanged words on December 7, 1870. Williams pulled a revolver but fled when Clarke stared him down. Later that morning, in Ben Dowell's saloon, Williams drew his pistol on Fountain. The senator lashed out with his cane as Williams got off five shots at point blank range. Fountain was struck in the scalp and left arm, but his pocket watch and a bundle of letters in his breast pocket broke the path of the fifth bullet, turning it away from his heart down to a rib. Williams fled.[1]

A posse cornered Williams in a nearby rooming house. They grabbed axes and shredded the front door while Police Lieutenant Albert French moved around to the back. His fate sealed, Williams burst outside holding a shotgun. Most of the posse took cover, but Williams fixed his aim on Judge Clark. "Don't you dare, Williams! Don't you dare!" were the jurist's last words as Williams blasted him. At the explosion, French rushed back to the front of the house. Finding Williams "running *amuck*," French fired his revolver, hitting Williams in the right side. French then grabbed hold of Clarke's long hair and struck him two or three times. The enraged lawman responded to policeman John Hale's calls to desist, but then "stuck his pistol in [Williams's] face, shooting him in the eye, the ball coming out the back of his head."[2]

Fountain told a different story. After the incident in Dowell's saloon, his wife Mariana dressed his wounds. He then grabbed a rifle and headed for Williams's rooming house. Fifty yards from the scene where Clarke was killed, he fired his rifle, striking Williams in the chest. Although most reports of the shootout do not include Fountain, his enemies spread the story that he had murdered his opponent. "Spurious tales about Williams' death dogged the Senator for the rest of his public life."[3]

When French resigned his lieutenancy in January 1872, he was succeeded by his unpleasant sergeant, John G. Atkinson, another California veteran firmly in Fountain's camp. Atkinson could rely upon policeman Pablo Mejia, a trusted friend. Between them, Atkinson and Mejia were perhaps the squad's nerviest officers. One or the other was involved in each of the detachment's ten successful arrests made of murder and attempted murder suspects.[4]

Atkinson's attention was not limited to criminal targets. In 1871, he arrested W. W. Mills and his brother Ethan for carrying arms in violation of state law, even though the law permitted such possession while traveling. Maximo Aranda, the justice of the peace at San Elizario, ordered the prisoners' release. That evening, Mills attended a dance, as did District Judge S. B. Newcomb, District Atty. James P. Hague, and Sheriff Charles Ellis, "all enemies" of Mills. At the dance, said Mills, "Atkinson stood before me, one hand on his six-shooter and the other in my face, and called me many pet names, the mildest of which were 'coward' and 'liar,' and threatened to shoot me if I spoke or moved." Mills kept his mouth shut until Atkinson strode away. The next day the grand jury, packed with Mills partisans, indicted Atkinson for aggravated assault. Hague refused to prosecute and, when Zabriskie stepped in to do it, Judge Newman lowered the charge to simple assault.[5]

In 1871, Radical Republicans lost control of both houses of the Texas legislature to a coalition of Conservative Republicans and Democrats. It spelled the beginning of the end to A. J. Fountain's senatorial career. Following a practice as old as the Roman Republic, Fountain's enemies, "Led by a grudge-bearing W. W. Mills," secured eighteen felony indictments against him. In November 1871, the senator was arrested at Franklin by Deputy U.S. Marshal William Treyer and troops from Fort Bliss. Austin's *Tri-Weekly Statesman*, a Democratic Party organ, gloated that Fountain was wanted "dead or alive," for "such little charges as swindling, embezzlement, fraud and forgery; personating Collectors and Assessors, signing their names, swearing to documents, and disposing of pay-rolls at half price to Dutch saloon-keepers." Comparing Fountain's corruption to New York's Tammany Ring, the *Statesman* asked, "What chance is there for reform in a party which through its central organ thus teaches disobedience and defiance to the laws?"[6]

The criminal case against Fountain went to trial in early 1872. Mills and Cardis were key witnesses for the prosecution, but other El Pasoans, including merchants Inocente Ochoa and Joseph and Samuel Schutz, had to be arrested and compelled to testify in Austin. In the end, the senator was acquit-

State Representative Louis Cardis. Courtesy of University of Texas at El Paso Library, Special Collections Department, C. L. Sonnichsen Papers, MS 141

ted of every charge. In Galveston, *Flake's Daily Bulletin,* initially convinced of Fountain's guilt, commented, "The trial of these cases shows that . . . Senator Fountain did not forge anyone's name, did not personate any other person, did not illegally collect a dime, and did not neglect his business either criminally or carelessly. . . . Senator Fountain has been the best abused man in Texas for the past six months. . . ."[7]

On April 1, 1872, Cardis, by now Fountain's chief nemesis, was brutally assaulted as he traveled home from Austin. The event provided the Liberal Republican boss the opportunity to tar his radical rival with an attempted assassination. Cardis was at the Head of Concho stagecoach stop when "a ruffian came into the room where I was, and, without saying a word, and from behind my back struck with a club a terrible blow on my head, and I fell senseless to the ground." The assailant was Sidney Elsworth, a stage line employee and, according to Cardis, Fountain's "emissary." Elsworth was arrested at Fort Concho, but the army let him go, citing lack of authority to keep civil prisoners. Cardis alleged that Fountain improbably met Elsworth at Head of Concho. Fountain was supposedly "overheard by Mr. — and another gentleman" to reassure Elsworth, "Never mind, I will clear you of all this, and also the indictment pending against you at El Paso." The virulently antiradical *Austin Daily Statesman* declared the "Attempted Assassination" to be the work of Fountain. His motive, said the *Statesman,* was revenge. Cardis had testified against Fountain "in the face of threats of violence and Radical official powers. [Now] he has paid the penalty of this temerity."[8]

After recovering from "pains . . . not much less than death," Cardis continued his journey to El Paso County. Soon after arriving, he was arrested as one of "certain evil disposed persons" who had conspired in 1870 to intimidate Dwight C. Marsh into quitting his post as collector of customs at El Paso. Fountain brought the federal charge before a crony, the Reverend Joseph W. Tays, an Episcopalian minister who doubled as U.S. commissioner for the Western District of Texas. No witnesses against Mills came forward, and the case was eventually dropped. The *Austin Daily Statesman* attacked the "Radical Deviltry" and "reign of terror" practiced by Fountain and "Parson Tays . . . a canting, praying hypocrite who steals the livery of heaven to serve Fountain in."[9]

In a letter to the *Statesman,* Cardis charged that "Senator A. J. Fountain has been planning and plotting my ruin and my death for the last three years, by all ways and means, and has used every tool, good and bad, clean and dirty, to accomplish this, and all because I have opposed his dirty politics, which I intend to do as long as I live, in a plain and open manner, as I have done heretofore." Fountain could manipulate the courts; Cardis would rely on the electorate. "Echo answers 'November,'" he warned.[10]

"Parson" Tays, as he was known, was a native of Nova Scotia trained as both minister and civil engineer. Although Fountain had secured his services as the county's first Protestant minister, he had few co-religionists to minister to, enabling Fountain to keep County Surveyor Tays otherwise occupied. The previous legislative session saw the enactment of three laws incorporating the cities of San Elizario, Socorro, and Ysleta. Each new town had the power to grant and sell real estate within the town limits. Land buyers were restricted to town citizens, future town citizens, and businesses erecting buildings. The Tigua Indians who previously owned the land in Ysleta were not citizens and could not become citizens. They could not vote in any incorporation referendum, nor could they enforce their rights in court! Tays surveyed the land and benefited from the scheme.[11]

Many Paseño families also took advantage of the opportunity to homestead the land. Two deeds, for parcels of sixteen and seventeen acres, were conveyed in August 1873 to Francisco and Miguela Barela, in consideration of their "desire to cultivate, use and occupy" the tracts. For Barela, who had at least two children and two domestic servants to feed, the farmland was not just a godsend, but a measure of his personal stake in the future of *la Isla*. When the Supreme Court ruled that Fountain law was unconstitutional, but "the damage had been done; the Tigua land was gone." The Tigua scheme illustrates why Paseños may have been reluctant to repose ownership of the Guadalupe Lakes in a county government controlled by Fountain. Such concerns evidently were of no concern to the Parson, who was soon joined by three brothers. Two played leading roles in the Salt War and its aftermath.[12]

While the Fountain-Cardis drama played itself out in the courts and newspaper columns, a different sort of conflict arose in El Paso County. Details are sketchy, but a popular uprising came within a whisker of ripping apart the county five years before the insurgency of 1877. Its root causes were clashing religious and cultural values rather than ambition and greed. The trigger was the law passed by the 12th legislature instituting statewide compulsory education of minors. The measure faced stiff opposition at the state level because of its cost and its provision for the education of black children.

Father Borrajo had a different concern, the lack of proper religious instruction, a subject dear to his heart. Padre Borrajo previously had insisted "upon having the entire control of the public schools [and] the selection of teachers." Fountain had refused on both points, but the issues remained. In January 1872, Borrajo warned Bishop Jose Vicente Salinas in Durango that the county had "hired the most corrupt males and females as teachers, the worst of which we have here." Borrajo complained that children were prohibited from bringing religious books to the "atheist" schools, which "propose to cultivate only knowledge, leaving the will to the mercy of the world, the devil and the flesh."[13]

When the school law took effect in 1872, the two men chiefly responsible for enforcing it, both California Column veterans, were simultaneously the county's school commissioners and its top lawmen. A third was both teacher and deputy sheriff. All three would play notable roles in the El Paso Salt War. All would pay for their opposition to Borrajo, two with their lives.[14]

Sheriff Charles E. Ellis had held office since 1869. He settled after the war in San Elizario, where he became a general merchant. In 1870, he wed Teodora Alarcon, described as "large and stately and gracious, like her house." The Ellis family lived in the old Casa Ronquillo, a home befitting the status that Charles Ellis had acquired in El Paso. According to the 1870 Census, the thirty-five-year-old merchant and twenty-eight-year-old Teodora were the wealthiest residents in San Elizario, with real and personal property worth fourteen thousand dollars. Their household included seven employees but no children. His businesses included a general merchandise store and a grist mill. Ellis was an affable man, but "too good hearted to be strict in enforcing his duties" as sheriff, county assessor, and tax collector.[15]

Ellis's state police counterpart, Lieutenant Atkinson, also doubled as a deputy sheriff. "Atkinson was very often too harsh," said El Paso attorney Edmund Stine, "and the odium which had but justly often attached to Atkinson was always shared by Ellis, because they not only belonged to the same party, but were bosom friends, and so intimate were their relations at that time that many looked upon them as brothers." It was said he was "a man of bright intellect, a good English, German and Latin scholar, wholly self-educated" after leaving school around the age of twelve. He apparently studied law before the Civil War broke out. He was either "long" or "small," but in either case lean and wiry. He sometimes used a cane as a result of being shot in the hip during action against Indians late in the war.[16]

El Paso historian C. L. Sonnichsen described Atkinson as "the village Mephistopheles . . . quarrelsome and dictatorial." He inflicted his peppery nature on a series of Mexicano women, possibly contributing to his selection for execution in the Salt War. The first of these was Jacoba Alvorado. The two wed on June 10, 1863, before his discharge, in Paraje, New Mexico. The couple settled in San Elizario, where he worked as a deputy customs collector. In December 1868 he sued for divorce, alleging that Jacoba had committed

Sheriff Charles Ellis. Courtesy of El Paso County Sheriff's Office

adultery with "diverse other persons." In response, Jacoba stated that she met Atkinson, "a young, handsome, intelligent soldier . . . utterly destitute of property or money except his monthly pay." Jacoba claimed it was Atkinson who was faithless with "divers persons." The jurors found for Jacoba. To add to Atkinson's woes, he was indicted by the grand jury for suborning the perjury of a witness in the divorce case. These troubles were only a bump along Atkinson's road to success. He set up shop as a merchant in San Elizario, building a business second only to that of Ellis. After February 1871, he also served as the town's postmaster.[17]

Ellis and Atkinson had come up against Father Borrajo once before. In June 1871 they had "procured his indictment" for burying the dead in the San Elizario church cemetery after it was determined this was polluting the water. Borrajo was acquitted by Judge Newcomb, but the clash "may have left rancor" among the pastor and his flock.[18]

The third lawman was born Friederick Sperfechter in 1835 in Wurttemberg, Germany. In May 1855, after emigrating to America, he enlisted in the 3rd Infantry Regiment and was posted to Fort Fillmore, just south of Mesilla. Sperfechter rose to sergeant, was busted back to corporal, and again put on sergeant's stripes before deserting on August 22, 1860. In June 1861, after resurfacing in Denver, Sperfechter enlisted as Charles Kerber in the 1st Colorado Volunteer Infantry Regiment. Promoted to 1st Lieutenant, Kerber soon took command of Company I. The twenty-seven-year-old was roughly five foot, eleven inches tall, with gray eyes, brown hair, a ruddy complexion, and a propensity for taking the bit between his teeth.[19]

Kerber's second military career, like his first, might have come to a premature end. In February 1862, during the march in freezing weather from Denver to New Mexico, Kerber led his company in mutiny, refusing to march further until they received another wagon to carry equipment and supplies. A "violent argument ensued" between Kerber and the regimental commander, Colonel Slough. When Slough pointed his revolver at Kerber, Company I "instantly leveled their rifles on Slough as a voice from the ranks warned, 'You shoot Kerber, and we'll put sixty holes through you.'" At that point, Maj. John Chivington arrived to calm down the men. No action was taken against Kerber. Fifteen years later, as sheriff of El Paso County, he would have different ideas about the respect owed to authority figures. Kerber fought well in the Union victory at Glorieta Pass, where his company battled aggressively on offense and stubbornly on defense. It was "the best stand the Federals made all day," costing the ninety-man company dearly (twenty killed, fifteen wounded, and five taken prisoner).[20]

In June 1862, Kerber's regiment was turned into cavalry. Captain Kerber returned with the regiment to Colorado to protect settlers against Indians. On November 22, 1864, he was wounded in the left forearm during an Indian fight at Zapato Creek. After mustering out in November 1865, many veterans in his company settled in what is now Saguache County, Colorado, at a place still called "Kerber Creek."[21]

In November 1869, Charles Kerber returned to El Paso, a place Sergeant Sperfechter once knew well. He settled in Ysleta, purchasing a small homestead and looking for opportunity. By 1872 he was a deputy sheriff, company clerk in Captain Garcia's Texas Ranger company, and a county school teacher. The German immigrant accepted a task to "try the stoutest heart, for many of the pupils were full grown men, some with beards, who made it a point to go after the teacher's 'nanny.'"[22]

Now educator-lawmen Ellis, Atkinson, and Kerber came into opposition with the people. Borrajo was certainly an instigator. The parish priest of San Elizario "hated everything American, and particularly hated the compulsory school law [which] prohibited the use of any and all religious books. This latter clause so incensed Padre Borrajo that he forbid his parishioners to send their children to school. . . . [H]is will was law, and the children were not sent."[23]

Paseños did not need a priest to tell them something was wrong with the Anglo-mandated schooling. In addition to the lack of religious instruction, they objected to sending their teenage daughters to school in small school rooms with older boys and "full grown men with beards." One "Mexican gentleman" explained another problem to the Las Cruces *Borderer:* "Their children are taught in a language of which they understand not a word and consequently make no progress except in wearing out their clothes, and when unable to purchase a new suit, the parents keep them at home, they are brought up and fined. If the fine is not promptly paid, the offending parents are sent to jail." The *Borderer* aptly commented, "This is a school system with a vengeance."[24]

The first to commit civil disobedience was one of San Elizario's foremost citizens, Telesforo Montes. Standing five feet, eight inches tall, with black eyes to match his hair, Montes was somewhere between forty-five and fifty-five years old, depending on the record. He and his wife, Maria Quirina Alderete, raised twelve children in San Elizario. Three married three of Gregorio Garcia's offspring. Like Garcia, Montes was farmer, warrior, and civic leader. From 1870 to 1871, he had served Texas and his community as lieutenant and second in command of Garcia's Rangers.[25]

Now, in the spring of 1872, Montes refused to send a teenage daughter to the county school. He and at least two other men were thrown in jail for not sending their sons. "An old blind fiddler of San Elizario," according to the *Borderer,* "whom we have often seen and who was led about by his little son, is now in the county jail because he does not send his child to school, where the child does not understand a word that is said to him."[26]

In jail "they enjoyed martyrdom for three weeks." The arrests brought the county to a crisis point during the first week of June. Four years after the event, Kerber alleged that, "Cardis in company with a fanatic catholic priest had organized a company to drive all Americans, not his friends, from the Country or to kill them." At the moment of crisis, Atkinson, wearing his state police hat, informed the state adjutant general that "the whole of the police in [my] charge have been and are busily engaged in suppressing an armed organization, who armed themselves for the purpose of resisting the collection of the school tax by force." Unfortunately, these sketchy reports are all that remain in the Texas State Archives.[27]

No other evidence has come to light explaining exactly how many Paseños armed themselves, who led them, or how the state police went about trying to keep a lid on an explosive situation; but whatever tactics they employed to enforce the school law, Sheriff Ellis and Lieutenant Atkinson "made themselves very obnoxious at that time to a large part of the Mexican population." The situation was finally resolved when Sheriff Ellis "turned them loose to save worse trouble." According to Edmund Stine, "ultimately the priest triumphed and the school law became a dead letter in El Paso County." The school law also became a dead letter throughout Texas when it was repealed after the Conservative Republican-Democratic coalition took control of the 13th legislature in 1872.[28]

If the popular insurgency against the retreating Confederates in 1862 was but a dim memory ten years later, the arming of the people in 1872 provided discerning men with fresh evidence that this population would not meekly recoil when its perceived rights were threatened. The memories and hard feelings of the school crisis carried forward on both sides through the years.

The battle over schools was the last waged by Father Borrajo as the parish priest of San Elizario. It was reported that his victory in this fight prompted the *Anglos* to secure his removal. "Soon afterwards a new diocese was formed by decree from Rome, including El Paso County, and when the new Bishop arrived, every American in San Elizario used his influence to have Padre Borrajo removed, and his most active enemies were Ellis and Atkinson, which Borrajo knew well." While the priest may have made himself especially *persona non grata* over the school issue, his removal was for administrative reasons and had been in the works since 1869.[29]

It was then that Jean Baptiste Salpointe was named vicar apostolic of New Mexico, an ecclesiastical territory that stretched from Arizona to El Paso County. Salpointe, a Frenchman, recruited newly ordained priests in France to minister to the faithful in the Arizona vicariate. But when Salpointe notified Bishop Salinas of Durango that the parishes on the American side of the Rio Grande were now in his jurisdiction, Salinas refused to deliver them without a direct order from Rome. Borrajo and the priests of Mesilla, Franklin, and Ysleta vigorously opposed turning their flocks over to the French. According to the Spaniard, the substitution of "Franco-American" rituals was leading to "apostasy" among the faithful. "In the shadow of the usurping government, they come as a devastating flood, erasing the footprints of the first apostles of this new world."[30]

By November 1872, Salpointe had had enough. He announced his intention to assume jurisdiction over the American parishes. Borrajo and two other padres pleaded that Bishop Salinas of Durango had ordered them not to give up their parishes. "If it is now permissible for us to resist for lack of notice from our prelate, his priests and the persons who follow them would have no reason to oppose the ministry of the Devil, who against your excellency's will tries to be their true pastor."[31]

It was Bishop Salinas, and not Salpointe, who at last backed down. In December 1872, the American parishes were delivered to Salpointe's representative. Borrajo still refused to budge from San Elizario. It may have been at this point that Salpointe made the journey to the lower El Paso valley. According to reports, Borrajo "tried to incite the people against the bishop and all the priests under him, and called them Protestants, *pelados,* and thieves who had come to rob him of the parish." Salpointe had been warned that it was dangerous for him to travel, but he "came nevertheless, and had to camp out, because the people were afraid to offer him hospitality." Borrajo still refused to budge, but Salpointe prevailed upon Salinas to expressly order the obstructionist priest to leave San Elizario. On May 23, 1873, Antonio Borrajo conducted his last marriage ceremony at the Chapel of San Elizario. By mid-June, another priest had taken his place.[32]

Still another pastor arrived in 1876. Peter Bourgade, born in 1845 in Vollore-Ville, France, had been recruited for American missionary work by Salpointe in 1869. After several years in Yuma, Arizona Territory, he was shifted to the parish of San Elizario. For forty years in the American Southwest, Bourgade proved that he knew how to work with—not against—both ecclesiastic and secular authorities. He was no mere politician in priestly garb. Father Bourgade left behind a reputation as a "laborious, patient, humble, and special friend of the poor and lowly, zealous for souls." His career was unmarked by the accusations of greed that followed Borrajo.[33]

According to Bourgade, Father Borrajo refused to accept his defeat. "I can also say from the time he left, even until now [i.e., 1878], he has used his influence to estrange the people of my parish from me, and the priests who preceded me as well. He tried to make the people believe that he would come back here again; but why he wanted to do this I do not know." In his efforts to undercut Bishop Salpointe and other priests, said Bourgade, Padre Borrajo "worked in perfect accord with Louis Cardis."[34]

CHAPTER 5

"Relieve us from the insolent oppression of those men."

Charles Henry Howard "considered it his duty to convert El Paso County to the true faith, or Democratic party." He was born on February 3, 1842, in Virginia, the son of William Henry, a physician, and Sarah Catherine (DuVal) Howard. The family moved to Texas in the 1850s, eventually settling in Lavaca County. The family's prominence is indicated by the wartime and postwar associations of several Howard children with eminent Texans.[1]

Charles, the third of eleven children, was one of five brothers who fought for the Confederacy. The eldest, Thomas Calthorpe Howard, passed through El Paso on his way to and from New Mexico while serving as a lieutenant on the staff of Col. William Steele. Charles served throughout the war in the 8th Texas Cavalry, known as "Terry's Texas Rangers." A San Antonio acquaintance recalled that he "was a natural born warrior, and his achievements during the dark days of the Confederacy border on the miraculous."[2] The tales of Howard's Civil War daring reveal both a devil-may-care attitude and ruthlessness towards foes. One such story told of Howard's capture by a gang of "plunderers and murderers," who operated between the lines: "They condemned him to die at once, and offered him a drink of whisky [*sic*], telling him it would be the last he would ever take on earth. He didn't want the whisky, because he didn't drink, but being forced into drinking it, he took a mouthful of it. He spit it out and told them no wonder they killed and plundered people after drinking such whisky as that. They then made him dance, and he did it so well and told so many funny stories that they finally got into excellent humor, and being careless on their watch over him, he escaped. In less than three months, he had killed off half of the weasels."[3]

After the war, Howard struck out on his own. He studied law in San Antonio under Thomas J. Devine (1820–90), one of the most prominent and influential lawyers and jurists in Texas. Howard "made quite a reputation at San Antonio, holding his own with much older and more experienced lawyers," and ultimately became a partner in the firm of Devine & Simpson. In

Judge Charles H. Howard. Courtesy of University of Texas at El Paso Library, Special Collections Department, C. L. Sonnichsen Papers, MS 141

the 1870s, Howard was "stout and portly, of dark complexion, his hair being straight and black like that of an Indian, his brow like that of Mar's [*sic*], to threaten and command. . . . Although very stout he was remarkably active, and was literally a dead shot." Even a decade after the war, Howard was allegedly "able to gallop past a tree, and to unload his six-shooter into the tree-trunk from a distance of fifty feet, and within a space that could be covered with the palm" of a boy's hand.[4]

Howard decided to make his mark in El Paso. The task of converting the county to "the true faith" should have seemed daunting to one who could not directly communicate with the voters. "[N]either . . . was he acquainted with the Mexican character sufficiently to become their political leader." Howard found his answer in Louis Cardis, a man who "spoke Spanish like a native, understood the Mexicans thoroughly, and who already enjoyed a popularity and influence among them greater than their spiritual advisers." The two men became political partners soon after Howard arrived in the county. "Cardis entered into the alliance expecting to use Howard, who was a good and fluent speaker, which Cardis was not. He was a brave man, or a bully, while Cardis was a coward. Both were alike ambitious, and alike unscrupulous. They worked together very harmoniously for a while, and profited by their alliance." Although prior accounts record that they fell out around 1875, evidence from the 1872 election campaign reveals there was trouble between them almost from the start.[5]

On the surface at least, their alliance was strong. According to Howard, "For months preceding the [1872] election, unassisted and alone, with an energy and perseverance absolutely astonishing, Mr. Cardis worked day and night, to rouse up the people to make an effort to escape from Radical misrule." Of course Cardis did have help and not just from Howard. Allies included George W. Wahl and George Kolhaus, and Paseños like Gregorio and Vidal Garcia, Jesus Cobos, and Telesforo Montes. The pair could also count on the cooperation of that inveterate Fountain hater, W. W. Mills, who threw in his support from Austin.[6]

The first opportunity to directly assault Fountain's power would come that November. Given the rampant fraud of prior elections, it was evident to all parties that the casting and counting of ballots would go more smoothly for whoever controlled the local lawmen. The senator had the advantage—both Sheriff Ellis and Police Lieutenant Atkinson were his men. The former was up for reelection but would hold office and control the deputies through the campaign. The state police, however, might be up for grabs.[7]

On October 2, with the election only five weeks away, Atkinson requested twenty additional policemen to keep the peace during the first ten days of November. He explained that Cardis and Howard were "assisted by a fanatic priest named Antonio S. Borajo [*sic*], who intends to take possession of the Polls in El Paso County by force." Atkinson's request was seconded by Fountain and by Allen Blacker, the "Acting Chairman of the Democratic Committee," a group reluctant to accept newcomer Howard as their party leader.[8]

Cardis and Howard entered the fray in mid-October with letters to Adjutant General Davidson, calling for Atkinson's removal and replacement by his former superior, Albert French. Howard's note accompanied a petition signed by over two hundred citizens. During their combined Liberal Republican and Democratic County Convention, held in Socorro on October 13, the supporters of Cardis and Howard resolved to abide by the election results but also declared, "we will not submit to fraud." Fountain fired back, alleging that "not 25" signatures were genuine. Further, Cardis, "who has been three times indicted & twice convicted other crimes in El Paso Co . . . is a grand scamp . . . now under indictment for perjury at Austin." The charge that Atkinson was unfit was alleged to be a campaign ploy by Cardis, a candidate for sheriff against the incumbent Ellis. "[His] action towards Lieut. Atkinson is entirely for political effect and is instigated and directed by one Charles Howard, a Ku Klux lawyer who is a candidate for District Attorney on the ticket with Cardis & leads the petition with his signature." Fountain recommended that the petition be sent right back to J. P. Hague, the sitting Radical Republican district attorney for El Paso, who would prosecute Cardis for forgery. The senator expressed his trust that no action would be taken against Atkinson until the charges could be refuted.[9]

Atkinson may have felt the heat from the recall campaign. According to Howard, the policeman drunkenly declared "that he was Lieutenant of Police, and Judge of the Election and he would be damned if there should be any election at all." Whatever malfeasance he had in mind, Atkinson missed his chance when he was replaced by French at the end of the first day's balloting.[10]

The county-wide voting lasted four days, from November 5 through 8. The Republican-Democratic candidates won every contest handily, scoring lopsided majorities of nearly 80 percent. With the offices of district attorney, sheriff, surveyor, and treasurer within their grasp, and the state police in El Paso already under their control, the Cardis-Howard alliance could look forward to controlling the county. Fountain's party had no intention of permitting this to happen. The last ballot had barely been cast when Charles Kerber, acting as federal supervisor of elections for the Radical ticket, protested to the board of election that all votes cast for Cardis were "null and void" under the Texas constitution. County Registrar Edmund Stine joined in the protest, along with Fountain, Hague, County Judge Maximo Aranda, and others. The signatories claimed that Cardis was ineligible because he had been convicted of an assault with a deadly weapon. Stine forwarded the election returns and the protest to Austin on November 11.[11]

The Radicals did not stop with Cardis. They also filed sworn affidavits with Secretary James P. Newcomb "proving mob violence, intimidation and undue influence," and calling for all El Paso election returns to be thrown out. Specifics included a charge that "the Mexican voters of this county were influenced by religious considerations," undoubtedly at the behest of Father Borrajo, still the parish priest in San Elizario. Joseph Tays swore that prior to the election, G. W. Wahl, the parson's opponent in the race for county surveyor, told him, "We intend to carry the election by force." A mob allegedly intimidated Radical voters from casting their ballots in Ysleta.[12]

El Paso's isolation quickly became a major factor in how the crisis played out. Three weeks after the elections, Howard was aware that some sort of protest had accompanied the election returns to Austin, but considered it "only 'Buncombe." On November 28, Mills wrote to Secretary Newcomb asking "whether there are any representations filed in your office touching the eligibility of Louis Cardis Sheriff-elect of El Paso County." Suspecting voter fraud on the order of the 1871 election (when eight hundred votes were cast by the three hundred and fifty electors), Mills asked for a list of the registered voters in Fountain's district and the vote totals for legislative candidates.[13]

Newcomb was slow in replying. In the meantime, the still triumphant Liberal Republican and Democratic leaders were already showing signs of tension in their relationship. Cardis and Howard each wrote to Mills complaining about the other. Money was the trigger, but underneath the determination of each to be the dominant partner is evident. Howard and G. W. Wahl, the alliance's candidate for county surveyor, had spent hundreds of dollars on their campaign (or on voting day), certainly more than Howard could afford ("I went through to the bedrock."). He asked Cardis for help, but Cardis refused, miffed that they had spent the money "without profit to the election" and without consulting with him first. Now he was stuck paying useless campaign bills.[14]

Each partner congratulated himself as the architect of victory. Cardis declared that he "faithfully complied with my promise to them [i.e., Howard and Wahl] by electing them as none would have received 100 votes without my support which I [gave] to the risk of my life." Howard boasted that, "We have carried the county gloriously, and I surprised myself, by the energy and interest I took in the campaign. However, I saw that it was absolutely necessary to support Cardis and strengthen his hands, if we at all hoped to succeed." While Cardis was supported by "the whole Mexican population," these voters were "weak kneed from former defeats, and thought it useless to struggle—So I came out and declared that I would stand by Cardis to the last, brought with me half the Americans, neutralized the other half, and gave much confidence to the Mexicans. The ring tried to coax me off the warpath, and that failing tried threats, but I made a full stand off with them at their own game." Cardis, Howard did acknowledge, "deserves infinite credit for his untiring energy and watchfulness."[15]

As the two ignorantly basked in glory, the secretary of state hurled a bombshell. The election returns were "to be thrown out, unless rebutting testimony is presented." The secretary warned Mills he had just twenty days to provide "counter statements." At a time when the telegraph did not reach El Paso and the advertised five-day stagecoach journey between there and Austin was problematic, twenty days was no gift.[16]

Mills replied on the ninth day of the month, pleading, "The time is too short to procure testimony from so remote a point as El Paso." He asked for copies of all complaints, adding, "I infer from your letter that there are no objections to Mr. Cardis on the ground of eligibility."[17]

Newcomb's same-day reply offered no relief. The charges Mills found "too general, vague and uncertain" were "set forth in the law." Mills could see the evidence, if he wished, in the secretary's office. And infer as Mills might, the secretary had not mentioned Cardis in his earlier reply only "because the rejection of the election of the county superseded the question of illegibility of Candidates." The evidence "is clear and positive," wrote Newcomb, "that Cardis has disqualified himself under the Constitution from holding office." The governor was withholding Cardis's commission as sheriff.[18]

While Newcomb stonewalled Mills, the anti-Radicals in El Paso were busy responding to Fountain's attempt to deny Cardis his victory. In a second "get out the vote" campaign, 429 signatures were collected on a petition supporting Cardis. First, said the petitioners, the protestors were "the avowed personal and political enemies of . . . Louis Cardis and any charges against him coming from them should be received with great caution." There was also no proof that Cardis was ineligible to hold the office of sheriff. The petitioners asked the secretary "that no vexation, frivolous, or illegal objection be permitted to hinder or delay [Cardis'] commission as sheriff of said county."[19]

On December 14, Mills submitted this document to the secretary of state.

He added his own assurance that the signatures were genuine. Though under no illusions that Newcomb would be moved by the petitioners' plea, Mills had an even greater worry. The December 27 deadline set on the real issue, the rejection of all votes cast for the entire Liberal Republican/Democratic slate, was fast approaching, and Mills still did not know what charges the victors had to disprove![20]

Though Mills did not know it, not until December 22—the fifteenth day of Newcomb's allotted twenty—did his letter alerting his allies to the Governor's rejection of the election results reach El Paso. It is not difficult to imagine the apoplectic feeling that must have surged through the anti-Radicals when they read this news. With few specifics to go on, they were consigned to taking their swings at the shadows of charges, the hints of their supposed misdeeds. As federal supervisor of elections, Howard swore that "said election was conducted in a fair, free and impartial manner, without any attempt at fraud, violence, intimidation or other corrupt influence." No elector was restrained from voting by "by force, threats, or any other undue influences . . . , or was forced in any way to vote for any person or ticket against his will."[21]

Three hundred persons signed affidavits supporting the Cardis ticket. Election judge Telesforo Montes and two election clerks swore to the election's validity. Police Lieutenant French affirmed that he "saw or heard of no riot, armed disturbances, intimidation, or other corrupt influences to impair the freedom or purity of the election." In dispassionate tones, the affidavits all echoed Howard's assurances of fairness. Outrage was reserved for a petition to Governor Davis prepared by Howard. He expressed the voters' "great mortification and astonishment that the Election returns from this county are to be rejected—for what reason they are at a loss to imagine." Even before their petition was dispatched to Austin, the Howard-Cardis forces received a sign of the reception their pleas would receive. On December 23, French received the order to turn in his police badge. Atkinson was again in charge of the El Paso detachment.[22]

With the prospect of victory evaporating, the underlying tension between Cardis and Howard threatened to create a breach. Cardis had seen too much of El Paso politics to think that elections depended on vote totals. He began, Howard warned Mills, "acting very strangely." If Howard's observations were correct, Cardis was looking for a way to forge an alliance with Fountain. First, he made off with some fifty or sixty affidavits secured in Ysleta supporting their position, refusing to let Howard forward them to Austin. He also worked "to keep the people from giving affidavits—and boasts openly that he can 'carry the county for the very men he has just beaten' [and] 'that he is the county of El Paso' and other such like things." Howard charged that Cardis was making overtures to the Fountain camp "to join hands with them against Wahl and myself [and] telling the people that we have gone back on them and turned our Jackets &c." Worse, Howard alleged, Cardis was "inciting people to riot and murder." Cardis was accused by his enemies of doing exactly this in the first stages of the 1877 troubles; but this 1872 accusation came from a perplexed ally and provides indirect evidence that similar charges against Cardis five years later may have been true. "He is doing great damage to the county," Howard complained. He feared Cardis's actions "will make it necessary for us to denounce him publicly as a liar—a thing I would not like to do." Howard, still new to the county, was not yet prepared to break with the man who controlled the votes, yet from this point he had reason to treat his ally with caution.[23]

Meanwhile, with time running out, on December 26 Mills again requested an extension of the time to appeal. The effort was futile. There would be no new election. The votes for the Liberal Republican-Democratic ticket were declared void. The election losers, not one having gained more than 21 percent of the vote, were named victors by the governor.[24]

The anti-Radicals considered their options. Taking the matter to court was out of the question. Howard wanted "a judge on the bench who has sense, honor, and decency enough to keep himself aloof from neighborhood and party strifes," in other words, to rule in favor of Democrats. Unfortunately, the district judge for El Paso County was Simon B. Newcomb, a Davis appointee. A gathering of El Paso lawyers endorsed him as "an able jurist . . . and an incorruptible judge." But Newcomb also was the brother of the stonewalling secretary of state. That automatically made him, in Howard's mind, "a miserable sneak, without courage, sense, principle, or shame—a tool of Fountain, venal, corrupt, and so far as I know, without a redeeming trait."[25]

With no recourse to the bench, Howard looked to the legislature for relief. The timing looked right for attacking Senator Fountain in his bastion of power. Statewide, the 1872 elections had gone well for the Democrats. They captured both houses, and Fountain was now the senate *minority* leader. Now Howard, Cardis, and 250 other petitioners invoked the legislature to "repair our injuries [or] we will be driven to despair." The document charged that ballot switching in 1869 and ballot box stuffing two years later had been "planned and executed" by Fountain, who "crept" into the Senate "by a fraud and trick." Now the Radicals were again cheating the electorate out of their preferred candidates. "[W]e have been misrepresented and calumniated by men who have thrust themselves into office, against our wishes and consent . . . we appeal to you to relieve us from the insolent oppression of those men, who still retain their offices, although four fifths of the registered voters of this County have declared at the ballot box that they wish to be freed from them." The petitioners also begged the legislature to replace Newcomb.[26]

The Democratic Legislature refused to act. The overturned election results remained overturned. Ellis, Hague, Tays, Blanchard, and others kept the jobs their opponents had won at the polls. Nevertheless, Fountain was as clearheaded as politicians came. He knew that a battle won was not the war. The Radicals were no longer a power. The legislative program he had pushed through the senate in 1870 to 1871 came undone after 1872. Not only would the senator's power and influence diminish, but the time and money he would spend defending against attacks from Mills, Cardis, and now Howard would only multiply. Fountain came to a decision. He would move back to Mesilla, New Mexico, to the home of his wife's family, and to a Republican-controlled territory where a still young and ambitious Republican could make his mark.[27]

The next elections, those of 1873, marked a fault line in El Paso's politics. Republicans of any stripe were now a spent force. Anyone wishing to grasp political power must do so as a Democrat. So many ambitious office seekers in West Texas donned the Democratic Party label that the alliance with moderate Republicans was dropped, and it was the democracy that now split. Not surprisingly, the rupture was along personal lines. Louis Cardis, still the undisputed political leader of the Paseño community, stood for the state representative from the 30th legislative district (comprising El Paso, Presidio, Pecos and Tom Green counties), winning the nomination at a Democratic convention meeting October 15 in Laredo. The news was a gut punch to longtime Democrats, such as Simeon Hart and Allen Blacker, who reviled Cardis for his Republican past.[28]

On Christmas Day of 1872, Hart and Dwight C. Marsh began publishing the El Paso *Sentinel,* written with the editorial help of Blacker. The November 8, 1873, edition was little more than an anti-Cardis screed. The Laredo convention, declared the *Sentinel,* "has been hoodwinked and deceived into nominating the notorious Radical Communist Louis Cardini, alias Cardis, as a democratic candidate." In what amounted to either vile slander or damning accusations, Hart's paper charged that Cardis had allegedly "murdered his mistress, a Mexican woman, while en route from Chihuahua to this place, in order that he might debauch her daughter, a girl of thirteen years of age. He did debauch the girl—Juliana—and kept her as his mistress for over two years. This fact every citizen of El Paso is acquainted with. The mother has never since been heard of. Perhaps our Democratic Nominee can tell where her body lies."[29]

It was only to "gratify his greed for plunder," said the *Sentinel,* that Cardis joined the Radical Republican "oppressors." He "publicly embrace[d] negro soldiers" and drank with them, while advising the hanging of Confederate sympathizers. In addition to these alleged sins, Cardis was an "infamous" confiscator. In an apparent reference to Cardis's 1865 purchase at auction of Hart's home, the *Sentinel* charged that, "While the wife and children of one of the best and purest of our citizens were without a roof they could call their own, this model 'Democrat' peddled whiskey from the parlor of their little homestead, to which he knew he had no title."

The *Sentinel* heaped crime after crime upon Cardis's name: "Four times has he been indicted for assaults with deadly weapons upon inoffensive, unarmed Mexicans. A dozen times has he been publicly whipped for lying. He committed perjury—falsely swearing he was a citizen of the United States—to register as a voter in 1869. . . . He is a social Israelite, a moral leper, a thing devoid of honor, principle or religion. God Almighty could only have created such a being as an example of how much mendacity, meanness, cowardice and general rascality could be crowded in one human skin." Cardis's supporters, said the *Sentinel,* amounted to "a class of ignorant Mexicans [and] some three or four office-seeking carpet-baggers who disgrace the Democratic party of this county." For good measure, Hart and Blacker added a venomous dose of campaign verse.

Our Democratic Candidate for the Legislature

Our artist's pencil could not miss / In painting portraits of "Cardis"
His every feature, nose and eye / Proclaim this man a foreign spy
His furtive air, and manners meek / all go to prove this man a sneak.
His stealthy step, his style, in brief / Advertise him a petit thief
His oily tongue, and labored wit, / Exhibit him a hypocrite;
Just give a painter such a face / And a true likeness he can trace.

But all descriptive power must fail / Sin, vice and infamy will pale
In presence of a man like this / Vile more leper, L. Cardis.
"Dickens" could not this man portray / He'd drop his pen in blank dismay.
No power that human genius gives / Concerns that such a monster lives.

Yet we are asked with serious face / To perpetuate the foul disgrace
In aiding others to elect / A villain; no one can respect.
His politics are all a sham / A bald pretense, not worth a damn;
His perjured soul won't hesitate / To sell his party or his State.

We know this man, and well have known / The seeds of evil he has sewn;
The deadly purpose of his life / Of murder, perjury and strife/
His craven spirit shrinks in fear / Wherever honest men appear.
We must oppose this deep disgrace; / This fraud upon the human race.
We will oppose, just as we please, / This sum of human villainies[30]

Howard, who wisely remained "in favor of Cardis," retaliated in the *Austin Democratic Statesman*. "We do not recognize the El Paso Sentinel as a Democratic newspaper. . . . Politically, Judge Blacker and the Sentinel are the cups tied to Fountain's tail to rattle as he moves." Eleven months after privately telling Mills that Cardis had tried to cut a deal with Fountain, Howard now publicly charged that the Radicals promised Cardis "any office in the district, and the absolute control and leadership of the party, if he would work for them. All this he refused." Howard called on true Democrats to disbelieve the *Sentinel's* "disgraceful and dirty abuse" of Cardis.[31]

Howard himself ran for district attorney, while George Wahl again stood for county clerk and Albert French for county surveyor. In the end, Cardis carried the Democratic ticket to victory in most races. The vote was down from the previous year due to the absence of many electors buying corn in Mexico to replace failed crops. But the Cardis win was even more overwhelming than in 1872, with some four hundred ballots of four hundred and sixty cast. The anti-Cardis forces would again charge that the Italian's election tactics were "eminently" foul. "The priest Borrajo went from house to house threatening that those who did not vote for Cardis should not be buried in consecrated ground, but on dung hills, like dogs." Borrajo, it was said, not only sprinkled holy water on the Cardis tickets, but on election day even stood on the roof of one house and similarly sprinkled any "unbribed and unbought suffagans on their pilgrimage to the ballot box with unwashed tickets." While the stories of the padre's election meddling were probably true, this time no one in Austin cared to listen.[32]

Sheriff Charles Ellis had decided not to seek reelection in 1873. The voters replaced him with Deputy Sheriff Kerber, the one-time army sergeant-deserter-officer-mutineer. When the Radicals ran things, Kerber touted his own Union allegiance and disparaged the Confederate past of men like Howard. Kerber had even less use for the largely Paseño electorate, now his constituency. Nonetheless, bucking the Democratic sweep, Kerber won the sheriff's job, defeating Cardis's man, William Treyer, by thirty votes. With the state police disbanded in March 1873, Sheriff Kerber became "the law" in El Paso County.[33]

As the year ended, Fountain packed up his belongings. By January 1874, the former state senator, his wife, and their five children were residents of Mesilla. Before he departed, Fountain had one last conversation with Howard. Their relations had ranged from "unfriendly" to "uncordial." Now, they achieved "a mutual understanding that by-gones should be by-gones." Fountain warned El Paso's new district attorney to be careful about his friends: "I told him that Cardis was a murderer, that he was a fugitive from his own country, Italy, that he had been a member of the Italian Cabonari, had betrayed his associates and fled to save his life from their vengeance, that he had no equal as an intriguer except Borajo, and that between them they would compel him, Howard, to yield to them or they would ruin him. . . . When they had no further use for him they would compel him to do as I was doing, leave the County and State." Howard expressed confidence he could control his allies, but he was glad Fountain had decided to leave El Paso. "You have been in great danger here from men whom I have heretofore been able to control; but I am losing my influence with them because I will not accede to some of their schemes, and the day may come when I shall have to do as you are now doing."[34]

CHAPTER 6
"Both were alike ambitious, and alike unscrupulous."

The year 1873 began well enough for the little community of Franklin, which incorporated as El Paso, Texas, on May 17. Saloon owner Ben Dowell was elected mayor. Enterprising El Pasoans such as Blacker, French, Hague, and Joseph Magoffin put aside political differences to form the El Paso Real Estate Trust and Immigration Company. Cardis actively bought and sold property along the projected railroad lines and in town. The railroads were a key economic driver throughout the state. Perhaps no corner of the reunited Union offered such bright prospects for the laying of new tracks as the vast expanse of Texas. Statewide, miles of usable track quadrupled between 1865 and 1873. El Paso's leaders could look upon Dallas as a precursor of their own goldmine-in-waiting. That city more than doubled its population soon after the arrival of the Houston & Texas Central (1872) and Texas & Pacific (1873) railroads. More people meant more business, more cash, and rising property values.[1]

Even as El Pasoans invested in the future, however, the nation's economic prospects turned bleak. The Panic of 1873 struck the banking firm of Jay Cooke & Co. on September 18, and much of the national economy soon followed. The resulting depression lasted five years, the longest of the century. As Texas-chartered railroad companies failed or slowed their track laying, visions of locomotive smoke over the Rio Grande Valley evaporated. El Paso's town government collapsed through disinterest.[2]

Local farmers, already isolated by the lack of railways, were somewhat insulated from the national decline in prices and wages. Still, manufactured goods were harder to come by, affecting all those who would buy, sell, or barter for them. Some of the few "Americans" at the Pass gave up and left. Perhaps only freighters like Price Cooper and stagecoach operators like Cardis found good news in the failure of the railroads. The delay in the anticipated economic explosion forced El Paso's leaders, men with large egos and expansive appetites, to contest one another year after year for only the promise of power and wealth, rather than any actual spoils. Such conditions do not make quarrelsome men cooperative.[3]

Most Paseños had other, greater concerns than who won or lost the last election. It is instructive that the Paseños did not rise up when their votes were scratched out, as they allegedly were in 1869, or tossed out, as they certainly were in 1872. The theft of their ballots did not directly hit their livelihood, as the retreating Confederates had done, or assault their culture, as the school law had done. Elections were relatively marginal to their lives. Perhaps they understood that Anglo officeholders would always act to benefit themselves, no matter what the party label.

What mattered to Paseños, besides their culture, was survival. Would weather and water cooperate, allowing crops to grow? Would they be safe beyond their doors, their livestock safe in corrals and pastures? Or would Apaches deprive them of what little wealth they owned? Would bad weather and Indian depredations assault them in the same year, leaving them nothing but salt from the lakes, salt to sell, barter, and survive on?

Recurring climate catastrophe in the 1870s illustrates how crucial the reliable Guadalupe Lake salt deposits were to the Paseños' shaky economy. Bad times started in the spring of 1872, when melting snows sent a "great force of high water" rushing down the Rio Grande. The dam protecting the *acequias madre* began giving way, threatening "an immense loss" to the lower valley. The river was still "constantly rising and booming along at a destructive rate" in early June. On January 27, 1873, six to eight inches of snow fell in the valley, "a circumstance unknown during the lifetime of the oldest of inhabitants." Continuing extreme cold claimed the lives of many animals. The peach, apple, and fig crops were "entirely ruined" and an estimated three-fourths of the wine crop was lost. Fortunately, the wheat crop survived and was declared "the only hope left the agriculture in this valley to meet the necessities for the present year." Later in 1873, drought appeared in place of the summer rains, causing the "the entire failure of water in the Rio Grande" and leading to widespread crop failures. As the *Sentinel* put it in November, "Water is at present dearer than whiskey."[4]

The weather in 1874 seemed tame enough at first, but in mid-May residents of the Mesilla Valley and at the Pass became aware of approaching calamity. Melting snows in the mountains to the north were already sending torrents of water south along the Rio Grande. The river rose rapidly throughout the month. By May 30, twenty-five homes in El Paso del Norte had been washed away. On the American side, the town of Ysleta was "in great danger."[5]

On June 6, the *Mesilla News* reported, "The Inundation" brought "tidings of disaster. The unprecedented rise in the river has baffled all calculations, towns have been destroyed, fields submerged, life lost." The river was still in flood stage in mid-June. Although the worst damage to homes and vineyards at the Pass came on the Mexican side, the American bank was also stricken, as "the river . . . submerged the bottom below town [i.e., El Paso]." The flood could not have come at a worse time. The *Las Cruces Borderer* reported from Ysleta that, "great destitution prevails in the county owing to the entire failure of the crops last year, and that several persons are on the verge of starvation, but no one will starve." Leading citizens distributed some 150 *fanegas* of corn "among the poorer class of that place." Ward Blanchard, the new county treasurer, "almost sacrificed himself to procure flour for the poor, he not having any corn on hand." Crop failures forced many Paseños to buy their corn and other food in Mexico, some undoubtedly by bartering salt from the Guadalupe Lakes. Conditions in El Paso were so harsh that nearly fifty families emigrated to the Mesilla Valley.[6]

Ultimately, the floodwaters receded, and farmers put their "spades, pick-axes and shovels" to the task of restoring their homes, fields, and *acequias*. The summer rains needed for autumn and winter crops came late, causing renewed anxiety for a time. It was not until the last week of August that the area saw significant rain, "a Godsend to our almost disheartened farmers."[7]

In the first days of 1874, the Cardis-Howard alliance had work to do, Fountain men to clean out. First on their list was District Judge Newcomb, vulnerable following the Radicals's statewide electoral defeat. On April 13, the Democratic Legislature removed him from office. Governor Coke appointed Charles Howard as District Judge, an elevation confirmed by a special election held June 2. Three candidates contested in the race for district attorney, two of whom, Fountain supporters James P. Hague and William M. Pierson, sued the third, Mills ally James A. Zabriskie, on the grounds that he was a New Mexico resident. Zabriskie won the election, but runner-up Hague pressed his lawsuit. In January 1875, Judge Howard confirmed Zabriskie's election.[8]

By 1875, virtually every political plum in El Paso County belonged to the Democrats. The federal diplomatic and revenue posts were still in Republican hands. Sheriff Kerber was a Republican but also a strong admirer of Howard. That feeling did not extend to Cardis. From this time forward we may be certain the two party chieftains were moving apart. At first, any tensions between them were mitigated by physical distances. Howard spent a third of each year traveling the four-county 25th Judicial District. Meanwhile, Cardis sat in legislative sessions in Austin and attended to his expanding stagecoach business. By 1876, he was the proprietor of the El Paso Stage Line, with service running from El Paso north to Mesilla, southeast to Presidio County, and east to Fort Concho. Business kept Cardis on the road at times.[9]

Sometime early in 1875, Howard and Cardis "became unfriendly, quarreled, and two parties sprang up." According to one who knew them both, Howard "considered himself the head of the party in El Paso County, and presumed to dictate to the man through whose influence he had been elected to the judicial bench; but there he made an error, because Cardis was not the man to play second fiddle, particularly as he well knew his influence among the voting population in El Paso County to be too strong to be ever shaken by Howard or anybody else. Their first misunderstanding soon grew to be a bitter feud."[10]

The two politicians battled on several fronts, most notably in the matter of El Paso's *acequias*. No sooner had Cardis taken his seat in the state legislature than he introduced a bill to subsidize the building of irrigation ditches in the county carrying free water "to all who will provide labor and keep them clean and in repair." The measure, approved April 25, 1874, authorized the issuance of land certificates to the towns or irrigation companies that constructed the ditches. For each mile of ditch constructed and operated, the responsible municipality or association of citizens would be entitled to eight sections of land, that is, eight square miles![11]

This and later enactments spurred Paseños and Anglos alike to get in on the construction. The towns of Ysleta, Socorro, and San Elizario organized their efforts under the leadership of trusted commissioners, while Cardis created an irrigation company to dig a five-mile *acequia* from Hart's Mill to Ysleta. Magoffin, the Schutz brothers, and Kerber incorporated a business to build *acequias* supplying free water to Ysleta. Magoffin, Hague, and others, acting as another company, undertook to dig over thirty miles of ditches from Magoffinsville to Socorro. Throughout the spring of 1875, letters poured into Governor Coke's office from the various towns and corporations, urging the appointment of an inspector to certify that the ditches met the legal specifications necessary for land certificates to be issued.[12]

Not surprisingly, everyone argued over who dug what ditch. After one inspector examined the Magoffin-Hague network of *acequias*, Cardis complained to Coke that "an act of injustice" was in the making. Some Magoffin ditches were said to be "one hundred years" old! Accused of monkeying with the ditch measurements, Howard protested the slur on his integrity and succeeded in frustrating the Cardis-backed town efforts. The legislator accused the judge of graft and of misrepresenting the people's case to the governor.[13]

A second issue divided the two camps. This was El Paso, where it helped greatly to have an armed force at one's command on election day. The politicians had battled for control of the state police in 1872. Now the warring factions were each determined to handpick a new Texas Ranger company. The 1874 legislation that created the famous Frontier Battalion of Texas Rangers also established special ranger detachments in certain counties, including El Paso. Cardis endorsed the appointment of Albert French as captain of the local company. The position went instead to Telesforo Montes, an experienced Apache fighter. Of Montes's twenty-four enlisted men, all but one were Paseños.[14]

Fiscal restraint forced the company to stand down at the end of 1875. In begging the governor to reform the company, Howard laid bare his motive: "[B]oth this and the San Antonio Senatorial Dist seem to have gone for the enemy. I can reclaim El Paso County if I can have the filling of the company that we talked of. Cardis has had two or three companies mustered in, as they think. The whole frontier appropriation would not pay his promises. If you will let us have a company, and let me name men and officers, it will do more to break down Cardis, than any other one thing. The people will then see that he has lied to them. They are so ignorant, that they require something tangible like this."[15]

While the judge looked for political advantage, Kerber was motivated by a palpable dread. The sheriff warned Coke that Cardis was already organizing a Ranger company "among his Mexican friends" in fulfillment of a campaign pledge. This was, "dangerous for all white men of this County, which he threatens to punish personally for not voting for him." Four years after the school law crisis, Kerber claimed, Cardis and Borrajo were still conspiring, and "every white man, not Cardis' friends, has to expect to be assassinated."[16]

El Paso's frontier company was remustered in late 1875, with Montes still in command. The sheriff's fear that Cardis would use the unit to kill "Americans" was baseless, but he was not entirely mistaken in the danger posed by a military unit composed of Paseños. A number of Montes's rangers joined the insurrection of 1877, including two who became leaders, Corporal Francisco Barela and Private Sisto Salcido. During the Salt War, the former would display a number of qualities desirable in a commander. It cannot be discounted that the Rangers' organized forays against the Apache enemy gave Barela a chance to demonstrate his martial qualities to his countrymen and to himself.[17]

Cardis and Howard also battled over control of the county's Democratic Party apparatus. Each sought selection during the summer of 1875 as a delegate to the state's latest constitutional convention. Three spots were open in each senatorial district, and both Howard and Cardis were initially elected in the 30th District conventions held in June. Cardis withdrew from the county precinct meeting, claiming fraud engineered by Howard's "tool," District Clerk G. W. Wahl. Running as an independent in the final round of voting, Cardis whipped Howard, the Democratic nominee, 370 votes to 104.[18]

The Convention met in Austin during from September through November 1875. Cardis "worked diligently to represent the sentiments of his constituents." Speaking up in opposition to a poll tax, he argued, "did all of the citizens of Texas have a right to vote, or only a portion of them. Which class was to be deprived of the sacred right? The [poll tax] article would indicate that it was the poorer class." He proposed (for home consumption) "that any citizen of the State too poor to pay his taxes should be allowed to vote twice."[19]

The Constitution of 1876 that Cardis helped create contained a clause of particular interest to El Pasoans. Found in Article XIV, Section 7, it read, "The State of Texas hereby releases to the owner or owners of the soil all mines and minerals that may be on the same, subject to taxation as other property." This language, nearly identical to similar statements in Texas' two previous constitutions, once more proclaimed the right of individuals to legally establish personal ownership of saline deposits, such as El Paso's Guadalupe Salt Lakes. It was language that Cardis affixed his name to as a signer of the Constitution. No evidence has surfaced indicating that *Don Luis* conveyed its import to his constituents.[20]

It may have been during the months of the Convention that Howard caned Cardis on the streets of Austin. At about this same time, Howard apparently began or continued a less public and more intimate relationship with Tiburcia Jurado of Socorro, who became pregnant. It is unlikely that Howard made any serious offer of marriage, for in late 1875 he embarked on a new alliance, one that promised both personal happiness and a stronger political position. It also led him directly to a firing squad.[21]

Dark haired, dark eyed and dark complexioned, George B. Zimpelman was driven by an attitude anything but dark. His life was marked by the ready seizure of opportunity, a quality that may have cost him the lives of his beloved daughter and trusted son-in-law. A native of Bavaria, Zimpelman was born in 1832 to a farming couple who ensured that he received an excellent primary education. At thirteen, in the company of an aunt and uncle, he joined the mass migration of Germans to America. In 1847 he moved from Galveston to Austin. Here he worked as a carpenter and then gunsmith before turning to "stock raising and agriculture." A veteran of "almost every engagement" between local settlers and Indians, Zimpelman volunteered for service in the Confederacy's 8th Texas Cavalry. He allegedly "took part in more than four hundred battles and skirmishes [and] had perhaps more horses killed and wounded than any other man." After serving as his company's "Master Armorer" for much of the war, he was eventually promoted to 1st Lieutenant. He was captured during the fighting for Atlanta, where he was wounded three times, bringing his total wounds during the war to nine.[22]

Zimpelman returned home after the war. There he found a community torn asunder by economic collapse, social upheaval, racial hatreds, an occupying army, and a flood of cheap surplus weapons. Texans had always relied upon the gun to defend their homes; now they used them with increasing frequency to settle personal disputes. Between 1866 and 1868, Travis County experienced eleven murders and attempted murders.[23]

Into this maelstrom stepped Zimpelman. He was elected Travis County sheriff in 1866, removed by Reconstruction authorities in November 1867, voted back into office in 1869, and returned in the Democratic sweep of November 1873. Within days of that election, he helped secure the removal of the intransigent ballot box loser, Gov. Edmund Davis, whose armed supporters occupied the lower floor of the capitol building. The one military force in Austin, the Travis Rifles, "fully armed and equipped," was on its way to bolster the Davis forces when Sheriff Zimpelman intervened. "[K]nown to be prompt and fearless in the discharge of his duty as a peace officer," Zimpelman halted the formation and deputized it to "preserve the peace" for Coke's benefit. The sheriff next facilitated a negotiated resolution. The armed forces on both sides dispersed, and the conflict rapidly ended.[24]

On December 31, 1875, the popular sheriff and his wife, the former Sarah Katherine Matthews, gave their first child in marriage to Charles Howard. Marie Louise Zimpelman, born in 1856 and known to all as Lou, was an accomplished and attractive young lady. She provided music for the library's entertainment program, and at the 1875 Capitol State Fair won first place in the "Young Ladies" riding competition. Charles loved Lou deeply (he reportedly wrote at least one "very tender poem" to her), and the courtship progressed "like a whirlwind." The couple was married at Austin's First Presbyterian Church and departed the following day in a wedding gift buggy. They honeymooned in Galveston before making the long trek to El Paso.[25]

The bond between Lou's husband and her father was also strong. The two had served in Terry's Texas Rangers. Howard had demonstrated horsemanship and courage, qualities his father-in-law appreciated. More important to the Zimpelmans would be his ability to provide. The lawyer must have seemed an excellent prospect. Only three years after arriving in El Paso, he had flourished in the rough-and-tumble of that county's politics, rapidly rising to district attorney and then judge. He had both ambition and the intelligence to back it up. If Howard lacked the sheriff's diplomacy, perhaps El Paso was a place where courage and perseverance were more useful qualities.

Before long, the bridegroom was deeply in debt to Zimpelman. Howard's financial troubles began as early as 1874. Robert M. Keating, who ran Howard's dry goods store in San Elizario, ran it into the ground. Keating allegedly gave away goods, left the store unlocked at night, and stole from the liquor shelves. Howard brought suit, Keating countersued, and their cases languished in court.[26]

Fountain thought Howard's considerable "pecuniary embarrassments" grew out of trying to settle the estate of Fort Davis merchant Pat Murphy. According to E. W. Bates, Pecos County Clerk in the mid-1870s, Howard lost his shirt in 1876 and 1877 trying to fulfill government contracts to supply Forts Davis and Stockton with hay, oats, and corn. While other army agents obtained grain from their own local farms, Howard had to buy and freight grain all the way from Mexico and El Paso, Texas. "Howard lost money on all these contracts," leaving him "cramped for money." He was forced to borrow repeatedly from his father-in-law, directly or through Zimpelman's bank. Acting as go-between, Bates received one thousand dollars from Zimpelman for Howard in July 1876. Only one month later, more money was transferred to the insolvent judge. Howard paid off a portion of his debts by deeding land he had located in Pecos and Presidio counties back to Zimpelman or the bank. Bates estimated that the value of all of Howard's land did not "equal the indebtedness due from him to George B. Zimpelman." Still, in 1876, the in-laws had reason to view Howard's future in west Texas as promising. At some point, they saw opportunity in the still unclaimed Guadalupe Salt Lakes.[27]

CHAPTER 7

"Leave this county within ten days."

On February 15, 1876, the new Texas State Constitution was overwhelmingly ratified by popular vote. Elections were held the same day. The campaign marked yet another political realignment. Nearly everyone was a Democrat now, though rivals attacked one another as being a "carpet-bagger" or "the worst of [Governor] Davis' gang." Cardis handily trounced Jesus Cobos in his bid for reelection to the legislature. Howard opted not to run for reelection as district judge. Instead, he backed attorney Allen Blacker, his rival for party leadership in 1873. Blacker's opponent was James P. Hague, formerly Fountain's ally and now Cardis's. Hague's decision to leave the DA race to his longtime rival, incumbent James P. Zabriskie, allowed that former ally of W. W. Mills to skate into office over El Paso's mayor, A. M. Jones. Cardis and Hague appeared on the same ticket with former Deputy U.S. Marshal William Treyer, candidate for sheriff in a rematch with the incumbent Charles Kerber.[1]

It would not be an El Paso election without fraud. Cobos took his allegations to the legislature, which tossed out the complaint. In the four-county race for district judge, Blacker carried the interior counties, Pecos and Tom Green, with a combined majority of 395 votes, while Hague won the Rio Grande counties of El Paso and Presidio, both Cardis strongholds, with majorities of 380. Precinct officials allied with Cardis tossed out enough Blacker votes to throw the election to Hague. Blacker charged that "two hundred or more" aliens voted for Hague and that Blacker supporters were interfered with at the polls. Cardis also allegedly delayed some vote counts until he knew how many more Hague needed. Cardis suffered an unlooked-for and telling blow from within his own camp. Treyer swore that "at least nineteen twentieths" of the electorate voting for Cardis and Hague were "Mexicans, and persons of Mexican origin [who] were neither naturalized nor had they declared their intentions to become citizens of the United States."[2]

Blacker appealed to Governor Coke. It cannot have hurt the Union veteran's cause that he was backed by ex-Confederate Charles Howard, the son-in-law of George B. Zimpelman, whose clear thinking and firm actions had proven decisive in Coke's battle with Davis just two years before. Coke awarded Blacker the district judge's commission. Like so many El Paso election victors before him, James Hague's moment of triumph vanished. He filed suit, to no avail.[3]

The election of 1876, the last in El Paso before the Salt War, was a turning point, one that moved the county closer to extra-legal resolution of the county's bitterest rivalry. Louis Cardis had tried repeatedly to gain control of local law enforcement through the offices of district judge, district attorney, and county sheriff. Howard's possession of the DA's office in 1873 and the judge's bench a few months later had been successes, but that alliance fell apart. Now the district judgeship was occupied by Howard's ally, Blacker. District Attorney Zabriskie was also independent of Cardis's machine; and in a baffling election anomaly, Cardis's influence was insufficient to defeat Kerber. With the law now exasperatingly on Howard's side, Cardis would be undone at every turn unless he wielded the only weapon still left to him, direct action by an angry citizenry, a political force as old as ancient Athens and Rome. Cardis reportedly had the people in his pocket. He knew their strength when aroused. The evidence indicates that he believed he could control that power to intimidate his enemies. He would discover, as others did over the centuries, that it was easier to unleash public anger than to bottle it back up.[4]

During the last decades of the nineteenth century, gunfighters and gamblers gave El Paso a well-deserved reputation as a "Six-Shooter Capital." Before the railroads came, the county was the scene of few gunfights. Most shots fired in anger seemed to be aimed by one politician at another. A remarkable exception took place on the normally quiet streets of San Elizario. The episode involved John Atkinson, arguably El Paso's most disagreeable citizen and, the gunfight demonstrated, one of its most courageous.[5]

The former state policeman repeatedly found himself on the other side of the law. In 1872 he was hauled into criminal court on a charge of assault and fined one hundred dollars. On October 9, 1873, he allegedly struck both the sheriff's wife, Teodora Alarcon Ellis, and her sister, Dolores Alarcon. The latter he allegedly "beat and cruelly abuse[d]" with an inch-thick walking stick. Mrs. Ellis was allegedly hit with a "buggy whip." Atkinson declared himself to be the injured party, countercharging that he never struck Ellis, although she "was assaulting, beating and abusing" him. The case was continued from term to term. Whatever transpired, it did not rupture the relationship of Atkinson and Charles Ellis. In 1877, they were described as "bosum friends," so close as to be considered "brothers." Atkinson successfully avoided trial on a variety of other violent crimes charged to him, including assault with intent to murder and disturbing the peace. These also were repeatedly continued, without resolution.[6]

The spring of 1876 found John Atkinson in yet another argument, this one caused by the unraveling of his marriage to John Campbell's sister, Juana. She had become the third Mrs. Atkinson in November 1875. Three months later the happy couple was anything but. It was then, according to Atkinson, he discovered that his young bride was already pregnant by some other man. She delivered the child on May 11, prompting Atkinson to file for an annulment. After husband and wife separated in February, Campbell warned his brother-in-law, "If you speak to your wife on the street I will kill you like a dog."[7]

Campbell also warned Pablo Mejia that he "had better draw out and have nothing to do with it." Mejia challenged Campbell to "come on"; he would find Mejia and Atkinson standing together. The most bizarre warning came in the home of a doctor named Smith, who read aloud from a piece of paper signed by Campbell, John P. Clark, and others cautioning Atkinson, "You are a dangerous character, and therefore are ordered to leave this county within ten days, or we will not be responsible for the consequences." If Atkinson stayed, he was warned, "Your doors will be broken down, and you will be taken to the chaparral and hanged." The deadline came and went, but the threats continued.

On May 28, Atkinson was walking away from the post office when he heard the noise of a bullet passing his head coupled with Campbell's voice calling, "Get out of the way, I am going to kill him." Reflexively, Atkinson drew his six-shooter. He saw Clark jump to the left, clearing Campbell's line of fire. Instead of shooting back, Atkinson called to Cleto and Maximo Aranda, "This man is trying to murder me, and I am not going to return his shot." Though Atkinson cocked his pistol, he left it by his side as he walked slowly toward Campbell. He moved forward about fifteen steps when Campbell fired a second shot. Atkinson raised his own pistol, aimed it at Campbell, and then lowered it to his side. Cleto Aranda called out in Spanish, "John, why don't you shoot?" "It is too far," replied Atkinson. "I cannot hit him."

A defective trigger affecting the range of Atkinson's pistol may have been the real reason he did not return fire at this point. He tried to bluff his way out of danger. He could not run. If he started he would have to clear the county. He could not hit the distant target. All he could do was unnerve Campbell into wasting his shots by walking steadily toward him, occasionally taking aim. By the time Campbell fired his third round, Atkinson had marched about fifteen yards with his cocked revolver in hand. He continued to close the distance to his adversary, who fired a fourth time and then a fifth. Campbell then disappeared. To no one in particular, Atkinson declared, "I suppose that finishes it." Turning around, he walked back toward the Arandas. He told them he would "show my pistol to the Justice of the Peace that he may see that I have not fired a shot."

Turning the corner onto the Socorro Road, Atkinson walked a few steps when he heard someone shout, "Cuidado, John!" Turning his head, he spotted Campbell at the corner of his home some thirty-five yards away, aiming to back shoot him. Instantly, Atkinson felt a bullet strike his leg, six inches above the knee. Maintaining his footing, Atkinson wheeled around and walked back toward Campbell. "That is your last shot," he crowed. Campbell disappeared again. Atkinson turned about and limped to see the JP. The whole episode took five minutes.[8]

Kerber arrested Campbell, who obtained a writ of habeas corpus from Judge Blacker. The case was postponed several times until Atkinson, a man who inspired women to lacerate him and men to attempt his murder, was executed by firing squad.

The explosive violence marking the end to John Atkinson's latest disputatious marriage to a Paseña was emblematic of El Paso County's slide from mere personal and social tensions toward bloody convulsion. In this more important drama, Charles Howard was at the storm center. Perhaps the first indication that he ought to "leave town" occurred in the autumn of 1876. He and Judge Blacker, riding a coach carrying them to the next court session in Presidio County, had been advised of a rumor that a mob planned to stop them. When the coach reached San Elizario, it was halted by an angry "committee" demanding that Howard explain his actions "in defeating an Acequia project in which . . . the people were interested." Howard refused to justify his stance, except to say that the governor could not approve their *acquias* without violating the law, "and all who put their money or labor into it would lose." He assured them "that time and reflection would satisfy the thinking portion of the people that he had really done them a great favor." With this, the crowd let Howard go on his way. Judge Blacker did not believe that any serious threat of violence had been made. The people were not yet angry enough.[9]

Meanwhile, Cardis's ally Albert French was going insane. A "queer change in his mental condition" was noticed in mid-1876. Suddenly in mid-November, he became "vicious and violent, so much so that it has been necessary to keep him chained, and to have him attended by four guards." The local army doctor pronounced him a "lunatic," and a local jury agreed. French was conveyed to an asylum in Austin, where he died on January 11, 1877.[10]

Howard, meanwhile, lost more than an ally. Lou Howard, the former Austin belle, did not flourish in the dusty border town of El Paso, Texas. She fell ill and remained in that condition "for a long while." In April 1877, her father came out from Austin to be with her. She died on Monday, June 25, 1877. The *Austin Statesman* noted that many in the city grieved "over the untimely end of so bright a flower." As for Howard, he became subject to "fits of depression" over his wife's death. The evidence suggests that he willed himself to keep busy, to focus on work and especially on the success of a new and risky business venture.[11]

By now, Zimpelman was no longer the sheriff of Travis County. He had joined the Austin banking firm of Forster, Ludlow & Co. Away from Lou's bedside during the spring of 1877, he and Howard discussed an exciting business opportunity presented by the Guadalupe Salt Lakes. Their saline treasure had attracted many a Gilded Age entrepreneur. In 1871, Hague advised Secretary of State Newcomb to invest in the Guadalupe beds, providing care was taken to cover *all* the lakes with certificates. "'Then,' Hague promised, 'it will be a fortune without end,' since the salt 'is inexhaustible.'" Six years later, the opportunity was still open. Howard would have explained the dangers posed by the opposition of Cardis, Borrajo, and the people. It would take men devoid of physical fear to attempt what Zimpelman and Howard were about to do. It was a quality the two shared.[12]

Zimpelman obtained possession of the old Maverick salt lakes and acquired "Memphis and El Paso Railroad certificates duly certified by the general land office . . . and located on the Texas Pacific Railroad reservation." All he needed to do was survey the remaining lakes, "locate" his certificates upon them, file the right papers in Austin, and the entire Guadalupe salt lake country would be patented in his name. All Howard had to do as Zimpelman's agent was extract payment from an angry population.[13]

Their plan leaked. Sometime in May or June, Cardis met with perhaps sixty Paseños at his home. According to Ben Dowell, "Every man that I saw was armed either with a pistol or a gun. The intention was to force Howard to relinquish his idea of locating the salt lakes, taking their instructions from Don Louis Cardis, but they didn't say anything about killing him." Nevertheless, Dowell alerted Howard, suggesting that Cardis "was trying to engage the Mexicans against him, so as to get them to kill him and get him out of the way."[14]

The idea that Cardis would encourage Howard's murder was not far fetched. El Paso County was next door to New Mexico Territory, where "[a] ssassination was condoned and encouraged." El Paso was also home to such a political culture. In 1870, Benjamin F. Williams attempted to murder Fountain and shot to death Gaylord Clarke after Fountain appointed the latter to a judgeship desired by Williams. In the years following, Cardis, Fountain, and Howard accused one another of assassination plots.[15]

Howard brushed off the threat, and Zimpelman engaged El Paso County's deputy surveyor, Ward Blanchard, to survey the lakes. In early summer, Blanchard set out from Ysleta with Howard, John E. McBride, three African Americans, and three Paseños, all unidentified. McBride, thirty-two years old, was another Canadian who preferred the warmth of the Rio Grande country. A former soldier, he tried his hand working as a civilian saddler and then as a freighter. In June 1871 he sold his wagons, oxen, and cattle in order to settle down and marry Louisa Lafayette. In 1873 he was elected a justice of the peace, and one year later his wife gave birth to a daughter, named after her mother. McBride's good fortune failed when his wife died, leaving John to raise little Louisa on his own. McBride was tough, reliable, and completely loyal to Charles Howard.[16]

No sooner did the party reach San Elizario than rumors of an ambush ahead caused the Paseños to back out. Blanchard's remaining party headed first to Fort Quitman, eighty miles southeast of El Paso. From this point, they intended to survey silver mines supposedly nearby. Fort Quitman had been vacated by the army the previous January, but Cardis operated a stagecoach station here. As coincidence would have it, Cardis was at the station inspecting the condition of his animals and equipment. Upon unexpectedly sighting his enemy, Howard assaulted him, "accusing him of trying to have us mobbed." The smaller man fled into the back room of the station house, with the burly, furious judge following. Howard had reportedly pounded Cardis twice before, in Austin and San Antonio. Now, gun in hand, he was prepared to do it again or do worse. Howard's friends waited, expecting the explosion of gunfire from within. A few moments later, Howard emerged, without the grim look of satisfaction that only Cardis's death could give him. Ben Dowell heard the reason: "Cardis had run under a table and acted so cowardly that he didn't have the heart to kill him."[17]

There were no further run-ins, and the party moved north to the salt lakes. Here Blanchard surveyed three sections of 640 acres each, adjoining Maverick's sections. Upon their return, the survey documents were dispatched east to be filed with the General Land Office in Austin.[18]

Before Zimpelman's patents were issued, Howard announced "that no more salt could be taken from the Lake, except with my permission, at the same time offering the same price per *fanega* for the freight that they had formerly received for sale of salt." At first, Howard received no takers. Paseños "refused to freight salt for [Howard], and said that they had been advised by Louis Cardis that the Lakes belonged to them." John Clark advised Paseños to buy their salt from Howard, "and afterward they could see who owned the salt and then get back their money if the salt lakes did not belong to Howard." The people would not budge. They would collect their salt as before, holding firm "that they were going to get salt whatever the consequences might be; that the lakes belonged to them by right of discovery, which they would use, law or no law; that their arms were the law."[19]

The people also responded politically to this direct challenge to their communal ownership of the lakes. Effectively cut off from Mexico City's centralized control, Paseños had long relied upon local institutions to govern themselves, such as *ayuntamientos*, or town councils, also known as *municipal juntas*. Ad hoc juntas were now called together. Several key leaders can be identified, notably Sisto Salcido, Leon Granillo, José María Juárez, and Luciano Fresquez, all of San Elizario, and Francisco Barela of Ysleta. Juan Mauro Lujan's home was used as a junta meeting place. Some participants reported that Cardis blamed Howard for stirring up trouble, "on account of his malice toward himself and the Mexican people." After listening to his own junta informants, Edmund Stine was sure that Cardis sought "the destruction of Howard." Stine was warned that Cardis "advised the Mexicans to kill Howard as the best means to recover the salt lakes to themselves." Not everyone spoke of taking arms against the law. "The only plan discussed," Blacker was told, "was to defeat [the Howard-Zimpelman title] in the courts in a regular proceeding, and by the laws of Texas."[20]

Subsequent events indicate that the juntas decided that summer to defend the peoples' salt rights. Circumstantial evidence indicates they established a system of rapidly communicating unfolding events and calls to armed action. The war that soon consumed El Paso County was not the result of mob violence but the product of a widely shared community decision to organize and act decisively, along legal lines if possible but militarily should Howard force the issue. In fact, what happened in El Paso strongly suggests similarities between American colonials who resisted the execution of British laws deemed unjust and those of the Paseño juntas. Like the Sons of Liberty and committees of correspondence or safety, the insurgent Paseños "acted with vigor, indeed with violence."[21]

Insurgent Headquarters (Juan Mauro Lujan home, San Elizario). Courtesy of the Center for American History, UT Austin, CN Number 07120, Webb (Walter Prescott) Papers, 1857–1966

The juntas continued as August gave way to September, amidst growing anxiety over the weather. The summer of 1877 came and went without rain, the Rio Grande was uncommonly low, and the *acequias* were running dry. In such hard times, local farmers trekked to the lakes for salt to barter for other necessary goods. But here was Howard, posting signs that no one could take salt without his permission. Though he had offered to pay the going rate, they knew it was not Howard's salt. With the season's drought deepening, "a restless, uneasy, disturbing appearance among the Mexicans was apparent."[22]

While in Mesilla to handle certain legal matters, Cardis was interviewed for the *Independent* by his former mortal enemy, A. J. Fountain. The conversation (not published until after Cardis's death) turned to the subject of Howard, who, said Cardis, had pledged to gain ownership of the lakes for the benefit of himself, Cardis, and Borrajo. The deal, said Cardis, was the same one proffered Fountain years before. Howard had reneged by securing the lakes solely for his father-in-law. When Cardis protested the betrayal, Howard cursed him. A bitter Cardis wanted no trouble, "but that there certainly would be very serious trouble in the event of Howard's attempting to assert his right to the lakes."[23]

Meanwhile, a few men willing to freight salt for Howard brought in five hundred *fanegas*. This break in the ranks "greatly exasperated the remainder of the Mexican population." Judge Blacker confronted Francisco Barela, said to be organizing an armed convoy to take the salt, "law or no law." Barela denied it. Though he could learn "nothing tangible," Blacker firmly believed "that there was some kind of effort toward an armed force." He and the ever-alarmed Sheriff Kerber wired the governor on September 14, "Bands of Mexicans of this county, organized to take salt from Lakes, located by George B. Zimpleman [*sic*] by force. Say they will resist civil authorities. Not enough Americans for posse. Mexicans unreliable. Troops needed. Answer by wire. A speedy answer may prevent riot."[24]

CHAPTER 8

"My men are all frontiersmen."

While El Paso's politicians and capitalists concentrated on schemes of riches and ruination, other men contended with a land of lawlessness and barbarity that lay in all directions beyond the Pass. Men such as Price Cooper, Gregorio Garcia, Francisco Barela, and Jim McDaniels were attracted to the harsh country beyond the Rio Grande by the possibilities of monetary reward, community defense, or personal independence. They knew little of luxury and a great deal about searing heat, bitter cold, the taste of dust, the lack of water, and the precariousness of life. The grassy bolsons, shrub-pocked limestone mountains, and dry heat of the Chihuahuan Desert shaped such men, toughened them, and inured them to killing their fellow men. Indian fighters, freighters, stagecoach drivers, and cowboys daily faced the possibility of sudden death, whether dealt by Apaches, climate, or bad companions. Keeping their eyes open, their wits sharp, their canteens filled, and their guns clean, loaded, and within easy reach, they learned how to stay alive and dish out death. For these "knights" of the Trans-Pecos, the land beyond the valley was a training ground, a military academy turning out the soldiers who escalated the personal squabbles of El Paso's politicians into a full-scale war.

Until the railroad came, people at the Pass were entirely dependent on grass- and grain-fed animals for travel and trade. Travelers, animals, and vehicles made attractive targets. Anything could happen during the long journey along the long miles from San Antonio or the stretch south from Santa Fe. The federal government's unwillingness to adequately fund either its Indian reservations or its army encouraged renegade bands and whole tribes to periodically strike out in search of adequate food, the dignity of independence, and blood revenge. For the stage line operator and the freighter, Indian attacks were a constant threat and an expected cost of doing business.

One hardy soul who delivered mail and freighted goods was William Price Cooper. "Old Price" was a boy when the opening of the Santa Fe Trail to Americans brought him to Paso del Norte with his brother Richard. Price learned the freighter's trade and by 1843 decided to make the Pass his per-

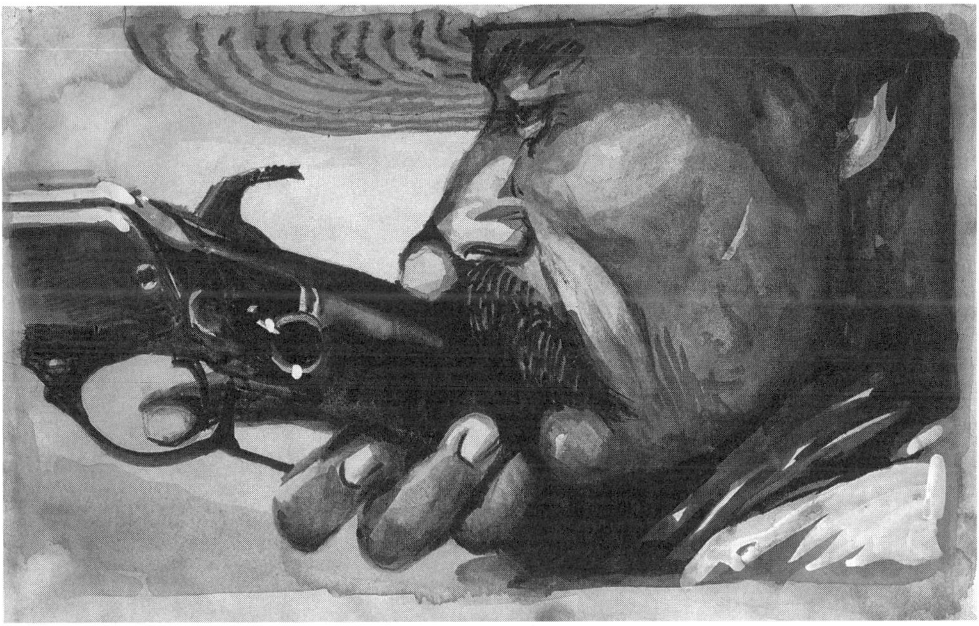

Paseño sharpshooter. *Courtesy of the artist, Bob Boze Bell*

manent home. He ultimately settled in Ysleta, marrying Maria Sanchez, and siring as many as seventeen children. His journeys took him north along the Santa Fe Trail, west to Arizona, and east to the Texas Hill Country. Though largely forgotten today, he was the subject of amazing stories during his life. According to one, Indians captured him twenty miles north of El Paso in 1848. His captors "sentenced him to death in that most horrible of all ways, roasting on the altar." For several days he sat, tied to a tree, watching a fire under his "altar" heat the stones. Finally, he was stripped of his clothing and led to the stone mound. Cooper had no intention of dying with heroic dignity. On the path to his grisly execution, he grabbed an infant from its mother, tossed it onto the altar and, in the moment's confusion, fled "at full speed" from the encampment. Pursuit began quickly, but "Cooper was the fastest man in the west." He ran naked across the Chihuahuan Desert until he ran smack into another band, unfriendly with the first. Though saved from immediate death, it took several days for his latest companions to let him go his way. Although his tales may have been embellished in the telling over the years, he was known as quiet and unassuming, not given to braggadocio. In a world of rough-and-ready frontiersmen, Cooper was considered the genuine article.[1]

Cooper worked for George Giddings's coach line from 1854 until Butterfield took over the route in 1858. Few businesses suffered greater operating losses or placed its employees more squarely in harm's way than the Jackass mail line, so named for the Spanish mules that drew the coaches. Cooper was kept busy replacing Giddings's stock and other supplies. The Indians "frequently stole from nearly every station on the route, horses and mules, and burned and destroyed considerable property at a good many stations."[2]

Cooper fought his way out of several roadside attacks. One such occasion took place in 1855 along the infamous *Jornada del Muerto*. This shortcut between Doña Ana and Paraje, just south of Fort Craig, bypassing a large bend in the Rio Grande, was well named: for ninety miles, travelers found no water, no firewood, no shelter, and no safety from Apaches. The attack in question resulted in the deaths of two teamsters, plus the loss of six mules, one (or more) wagon, and the mail. In November 1856, Cooper fought another running battle along the *Jornada*. Several passengers were wounded, but Cooper and others managed to shoot their way clear.[3]

On a spring day in 1866, Cooper, by then a freighter, lost his entire outfit in a Comanche attack. It occurred at the old Butterfield Head of Concho station, the first reliable water supply since he departed Horsehead Crossing some seventy-five miles to the southwest. Earlier that day, he had drawn his train up at the crumbling rock and adobe buildings that once served as the station. In 1866, the building was still an impressive fortress, with walls eleven feet high. At day's end, Cooper and one other man, a Paseño herder nicknamed "Old Joe," were still alive behind those walls. Beyond them were Cooper's teamsters, five or six Paseños, all dead or dying. Somewhere in the darkness, Comanches waited. Cooper had taken cover at the outset of the attack and stood them off all day, killing two. The dead men had been driving Cooper's wagons, each one loaded with salt from the Pecos salt flats. Each wagon had been pulled by six yoke of oxen. Everything of value in Price's business, his wagons, equipage, oxen, freight, horse, and silver-mounted Mexican saddle, was out there, unreachable and expendable.[4]

With darkness as cover, Price and "Joe" slipped out "and got away by going down the timber on the creek." Though not followed, more trouble lay ahead. They missed the nearest settlement, a ranch thirty miles away, and proceeded on foot over sixty more miles of creek bed and open prairie to the next ranch, belonging to Jim Ketchum. As long as they stuck to the creek bed, the imperiled men had fresh water to drink. Their route was filled with game, large and small, but they may have caught little. They were hungry, tired, and suffering from exposure when they finally reached Ketchum's place.

Before long, Price's fight at Head of Concho became locally known as "Cooper's Massacre." Financially, Cooper "went to the bottom, of course; they took everything he had, first and last and afterwards." He did own a home three hundred miles away in Ysleta, where his family waited while he worked the road, but he had no means to get back, not even money for a stagecoach ticket. He would return, but not in the spring of 1866.

Nearly a dozen years after the massacre, Price Cooper, by then a still-tough frontiersman in his sixties, was one of many Salt War Texas Rangers who rebut the charge that the unit was composed of "mediocrities or worse."[5]

By the time of Cooper's massacre, General Carleton's solution to the Indian problem, dumping Navajo and Mescalero Apache together in the Bosque Redondo, had already failed. The year 1865 had been "the worst year New Mexico had known for generations. Frost, hail, blight, drought, destroying insects—all seemed reserved for this particular time." The government responded to the crisis by cutting the Indians' rations. Convinced that winter on the reservation would be devastating, "every Apache who could travel arose and vanished" one night in early November. For more than seven years, the Mescaleros remained at large, within reach of any attractive target and just beyond the reach of punishment.[6]

The people and livestock of *la Isla* presented prime targets for the Apaches. Raiders struck the valley nearly every month. Martin Lujan of Socorro was hard hit in December 1865, when Indians made off with sixty-six horses and mules, together valued at nearly two thousand dollars. Though others lost far less, the loss of any beast could be a significant injury. The first deaths occurred in February 1866, when four citizens of San Elizario were killed and twelve oxen taken. None of this prepared the Paseños for what was to come. On March 14, 1866, Lujan's ox train was attacked. The wagons were burned, 103 oxen were driven off or slaughtered on the spot, while thirteen men were killed. One day later, San Elizario lost another ten men and nearly a thousand goats and sheep, along the road south to Presidio del Norte. In two days, the citizens of *la Isla*, most of whom lived their lives on a thin margin, lost over five thousand dollars in livestock and other property and twenty-three relatives and neighbors were dead. Not since the 1862 insurgency against the Confederates had Paseños suffered so grievously. The raids continued. Between March 1865 and January 1867, the people of *la Isla* reported the loss of more than twelve hundred animals, plus wagons and equipment, a staggering loss. Even worse, thirty-five Paseños were reported to have lost their lives in Indian depredations in less than two years.[7]

To counter continuing Indian depredations, in the spring of 1870, Governor Davis signed the Frontier Force bill. El Paso County was allotted a sixty-man Ranger company (soon reduced to fifty). Its payroll promised three thousand per month in wages to a poor county. El Paso's Company N was mustered into service on August 12. Its captain was an obvious choice, Gregorio Garcia, the veteran Indian campaigner who had saved Lieutenant Lazelle's column at Dog Cañon. Fifty-eight Paseños plus Civil War veterans Caleb Miller and Charles Kerber (company clerk) completed the roster. Though many Texas Rangers were related by blood and marriage, the presence of forty-two Mexicano surnames on the rolls testified to a community-wide response to the needs for soldiers *and* income.[8]

Putting Texas Rangers into the field—armed, equipped, and mounted—required money. When the army reneged on its agreement to provide the Rangers with weapons and supplies, Garcia was forced to campaign with "borought [*sic*] carbines and shoulder guns." Garcia stressed his need for arms, ammunition, and equipment "so I am able to chastise Indians," but did not wait to take the field. On December 6, he received word that Apaches had stolen cattle at a ranch some twenty miles southeast of San Elizario. A sergeant and ten men pursued, but the Indians escaped. An unexpected call for help came from El Paso on the night of December 7. Albert French, at that time the local state police chief and not yet insane, requested the Rangers's help to ensure order following the killings of Judge Clarke and Benjamin Williams. Garcia and his men saddled up at once, riding "25 miles in 2 hours and 15 minutes" from San Elizario to El Paso. The town remained quiet, and Garcia's men returned to camp two days later.[9]

In February 1871, Garcia's command cooperated in a winter campaign with four companies of the 9th Cavalry under Maj. Albert P. Morrow. The Rangers still lacked arms, ammunition, and sufficient pack mules, but the experienced Garcia still hoped to "make good record for myself and Company." The formations linked at Independence Spring, near the Guadalupe Pass. Here Morrow lent the Texas Rangers mules and armed them with Springfield needle guns. Though the ammunition was "quite unserviceable, not over half of them giving fire," at least the men were on campaign. While the cavalry scouted the center and east side of the Guadalupe and Sacramento Mountains, the Rangers scouted the west side. Garcia spotted the trail of a dozen or so Indians estimated to be eight days old. The Rangers followed the tracks as far as

Dog Cañon. Here the trail split in two, some Indians going to San Andres, and others headed to the safety of Fort Stanton. Finding no more signs, the Rangers returned to Independence Spring, "having marched an average of 26 miles per day over rocky and steep mountains, with scarcity of water, being 78 full hours without a drop of water for man or beast, also under very cold weather." Another seven-day patrol east to the Pecos river country also failed to turn up Indians. Garcia made the best of his obvious disappointment. "The reports of the murdering of 8 Americans & one white lady about 90 miles west of the Rio Grande, also of the many murders in New Mexico and Arizona makes all believe that the Indians are aware of our large command or fled to other parts."[10]

In April, nearly eight months after they were mustered in, Company D finally received arms, Winchester carbines instead of the promised rifles. Garcia also received fifteen thousand rounds of Winchester ammunition and five thousand rounds of ammunition for pistols, but no pistols. Unfortunately, the Frontier Force was disbanded in June 1871. Captain Garcia's final report eloquently expressed his pride in the company's service as well as his many frustrations. Though his men had not met the Indians in battle, Garcia was confident the unit had performed a great service to Texas. "[T]housands of miles have been traveled by my Company under great wants of water and great cold so we could make drive [*sic*] the Indians for Arizona and Mexico. We succeeded in that." His boast was not idle. As Texas Ranger historian Frederick Wilkins noted, "Indians tended to avoid regions in which they observed the tracks of shod horses, and there is little doubt that Indian raids decreased in the districts patrolled by the Rangers." Indeed, the Apaches resumed their raids along the river once his unit disbanded.[11]

The company was discharged without having received any pay. Garcia advised the adjutant general, "[T]he majority of the Company are poor men and heads of families; for most their horses and pistols I myself had to go security with the promise to pay them at the first payday." Garcia's warning that the Democratic Party would make hay with this nonpayment of sixty Paseños at the next election had its effect. The Rangers finally received their full back pay, a small fortune for some. For others, perhaps even more valuable was the military experience gained.[12]

In the spring of 1874, the state government raised special companies of Texas Rangers called "Frontier Men" or "Minute Men" to meet emergencies on the frontier. The need was great, and there was no difficulty finding officers and recruits for the El Paso company. On May 27, the twenty-four Paseños and one Anglo sworn in as Frontier Men voted in their officers and non-coms. Telesforo Montes, a leader in the community's opposition to the recent Texas school law and one of Captain Garcia's sergeants in 1870 to 1871, was elected lieutenant commanding the company. Rangers that would figure prominently in the Salt War included Sgt. Tomas Garcia, Cpl. Francisco Barela, and Pvt. Sisto Salcido, another veteran of Garcia's company. The outfit was something of a family affair. In addition to Tomas Garcia, Montes commanded son-in-law Carlos Garcia, their brother Secundio, and the lieutenant's own sons, Jesus and Severo, a fact that later caused complaint.[13]

Again, the El Paso Rangers lacked sufficient arms and ammunition. Echoing Garcia's earlier pleas, Montes prodded Adj. Gen. William Steele (the same man who had led the Confederate rear guard through El Paso in 1862), "The Company are all anxious to go on a scout to prevent depredations from these marauding Indians, & to drive them from the Country or secure their capture." Montes's Frontier Men did not wait long. In July the Rangers received twenty Sharps rifles. Texas was stingier with ammunition, furnishing only two boxes of a thousand rounds each. Supply snafus were repeated. The pistol cartridges were unsuitable for the Rangers' old Colt revolvers. Promised cartridge belts and boxes never arrived, prompting Montes to hire a local saddler to make them.[14]

In mid-September the Montes company rescued an eight-year-old boy. The Rangers first scouted to the Guadalupe Mountains, then struck a trail headed back toward the river. The Rangers came upon their quarry on the third day out, camped in a canyon inaccessible to horses. It was a party of seven, one of whom was the captive boy. Montes dismounted his men, left half with the horses, and took twelve with him into the canyon. The Rangers attacked, killing two men, scalping one. The boy, unharmed in the action, had been stolen from his parents near Presidio del Norte eight months earlier. The Rangers also captured five horses, gunpowder, lead, saddles, and "camp paraphernalia," all "divided among the men." Evidence of the Apache band's recent success was found in a pile of "dry goods, such as muslin, calico & evidently recently acquired which had not been unfolded." The captured horses were in poor condition, but one was in better shape than Montes's "exhausted" mount. With General Steele's permission, Montes later exchanged his horse for the captured animal.[15]

Just as it proved its worth, the El Paso company was ordered to disband. Montes warned that Indian raids would increase, leaving citizens defenseless. Opting to reinforce success, Steele kept the company in force for two more months. Unfortunately, the Rangers failed to catch the Indian parties they tracked to the Sierra Blanca Mountains in late October and to the Guadalupe Mountains in November.[16]

Budget cutbacks forced the disbandment of the emergency Ranger companies that winter. The Apaches thereupon resumed their attacks upon travelers and ranchers. The citizens of *la Isla* were hit in mid-September 1875 by Indians who ran off four horses from San Elizario and then attacked a salt train, capturing six mules. Alerting General Steele to the newest depredations, Montes asserted, "The U.S. troops don't understand the character of these Indians, nor are they acquainted with the character of the country. My men are all frontiersmen, thoroughly acquainted [with] this whole country and we [are] well versed in Apache tricks and tactics."[17]

The Rangers were mustered back in on September 27, 1875. Again, they were indifferently armed (apparently the Sharps rifles used in 1874 were unavailable, though the correspondence is silent on this fact), and ammunition was again a problem. A raiding party hit Socorro on November 11, taking nine horses. The Rangers followed the tracks southeast until they reached the Eagle Mountains, where they came upon a campsite with the fires still burning. Montes followed a new trail south, though the Ranger horses "were jaded for want of water." Six hours later, the Frontier Men spotted three Indians on horseback fleeing up canyon walls. The Rangers wounded one warrior, but his comrades carried him to safety. The Indians left behind most of the stolen horses and three ponies. Montes felt that, "Great praise is due to the men under my command, in the manner they behaved, being without food or water for 36 hours."[18]

Following this foray, the company ceased campaigning until spring. Again the Indians raided *la Isla* in relative safety and with some success. The unit mustered back in on March 25, 1876. By now there was bitter dissension between the lieutenant and one private. The first flare up occurred during a nine-day, 215-mile scout launched on April 1. The company marched from San Elizario to the salt lakes, where they found the first water in two days. From there they rode eighty miles to the Rio Grande and then back to camp, without having spotted Indians. Private Miguel Garcia (a cousin of the captain's family) sent his own report to General Steele, complaining that men and horses had been mistreated by Montes's manner of marching them for hours on end without food or water. "All of us knew that the expedition was bad, but none of us could say anything." Garcia was particularly critical of one militarily useless excursion. At one point Montes "ordered the soldiers to climb a hill and cut Sabina (a Juniper bush), which was loaded on the pack mules and brought down, to help celebrate Palm Sunday at church." Finally,

Garcia noted, "most" of the company were members of Montes's family, "and that's why he has lasted this long as a Lieutenant."[19]

The next patrol caused an even greater tempest. On the night of April 16, Indians stole some mares from San Elizario and cattle from the tableland above Socorro. Montes had the company on the road at dawn on the April 17. Upon reaching the Hueco Tanks, the Rangers discovered the remains of two butchered steers. Montes watered the horses, filled the canteens, and pushed on in pursuit of the Apaches, only two of whom were mounted. Miguel Garcia told a different and troubling story. Having reached the Hueco Mountains, where the *tinaja* (basin) was "so big the water lasts all year long," Montes took the company in search of another pool, "but he did not find any, got lost, and finally found his way back to the same *tinaja* in Cerro Hueco. We were unable to overtake the Indians that day." Garcia complained that the Rangers did not leave the Tanks until the following day "at mid-day."[20]

The pursuit resumed. Following the first night, the men found "a little bit of water" by digging for it. Garcia charged that Montes's horse was given "six big glasses of water and after that, water was also given to the horses of the two sergeants who are his sons in-laws. Water was also given to the horses of the corporals, the water was [then] gone and none was left, not even to fill one canteen." Nevertheless, the chase continued until the men reached a small gorge, where they again found a small amount of water. The command then rode on to the familiar Dog Cañon. The Rangers moved into the steep gorge to water the horses and rest. According to Montes, within moments they were struck by a ragged volley from above. The Rangers "looked up and saw the mountains thick with Indians shooting from all sides." Garcia said they camped "without any precautions," an assessment implicit in Montes's own report. Garcia obtained permission to scout the canyon. He, a corporal (possibly Barela), and five others scouted deep into the canyon, when they were met by the report of a dozen guns. According to Garcia, "they were coming towards us and I told them to stop, but they did not pay me heed. They took some of my companions. [Montes mentions no casualties.] One was coming after me, but could not get me because I give him a terrible blow." Garcia reached the main party, where he found the Rangers had saddled their horses under fire. Montes judged that two hundred or more Indians were shooting at them, armed with needle guns and pistols. Collecting the horses as quickly as possible, the Rangers raced out of the canyon. From the opening they challenged the Apaches "to come out where we was and fight and they said for us to come in there and get the horses if we wanted them."[21]

At dark, the Rangers retraced their steps and then cut toward the San Augustine Pass over the Organ Mountains. The bedraggled column—they had not eaten in twenty-four hours and some men had not had time to fill their canteens—reached the Shedd Ranch, notorious as "a known and welcoming haven for every horse and cattle thief in southern New Mexico." There Montes bought a cow for his men to eat. They may have paid Warren Shedd's standard rate of one dollar for a bucket of water, a bargain for thirsty men. After four days on the trail, the company reached home on the morning of April 21.[22]

A month later, Juan N. Garcia wrote his own letter to General Steele in support of Miguel (who may have been his older brother). Miguel's reports, said Juan, were "the unvarnished truth." Juan Garcia gibed, "The battalion of Senor Montes is not solely made of his family because he does not have enough sons to . . . make 25 that are needed." Montes, he added, was motivated by his "vile interest" in making money off the government. As to Montes's soldiering, "[O]ne can say that he does not know even one military tactic to guide his soldiers. Pure favoritism alone has made him lieutenant."[23]

Perhaps coming to his point, Garcia added, "There are men here that are more skilled in every way than he." Who were these men? Garcia noted that both Captain French and Cardis would back up his charges. (Cardis had previously lobbied to secure the lieutenancy for French.) Juan Garcia claimed not to want the job for himself, but he coyly begged Steele "to permit me become one of your servants. Obediently fighting . . . under your orders."[24]

Montes's report of the last patrol, though vague on details, appears to support the conclusion that his company failed to exercise ordinary caution entering Dog Cañon. While they could not have known that scores, even hundreds of Apaches were lying in wait for them along the canyon walls, the lieutenant should have anticipated some trap might be set in this place. The charges leveled by the Garcias either indict Montes as inept (on this occasion) or only provide evidence that the Paseño community, like any other, was divided by party politics, personal jealousies, and strong opinions about how things should be done.

Indian raids increased after the Rangers stood down. People were murdered, cattle stolen, wagon trains attacked. The army could do nothing. The only troops close to the settlements were a few companies of the slow-moving 25th Infantry posted to Fort Bliss. Even their regimental history described their ten year record in Texas as largely devoid of contact with Indians, except for "a few unimportant skirmishes."[25]

It would be two more years before Texas Rangers based in El Paso would take the field against hostile Indians. When that happened, they would be different Rangers. Instead of Paseños, who provided seventy-eight of the eight-one men identified in Captain Garcia's and Lieutenant Montes's companies, the El Paso Rangers of 1878 and for years afterward would be almost entirely Anglos. The Salt War served as the dividing line, the event that made it impossible for the county's Anglo power structure to trust its fate, or its guns, to "Mexicans." But if the Salt War put an end to the tradition of Paseño Rangers in the service of Texas, it was one more chapter in the history of Paseños military service to their community. For no fewer than ten veterans of Garcia's and Montes's companies fought with the insurgents against a new, and largely outsider Texas Ranger company in 1877. Two of them, Salcido (a veteran of both companies) and Corporal Barela, were leaders of the popular rebellion. Miguel Garcia, once more a Texas Ranger, stood against them, as did private citizen Gregorio Garcia and his sons, once more defending their community against a perceived threat to its present and future.[26]

While El Paso's Texas Rangers scouted in search of Apaches and stolen livestock, other men rode across the same landscape in search of cattle to steal and competing cattle thieves to put out of business. Such men would arrive in El Paso during the climax of the coming Salt War, some as Texas Rangers, others as "volunteers" committing murder and mayhem at the behest of the overwhelmed county sheriff.

Five of that type, dirty and thirsty after sixty miles in the saddle over the roads from Mesilla, arrived in Ysleta on a hot summer day in 1875. Shaking off the dust, they went looking for a judge. It was the first of August or thereabouts, and they were hard on the trail of rustlers whose path led across the river into Mexico. Their quarry was the Maese (or Mes) gang, "banditti of the worst class." The five men, cowboys from Lincoln County, were determined to retrieve their livestock and finish the gang's criminal career. To do that, they needed to cross the Rio Grande, and for that they wanted the law on their side. And so District Judge Charles Howard first made the acquaintance of the cowboys' leader, an amiable rogue, standing five foot eight, with auburn hair and blue eyes, named Jim McDaniels. Only two years later, the cowboy would daringly expose himself to the aimed rifles of a Paseño army in an attempt to save the lives of Judge Howard and his fellow Texas Rangers.[27]

McDaniels and his men swore in affidavits that the Maese gang had stolen a score or more of horses and mules in Lincoln County. Howard checked the men's story, and confirmed "that Mr. McDaniel was a respectable citizen of the Territory of New Mexico, and that the persons accused, for a long time had been known to belong to an organized band of thieves and murderers, infesting the lower part of New Mexico, and committing almost weekly robberies." The judge issued a writ of extradition and sent it to the *jefe politico* at Paso del Norte.[28]

McDaniels presented Howard's writ to the *jefe,* who had five members of the gang arrested in San Ignacio. One escaped extradition, but four were turned over to El Paso alderman John S. Gillette. Without giving notice to Howard or Sheriff Kerber, Gillette delivered them to McDaniels. The cowboy, "having previously [given] his word of honor, that he would turn the murderers and thieves over to the civil authorities of Lincoln county and not kill them on the road," quickly departed with his prisoners in tow.[29]

On August 8, the McDaniels party reached the San Augustin Pass. From here he sent word to Las Cruces "that he heard friends of the thieves were coming in a band to rescue them." At about 11 P.M., according to McDaniels, "a party of masked horsemen rode up, demanded them to halt, and with their guns pointed at them marched McDaniels and guard one side of the road 75 yards, then returned and shot the prisoners." McDaniels could not identify the masked assassins. A coroner's jury determined "the men came to their death by gunshot wounds in the hands of unknown parties." The *Mesilla News* declared the killings "unjustifiable," adding "this may be a warning to thieves and murderers that it is safer to operate in some other section of the country."[30]

Open to criticism for having facilitated the prisoners' murder, Judge Howard declared that state and local authorities were neither careless nor to blame for the murders. In any case, he told the governor, "There is no doubt that the men killed were robbers and murderers in fact banditti of the worst class."[31]

Jim McDaniels was likely responsible for the Maese gang's demise. He drew his gun indiscriminately, against Indians, rustlers, soldiers, Lincoln County "Regulators," personal enemies, and Paseño insurrectionists. He seems to have been amiable but not above playing the bully. Described as a man who avoided fights, he was in one scrape after another. His courage is unquestioned, yet he shot unarmed or unsuspecting men more than once. He would give a horse to a widow in need, but recoup the loss with the theft of three more.[32]

He first enters the historical record as a young foreman for John Chisum. Perhaps McDaniels was, like saddle pal Jessie Evans, more useful as a nervy gunman. His fearlessness in a scrap cannot be disputed. In 1872, McDaniels and ten or eleven companions fought off some fifty to seventy-five Indians in a day-long battle near Horsehead Crossing. They lost Chisum's herd of horses and mules and were forced to walk back alongside their wagons for 250 miles. In 1874, McDaniels left Chisum's employ "because of a dispute over an alleged indebtedness."[33]

In Las Cruces on New Year's Night 1876, he committed wanton murder. Five others joined him, including three of the worst desperadoes in New Mexico Territory: Jessie Evans, John Kinney, and Charles Ray, the latter also called "Pony Diehl." It began at a New Year's *baile,* or dance, when a row between troopers of the 8th Cavalry and "Texas cow boys" ended with Kinney getting pummeled. The cowboys retreated from the building. While the soldiers celebrated their triumph at the bar, Kinney, McDaniels, and their pals slinked up to the windows and began firing from the outside. According to the *Mesilla News,* "the fight *wagged* hot and lively." When it was over, six revelers had been shot, including three men killed or mortally wounded. In his report, Lieut. Col. Thomas C. Devin bitterly noted that "through the supineness of those charged with the administration of the civil laws not a murderer has been punished by law."[34]

John Kinney, a native of Massachusetts, born in 1853, spent three years in the 3rd Cavalry Regiment, mustering out as a sergeant in 1873. It is possible he and Ray were associates of McDaniels and Evans as early as the murder of the Maese gang. Within a short period of time, these four would all become infamous members of what Lincoln County historian Frederick Nolan calls "the rustler chain," stretching from the Colorado to the Pecos on both sides of the Mexican border. They shared one profession besides rustling—"lawman." Whenever a county sheriff needed gunmen, they were among the first to hire on, bringing mayhem, not order. Between them, these four participated in virtually every major "war" along the border between 1877 and 1883. The first of these conflicts would take place in the towns of *la Isla.*[35]

CHAPTER 9

"Invasion of our territory should no longer be endured."

The El Paso Salt War did not happen in a vacuum. Tensions between Mexico and the United States were already high in 1877. Relations between Mexico and the Anglo-dominated state of Texas were beyond bad. The Texian rebellion, Alamo, Goliad, and San Jacinto had been succeeded by decades of border violence committed by both sides. Mutual distrust, born of racial and ethnic enmity, cultural differences, war, and real or imagined territorial designs, was fed by intractable border chaos. Indian bands took full advantage of international complications to raid back and forth across the Rio Grande. Rustling by bandits "had become an organized business probably overshadowing all other forms of border disorder." Filibustering expeditions bent on plunder added to the grievances. Tax-cheating smugglers, a problem encouraged by the *Zone Libre,*[1] and *prestamos* (forced "loans" to rapacious government agents) exacerbated the cost and dangers of doing borderland business. Each side was culpable for the troubles; each side saw itself as the victim. The political weaknesses of the two nations' presidents prevented cooperation toward mutually acceptable solutions.[2]

Mexico was torn by political unrest and civil war following the defeat in 1867 of the French puppet Maximilian. Relative peace was shattered in January 1876 by renewed civil war resulting in the presidency of Gen. Porfirio Diaz. Consolidation of power in Mexico City was one thing; assertion of authority in the northern borderlands was another matter. Followers of deposed President Lerdo de Tejada established themselves in Texas, using her soil to stage their attempts at military counterrevolution. While the *Lerdistas* never seriously challenged the Diaz presidency, they contributed to border instability. Their filibustering gave Americans second thoughts about the new regime's stability and contributed to President Rutherford B. Hayes's delay in recognizing Diaz's government.

Paso del Norte escaped serious upheaval during 1876, but the relative quiet here had its downside for El Paso County. Persuaded by apparent peace and budget cuts, the U.S. Army closed both Fort Bliss and Fort Quitman. The nearest troops were then at Fort Davis, two hundred miles south and east, and Fort Stanton, New Mexico, some one hundred fifty miles to the north. Citizen anxieties raised by the army's departure were expressed by Sherman C. Slade, the U.S. special deputy customs collector at El Paso, Texas: "Northern Mexico is now in a state of revolution and anarchy, filled with desperadoes and outlaws only waiting an opportunity for general plunder. The absence of troops here will embolden them, and leave us at their mercy on the one side and the Indians on the other."[3]

North of the border, Republican Hayes was the beneficiary of the disputed 1876 presidential election over Democrat Samuel J. Tilden. As his party's compromise, presidential nominee Hayes had only lukewarm support from both the moderate "Half Breed" wing of the Republican Party and the "Stalwart" wing led by New York Senator Roscoe Conkling. Early in his term, he alienated both factions. Hayes paid the price of his electoral victory by withdrawing troops from Louisiana and South Carolina, turning those states over to the Democrats. His actions served to undermine equal rights for African American citizens recently gained by the hard-won Union victory. As a result, the "Half Breeds," his natural allies in the civil service reform fight, fell out with their president. His dedication to ending the spoils system also earned him the bitter opposition of the "Stalwarts."[4]

Hayes, who had visited Texas for several months in 1849, found friends among the Democrats in the Texas congressional delegation, whose goals included an end to border depredations. Gen. William Tecumseh Sherman, commanding general of the army, credited the Texas delegation with preventing the Democratic-controlled Congress from slashing the army's strength during the 1877 term. The most powerful Texan voice in Congress was that of the mountainous Gustave Schleicher, who represented San Antonio and points west. Weighing 350 pounds and standing six feet eight when not seated in a specially made chair he used in the House, this fifty-three-year-old German immigrant was chairman of two House committees important to those interested in Mexico, Foreign Affairs and Railways and Canals. Schleicher favored unilateral U.S. military action, including hot pursuit and punitive expeditions, to halt Indian and outlaw raids launched from Mexico into Texas.[5]

Another Texan favoring hot pursuit influenced the president's views on border policy, Guy M. Bryan, a classmate of Hayes at Kenyon College. Hayes had stayed with the Bryan family on that extended Texas visit, and they remained warm friends through the years. Bryan, speaker of the Texas House of

Representatives in 1874–76, visited his old college chum at the White House for three weeks in April 1877. There he advanced his support of an aggressive border policy, namely, "that the best course for our government to pursue to prevent raids, is to follow raiders into Mexico, catch and punish them until Mexico can control her raiders; if war follows this course and it cannot be prevented, then let it come."[6]

Both Schleicher and Bryan were strong proponents of railroad commerce, and it was economic rather than territorial expansion they saw stymied by endless border troubles. Some Texans favored the permanent occupation of northern Mexico and the establishment of a protectorate over a new "sister Republic," wishfully called by some the "Republic of Sierra Madre." An American protectorate, said the *San Antonio Daily Express,* would secure Texans in their lives and property and be a boon to American commerce. Bryan warned "Rud" Hayes against such action, advising, "Efforts will be made to drag you into a War with Mexico[.] *avoid it if you can.* We want no more territory or semi-barbarians now. We have enough of both."[7]

Just ten weeks after Inauguration Day, the Hayes administration expressed its frustration with the breakdown in law along the Rio Grande. Kickapoo and Apache raiders, rather than *Lerdistas,* were the chief cause of this chaos. In response, U.S. Army columns had many times crossed the Rio Grande in pursuit or to retaliate. No Mexican president dared to openly sanction these incursions. On June 1, 1877, Secretary of War George Washington McCrary took American aggressiveness and Mexican anger to new levels, instructing Commanding General Sherman to, "direct General Ord that in case the lawless incursions continue he will be at liberty [to pursue] marauders . . . across the Rio Grande, and to overtake and punish them, as well as retake stolen property."[8]

McCrary's order caused indignation in both countries. Maine Senator and presidential hopeful James G. Blaine, leader of the "Half Breeds," suspected that Democrats, already running the House of Representatives, wanted war to gain the states that would give them control of the Senate as well. These suspicions prompted Blaine to denounce "any policy which contemplates or by any possibility involves the annexation of Mexican territory or any other addition to the disturbed element of the South."[9]

The order "caused an extraordinary outburst of rage in Mexico." Jealous of their sovereignty, Mexicans feared that the order was a prelude to annexation, whether by incorporation or creation of a protectorate. An active Mexican press was watchful for any signs that a weak Diaz might countenance any encroachment of territory. Even the Mexican Catholic Church, through one journal, suggested that any invasion would lead to the guerrilla warfare that so recently ground down the French occupiers. Diaz dispatched soldiers under Gen. Geronimo Trevino to Piedras Negras, opposite American troops near Eagle Pass, with orders to "repel force by force." Fortunately, he and General Ord quickly established a trust that each would step up patrols to stop raiding parties.[10]

An attempt to raise the *Lerdista* banner at Paso del Norte in May served, in some ways, as a dry run of later Salt War violence. The episode highlighted the fragility of civil order at the Pass, the porous nature of the border here, the courage of the ordinary Paseño, and the need for U.S. troops to ensure stability on the American side. In the predawn hours of May 2, a band of twenty-eight Lerdo men under the command of Colonel Paulino J. Machorro silently crossed the Rio Grande from the American side to capture Paso del Norte, including the *jefe politico,* a collector of customs, and the city guard. "This was done without firing a gun & by breakfast time he was in possession of the whole town."[11]

Machorro needed men and money to capture the state capital at Chihuahua. Rumor spread that thousands were answering his call for volunteers, prompting the *Porfirista* governor to decamp from Chihuahua City, "without saying goodbye." In reality, Machorro was forced to dragoon citizens into service. Cash was even harder to obtain. Machorro's extortion of fourteen Paso del Norte merchants, including ten American citizens, was ham-handed. The Americans were further outraged by the new *Lerdista* mayor's repeated public threats "to run every *Gringo* out of the place." Schutz feared that would come to pass unless "the strong arm of the military" returned to Fort Bliss.[12]

The citizens of Paso del Norte were less impressed with the adventurer. When the colonel levied a *prestamo* of 1.5 percent against those on the tax rolls, "the poorer people complained and demurred." Machorro's threats to shoot resisters only stiffened the people's resolve. On Monday, June 3, a junta began to organize resistance. On Tuesday, a "fierce fight" between Machorro's followers and a junta guard left three citizens dead and one or more *soldados* killed or wounded. "This lent the citizens courage and they advanced and soon all of the hills near the barracks were covered with riders who carried weapons of the previous century. There was even a body of 40 Pueblo Indians armed only with bow and arrows." When a growing force of citizens gathered around the army barracks, the colonel "held his life more dear than his honor and fled on his horse to the American side of the river, not bothering about the fate of his followers." By midnight, the citizens were again in control of Paso del Norte, ending what Kohlberg called one of Mexico's "7 day revolutions."[13]

Machorro's defeat threatened peace on the American side. New *politico jefe* Jesus Telles threatened to forcibly abduct *Lerdistas* from Franklin if the Americans failed to hand them over. The citizenry "at once assembled and organized themselves for protection, appointing committees of safety and defense." Slade informed the treasury secretary, "A sharp conflict is anticipated." Too few in number to withstand any determined assault, the Americans prepared to sell their lives dearly. "How this may finally end is hard to tell," U.S. Commercial Agent Solomon Schutz advised the State Department, "but at present everything is excitement, and everybody seems to feel that the worst is yet to come."[14]

Cardis and others telegraphed a plea to General Ord, "as we are . . . now more than ever, exposed to the insult and depredations of the Mexican revolutionists." Acting as he would in response to El Paso's later crisis over salt, Ord referred the problem to Brig. Gen. John Pope, commanding general of the department of Missouri, "as having troops more convenient than those of Texas." Recognizing a buck passed, Pope complained to Lieutenant General Sheridan, commanding general of the military division of Missouri, "I have . . . no jurisdiction over [El Paso], nor am I sufficiently acquainted with the facts and orders on the subject of the Mexican border troubles to take any action. Even if I were, I have not the troops to send without exposing the people of New Mexico to the Indians in the southern part of the Territory." Sheridan forwarded the message to Sherman. It was Pope's job, Sherman replied, to ensure that U.S. soil not be used as a base for carrying on a personal war with Mexico. Pope was instructed to send an officer to report the facts "and, if need be, a company of troops." That settled the matter. On July 12, 1877, 1st Lt. Louis Henry Rucker was dispatched to investigate matters in El Paso County.[15]

Rucker's junior grade belied his military experience. A native of Chicago, born in 1842, he distinguished himself during the Civil War as a member of the excellent 8th Illinois Cavalry, notably under Buford in the Gettysburg campaign and in antiguerrilla operations against John Singleton Mosby in northern Virginia. He mustered out as a First Lieutenant in April 1865.[16]

In December 1865, Rucker applied for one of the Regular Army vacancies reserved for "Young men who served at least two years in the volunteers." Letters of recommendation from superiors described him as "a man of education, great energy of character, and of sound morals and sober habits . . . an excellent officer," and as "faithful and brave . . . one of our best officers." Rucker was appointed Second Lieutenant, 9th U.S. Cavalry on July 28, 1866.[17]

The 9th Cavalry was one of four regiments of "Colored" troops established by Congress on July 28, 1866.[18] Its colonel, also a brevet major general, was Edward S. Hatch, whom Grant called "one of the most active and efficient Cavalry Officers in service." Hatch now relied upon Rucker, his regimental quartermaster, to discern the situation at El Paso.[19]

Rucker reached Franklin on July 18 and immediately began gathering intelligence. He observed that the Mexican population was calming down, but Mexican authorities were still "very much excited" by the *Lerdista* threat. While Colonel Machorro had apparently departed for San Antonio, leaving a band of supporters in Ysleta, both Sheriff Kerber and *Jefe* Padilla warned Rucker that the colonel was expected to return on July 23 for another assault on Paso del Norte. The Mexicans had posted guards up and down the Rio Grande for forty miles. Upon further digging, Rucker discovered . . . nothing. If Machorro's followers were reorganizing, "their meetings are held so secretly that the fact is not known." Slade's customs inspectors were "on the lookout" for the *Lerdistas* but did not spot any. Nevertheless, Rucker found among the Americans "a general feeling of insecurity prevailing here on account of the absence of troops and inability of civil authorities to enforce the laws when necessary."[20]

As summer progressed, Machorro's forces moved about as fast as the latest rumor. On July 24, Rucker reported the colonel on his way to Ysleta. Other information had *Lerdistas* organizing at Las Cruces. "The citizens of El Paso [Mexico] are preparing for defense—have barricaded the streets, and taken other necessary precautions to prevent the capture of their city." Based on these messages, Hatch advised Pope, "It is evident . . . that the garrison of El Paso should have been retained." Pope disagreed but authorized Hatch to send a company down to El Paso "to prevent any party from invading Mexico from our territory."[21]

Rucker reported "all quiet" until August 8, when Jesus Padilla, the latest Paso del Norte *jefe*, advised the lieutenant that Machorro was now organizing eighty men at the Hueco Tanks and planning a move into Mexico on August 10. The predicted *Lerdista* attack never materialized. Machorro's men apparently left the region. Still, Hatch cautioned Pope, "It will be well to keep an officer at El Paso, however, for some months to come."[22]

In early August, a border incident occurred at Rio Grande City, situated roughly half way between Laredo and the Gulf of Mexico, which provided ammunition to those who sought war with Mexico. It also illustrated the difficulties in applying military force in response to civil disturbances, even before the passage of a *Posse Comitatus* law. A dozen armed men from Camargo, Mexico, broke open the city jail and liberated two criminals, one a *Lerdista* officer, the other a notorious killer. In the breakout, the gang shot the county attorney, a jailer, and his wife. Troops from nearby Ringgold Barracks arrived too late to prevent the Mexicans from recrossing the Rio Grande.[23]

The soldiers had little incentive to catch up with the gang. In testimony that helps explain the army's later inaction up the Rio Grande at El Paso, Ord stated, "I should be very glad to have some State troops [i.e., Texas Rangers] or deputy sheriffs along with us on occasions when we pursue these raiders," just in case anyone shot by troops turned out to be a Texas citizen. "We have been already placed under serious embarrassment because pettifogging attorneys take up such cases and cause the arrest of the officers and men whenever we attempt to apprehend these raiders. . . . State troops can commence firing upon them; whereas if we did so, and shot any of them, their friends would have us arrested immediately and tried for murder."[24]

Governor Hubbard requested extradition of the two prisoners and their rescuers. On orders from Mexico City, local authorities did extradite to Texas the *Lerdista* prisoner and two of his attempted rescuers, but failed to assist further. Still, the delivery of three men to the Americans "provoked a fiery controversy throughout Mexico." When Mexican authorities released five jailed criminals wanted for extradition by Texas, the Extradition Treaty was, for all purposes, dead. The lack of an effective extradition agreement would later bedevil American authorities wishing to lay hands on Salt War insurgents.[25]

Texans were just as enraged over the Rio Grande City affair. The *San Antonio Express* thought Texas should give the Mexicans "a good whipping, just to let them know we are in earnest, and then talk about what they should do afterwards." Despite the outcry, Adjutant General Steele forbade any Texas Ranger forays across the border to secure the wanted men. Upon the advice of Congressman Schleicher, Hubbard did authorize the organization of mounted regiments. Texans driven by patriotism, ambition, and a desire for adventure bombarded the governor with offers to raise a regiment or company of Texas Volunteers to serve in the expected war.[26]

While events along the lower Rio Grande were pulling Texans to the brink of precipitating a war with Mexico, El Paso appeared to be one border county that had at last quieted down. As the *Galveston Weekly News* reported on October 1, "The District Court of El Paso county is in session. There were no criminal cases for trial this term of court, and but one indictment found by the grand jury. The grand jury concluded their labors in a little over two days, having had comparatively nothing to do, which is a gratification to all good citizens of the country, and an evidence to those living at a distance that the population of El Paso county is composed of good peaceable and law-abiding citizens." The item was already out of date. The Paseños had had quite enough of Charles Howard.

PART THREE

Chaos

CHAPTER 10
"Our county is in open insurrection."

Despite the drought, 1877 was a profitable year for some local merchants. Charles Ellis expanded his operations in September, buying John Campbell's large flouring mill and reportedly "doing a smashing business milling and merchandizing." John Atkinson was likewise "doing a good business," and so was El Paso merchant Don Ysmael Ochoa. By no means was the region booming. Ernst Kohlberg was vehement in his advice to younger brother Moritz to look elsewhere in America for his fortune, for "[c]onditions here are not promising at present." El Paso before the railroads was no place for a callow newcomer to elbow out the established merchants.[1]

That year Father Bourgade's parishioners broke ground on a new church, the fourth in San Elizario's history. Flooding had seriously damaged the third structure, so the citizens first elevated the ground beneath the new structure's foundation with dirt hauled to the site in their reliable *carretas*. In 1882, the graceful new chapel was consecrated. It stands today, a testament to the mound building.[2]

Louis Cardis had big plans. In mid-September he stopped in Mesilla to complete arrangements for a coach line between that town and Paso del Norte, Mexico. The stretch represented a gap in direct stagecoach service between San Diego and San Antonio that Cardis intended to fill. He also made arrangements for his customers to be able to purchase tickets as far as St. Louis "at remarkably low rates." According to the *Independent*, "Mr. Cardis has met and overcome many obstacles in bringing about this happy result, obstacles that would have dismayed and discouraged a man of less energy; and he is not only entitled to great credit, but also deserves a liberal support from the traveling public." Both his Texas & California Stage Co. and his El Paso Stage Line boasted four-horse Concord coaches and "good meals" at fifty cents each. In other good news, a telegraph construction crew guarded by the Army had brought the wires down from Mesilla to within twenty-seven miles of El Paso.[3]

SATURDAY, SEPTEMBER 29

Others might take Cardis's stagecoach, but not Howard. He stepped aboard his carriage for the two-hundred-mile journey from Franklin to Fort Davis, where court business demanded his presence. He had two stops to make. José María Juárez and Macedonia Gandara had threatened to gather Guadalupe salt without paying, and Howard was finished with threats. In Ysleta he picked up Deputy U.S. Marshal Ward Blanchard and Sheriff Kerber. They continued on to San Elizario, where they found County Judge Gregorio N. Garcia (not the captain). Howard swore out an affidavit against Juarez and Gandara, charging them with intending "to go to the salt lakes . . . and take salt without the proper permission." Howard sought a restraining order with teeth. He asked that the defendants "be put under bond, that they would not interfere with the Lakes." Judge Garcia, rumored to be in Howard's pocket, issued an arrest warrant for Gandara, Juárez, Mauro Lujan, and four others. Howard waited for the sheriff's action.[4]

Later that day, Kerber hauled Gandara and Juárez before the judge. Faced with fine or incarceration, Gandara testified "that he had no intention of going to the salt lakes, nor of breaking the law in any way." At Howard's request, the charge against him was dismissed. When no convincing evidence was introduced against Juárez, Howard laughingly said, "I reckon we will have to dismiss the case against him too." Not understanding English, Juárez "jumped up and said, in a very threatening and insulting manner to the court," that he would go to the salt lakes. He added, "in a vociferous manner that he did not care about the law; that if others went to the salt lakes, he would go also." Now really worked up, Juárez proclaimed that he would go to the lakes whether or not anyone else went. Garcia ordered the sheriff to arrest Juárez, pending payment of a two hundred dollar bond. Kerber returned to Ysleta, presumably with his prisoner in tow.[5]

At about 10 P.M., Judge Garcia was rousted from bed by armed men demanding a warrant for Howard's arrest. He told them to file a complaint first. Grumbling, they left. A few minutes later, Garcia went out to investigate the clamor in the town plaza. He was stopped by Desidario Apodaca, who pointed a rifle in his direction. Forty-five years old in 1877, Apodaca was like many in his community: a farmer who answered the call when danger threatened. He had done so as one of Captain Garcia's Texas Rangers. Now, taking aim at Judge Garcia, Apodaca marched him down to the home of Leon Granillo, where "a great many men fully armed" gathered. Somewhere else in town, Judge Garcia's older brother, Justice of the Peace Porfirio Garcia was being held prisoner. He too had refused to sign a warrant for Howard's arrest.[6]

These stillborn attempts to secure a warrant illustrate a practical notion of Anglo law. Seeking action, Apodaca and the others had been presented with demands for paperwork, the first and last refuge of lawyers. As parties, witnesses, and jurors, Paseños had watched pettifogging Anglo attorneys grind away at one another with little result. Charges filed, charges quashed; motions made, cases continued; justice delayed, justice denied. They accepted the desirability of cloaking their actions in the law, but what they really desired was justice granted.[7]

Holding two reluctant jurists hostage assuaged no one's anger. It was now that the junta discussions and decisions of the past summer bore fruit. The "fully armed" men collecting in town knew what had to be done. "Soon after Howard passed San Elizario, the Mexicans began to boast that the party gone ahead to play jimjam on Howard and would soon have him so." Meanwhile, his business completed, Howard had climbed aboard his carriage that evening and with employee Wesley Owens headed down the road to Fort Davis. They stopped for the night six miles down the road at Quadrilla, ignorant that behind them, Howard's world was collapsing.[8]

SUNDAY, SEPTEMBER 30

John McBride, the salt agent's own agent, had escaped the insurgents' attention and, "getting wind of the plot, went full chase to overtake Howard before he was ambushed." He reached Howard around midnight. The latter fumed, "if Cardis don't let me alone, I'm going to kill him. I'm going to kill him anyway, for he has been bothering me long enough." Howard led his party back up the Rio Grande along the Mexican side before crossing back to the United States at Ysleta. Early in the morning, they reached Kerber's home. From him Howard learned that the two Garcias had been "roughly treated" and that bands were scouring the county in search of him. The sheriff and salt agent, along with Owens and McBride, took to the safety of Kerber's roof. Meanwhile, word that Howard was now in Ysleta reached San Elizario. A contingent of several dozen armed Paseños marched that evening for the county seat, vowing to bring him back.[9]

The names of the determined men seeking Charles Howard are mostly unknown. Their leaders were solid citizens, not young firebrands. Lino (Leon) Granillo was sixty-two years old. His sons Luis (age twenty-five in 1877) and Pedro (age twenty-three) may have trod beside or behind their father. Perhaps no Paseño was more resolute that night than Granillo's co-leader, fifty-year-old Francisco "Chico" Barela. We have met Barela before, as a corporal in Lieutenant Montes's company of Texas Rangers. His thirty-three acres of

Ysleta farmland may have been modest, but he was well off enough in 1870 to have two domestic servants. With a wife, several still small children, and a property stake in the community, Barela had much to consider beyond anger at Howard. Overturning established legal authority was not something he could take lightly. The thoughts of these men as they marched through the night are not recorded, but they must have felt both the gravity of their actions and the lifting of a heavy burden that often accompanies the resolution to put aside talk and do something. It is likely that the men echoed the jumbled thoughts of their leaders with grim remarks, encouraging shouts, and grisly jokes rising above the steady rhythm of their leather soled and bare-footed search for Howard.[10]

MONDAY, OCTOBER I

What happened next is uncertain. According to Howard, Kerber thought it best to get his wife Julia away from their endangered home. The sheriff and Mrs. Kerber took off down the road for the safety of a neighbor's house. On the way back, Kerber "was surrounded by a band of armed Mexicans" and held captive. Kerber recalled a more heroic scene. Sometime after midnight, the contingent from San Elizario united with Ysleta's "militia" and marched in the darkness, some fifty strong, on the sheriff's house. According to Kerber, he was there to meet them. If so, he had not changed in the fifteen years since Glorieta Pass. Acting with the same bull-headed fearlessness Captain Kerber had displayed in battle, the sheriff attempted to seize the initiative from the angry throng. Leaving the roof, he stepped outside and ordered his unwelcome visitors to disperse. The tactic failed. Kerber was seized, disarmed, and taken a mile out of town to the insurgents' local headquarters. As they passed the home of County Clerk G. W. Wahl, the sheriff called for him to join the parade, which Wahl did.[11]

Kerber and Wahl were taken by sixty men, "some of them from the other side of the river," back to the sheriff's house. There the growing insurgent force called for Howard to surrender, promising not to kill him. After twenty minutes of fruitless dialogue, Barela and Granillo arrived to take command. Howard asked why they wanted him prisoner. "The people wish it," was the simple reply. Howard refused to submit, claiming the sheriff's protection. As Kerber described the scene, "Judge Howard spoke to them mildly, explaining to them the law about mobs, &c., but they did not care." Howard put it more bluntly: "They said . . . they knew no law excepting their guns."[12]

Barela was finished with negotiation. As Howard later described his ordeal, "I was seized by two of the men and dragged through the streets of Ysleta, surrounded by forty armed men, hooting and jeering at me. Near the plaza they were joined by twenty-five or thirty more and I was forced on horseback and taken to San Elizario. The whole mob followed, and yelled, hooted and jeered the whole way. . . . On reaching San Elizario, I was taken to the house of Dona Apolonia Lujan. There I found from 200 to 250 more Mexicans under arms—a more sullen, ferocious looking body of men, I never saw—I was taken to a room in the house, and immediately placed under a heavy guard."[13]

Wahl was immediately set free, but McBride, also herded back to San Elizario, was tossed into Lujan's home with Howard and the unhappy Garcia brothers. For some reason, the Paseños let the sheriff go, though his attitude toward all "Mexicans" cannot have been a secret to them. In his report to the treasury secretary, Slade explained, "In the morning, Sheriff told [his captors] he wished to go get his breakfast, after which he would return to them; but he came to El Paso instead." In his own report to Adjutant General Steele, Kerber does not mention the breakfast ruse. He simply explained, "Not having more than four white men in my town, and the Mexicans all in favor of killing us gringos, I started to El Paso."[14]

TUESDAY, OCTOBER 2

Not until Tuesday did Cardis learn of the arrests from J. R. Mariani, who "asked me to go to San Elizario and use my influence to pacify the excited people, which I did." Cardis received a more confrontational visit from Kerber, who suspected Cardis to be "the instigator of all these troubles." Kerber "told him plainly what I would do if Howard was injured." Cardis got the message. Promising the Anglo community to secure the release of all prisoners, he set out for San Elizario.[15]

There the scene was ugly. When not milling about, many of the hundreds there paraded through the streets proclaiming "Death to the Gringos!" and "Long Live Mexico!" Meanwhile, the Paseño leaders were discovering an age-old truth. Starting a revolution was far easier than realizing its purpose. Howard remained defiant. "[B]ecoming alarmed at what they had done" and at what might ensue, several of the leaders, including Sisto Salcido, the elected "commander" of the September uprising, Macedonia Gandara, Ambrosio Orgino, and Desidario Apodaca called on their parish priest. According to Father Bourgade, "the leaders came to me and wished me to help get them out of trouble." Admitted to the makeshift cell, Bourgade spoke with Howard, "who said that the mob were ignorant men, and that he would not prosecute them for what they had done to him, but that some of them might be prosecuted for what they had done to the judges." Howard's refusal to cooperate unnerved some Paseños and fortified others in their resolve. As Bourgade testified, the leaders "wanted Louis Cardis to come down and get them out of trouble, as he was more or less the cause of their getting in." The crowd outside, however, "intended to kill Howard."[16]

Kerber, meanwhile, began telegraphing for help, including Judge Blacker at Fort Davis and General Steele and George Zimpelman in Austin. Kerber and Deputy U.S. Marshal Blanchard then went to Ysleta to be closer to the action. They remained only two hours. Finding the place too dangerous, they "charged through their lines of sentinels and got safe to El Paso on Wednesday."[17]

Back in Franklin, Rucker had telegraphed the news of civil disorder to Santa Fe. The situation was serious. The sheriff, said Rucker, wanted the army "to disperse the mob and preserve the peace; . . . there are not enough Americans living in the county to form a force sufficient for the purpose. The Mexicans who are not with the rioters sympathize with them and cannot be relied upon nor obey his summons to act as a posse." Kerber added that "the mob numbers over three hundred and have sent to El Paso, Mexico for aid."[18]

When Cardis finally arrived in San Elizario, he "[f]ound the people very much excited against Howard only." He and Bourgade held discussions with one of the leaders in the priest's home. The man was bent on Howard's death. According to Cardis, "I begged for his life with all my might." Personal business rescued Cardis from his awkward position around midnight. His coach station at Eagle Pass had just been attacked by Indians, the hay cutter killed, and the mules run off. His drivers were now refusing to carry the mail without an escort. Cardis remained in talks until "satisfied that the people had taken my advice to let Howard and all the rest free." The legislator left San Elizario at 3 A.M., removing himself from events he could not control. Bourgade remained to learn what Cardis missed, or perhaps already knew. The Paseños had not agreed upon any deal, but continued to argue among themselves. That night, one of the leaders had "to push some of the mob away to keep them from killing Howard." One leader admitted "it looked very bad."[19]

WEDNESDAY, OCTOBER 3

The first reports reaching Austin were ominous but confusing. General Steele wired Kerber, "Send statement of mob action. Was it composed of citizens of your county or of Texas and where and by whom organized[?] Was Howard your prisoner and under your charge?"[20]

As the sun rose on the third day of his captivity, Howard's captors continued to demand that he sign over all rights to the salt lakes and leave the county immediately, never to return. Though he remained bound hand and foot, cut off from the world, and "at all times surrounded by three or four hundred armed *pelados,* raving and raging like hungry coyotes," Howard "refused to subscribe to the conditions required or in any way treat with the mob." Cardis, probably the only optimist around, returned to Franklin at 9 A.M. claiming that "all was settled." Two hours later, the sighs of relief abruptly stopped when Sheriff Kerber received word that Howard would be shot at 3 P.M. Kerber immediately "tried to summon a posse of white men," but could not find sufficient arms for the fifteen men in town. Merchant Joseph Schutz gave a more compelling reason: "Everybody concluded that this force was not enough, and did not go."[21]

At this point, Lieutenant Rucker decided to personally investigate the barrage of rumors of "riot at San Elizario." With Joseph Magoffin at his side to translate, Rucker reached Ysleta, where some citizens advised him to halt "as the rioters were greatly excited and had threatened to kill the Americans; that our arrival there would cause the death of Judge Howard." Veiled threats had little effect on this veteran of countless occasions of immediate danger and imminent death. Anxious that his reports "might be authentic," Rucker proceeded under the Ysleta priest's protection.[22]

Meanwhile, Howard's resistance was angering the increasingly restive assembly in San Elizario. Many Paseños had expected quick results. Now, the more impatient men just wanted to kill their prisoners, go home, and be done with it. In Howard's mind, "The mob was growing ferocious every moment." The prisoners, overhearing the feverish talk, were aware of fractures in the insurgents' ranks.[23]

Sometime around 6 P.M. an exasperated Kerber dispatched a plea to Second Lieutenant Simon Vedder, commanding the twenty-man detachment guarding the telegraph construction party near Mesilla. Kerber was in turn complaining, genteel, avuncular, desperate and patriotic: ". . . be so kind as to send those us those men, do not depend too much on your regulation when the lives of all the white men of this county are endangered. We are about 25 white men but we cannot fight 300, excuse my remarks but for the honor of the Old Army help me and help your countrymen." Kerber signed the letter and added, "An Old Soldier." Vedder forwarded the message to Hatch at Fort Craig.[24]

By Wednesday evening, the Garcia brothers had resigned their judicial offices. Howard alone remained defiant. Earlier that day, Father Bourgade was admitted to the prisoner's "cell" in order to administer the last rites, should Howard wish it. The priest then joined a meeting of the popular leaders, who "told me they were going to settle it." Bourgade worked with the insurgent captains to hammer out a deal that might end the crisis without bloodshed. The main provisions addressed their chief goals. First, Howard would pledge not to prosecute anyone for imprisoning him. Second, he would declare that he had no right to the salt lakes. Third, he would leave the county within twenty-four hours, never to return. Fourth, he would give sizeable bonds to ensure his compliance. The fifth clause had Howard confessing "that the prosecution that has been commenced against any of the people is unjust, improper and without cause and that the people of the county aforesaid have had just cause to raise against him." These details and arrangement of a twelve thousand dollar bond took up most of the day. Four local men volunteered as bondsmen: John Atkinson, Jesus Cobos, Charles Ellis, and Tomas Garcia. Some Paseños wanted a provision that the bondsmen's lives "should be forfeited" if Howard returned. The clause was not added, but the four understood the danger in the bargain they agreed to seal.[25]

Mauro Lujan prepared the document and handed it to Bourgade. The priest had his doubts Howard would sign. Salcido warned that the people's impatience to disband was not good news for their captive. "Tell Howard to sign anything, whether it is binding or not," Salcido advised, "for the mob [Bourgade's word] may have trouble among themselves." To the priest's surprise, Howard "was willing to sign anything, because on the night before, from what he had heard, they intended to kill him." Howard also realized there would be no rescue. And so he signed, as did his bondsmen. With that, he and McBride were released. They stepped outside to face an edgy crowd. A number of rifles were raised and aimed at Howard. Proving himself "a hero in canonicals, [Bourgade] threw his arms around Howard and marched through the mob, telling them they would have to shoot him too." Kerber agreed that "only the presence of the priest saved [Howard's] life."[26]

As Howard, Bourgade, and twenty armed insurgents reached the outskirts of San Elizario, they were met by Rucker and Magoffin. The officer was still bent on observing conditions in San Elizario, but Howard and the pastor strongly advised against it. Together the party rode to Ysleta and then on to Franklin.[27]

And so the insurgents achieved their goals — on paper. Howard had agreed to leave forever, taking with him his father-in-law's claims to their salt. Howard's oath to give up the fight was guaranteed by four friends whose fortunes and lives he was not likely to risk. Howard's puppet judges had abandoned their seats on the bench. Though Sheriff Kerber remained in office, he was powerless to arrest anyone, let alone see them safely to trial. No one was likely to suffer for their part in the uprising. All had been achieved without any bloodshed. As Howard's party rode away from San Elizario, they left behind a people who had much to celebrate.

THURSDAY, OCTOBER 4

Kerber escorted Howard and his wife out of the county. Armed men followed his carriage, but the party reached Mesilla in safety. The sheriff returned to his home in Ysleta. There was, for the moment, peace in the valley.[28]

Lieutenant Rucker reported to Hatch that "every American in the county would have been killed had not their terms in Howard's case been complied with." Though he only saw forty armed Paseños, he had been informed some 350 were under arms, joined by nearly a hundred Mexican citizens. "They appear to be well organized, and had been preparing this event for some time; yet their meetings were so secretly conducted that the civil authorities did not know anything about their movements." Rucker reported "that the presence of troops is necessary here, to protect life and property, and believe that in case this mob is not put down by force, the Mexicans will either kill or drive every American out of the country."[29]

So far, only one man had been driven from the county. Charles Howard had escaped death, but he had suffered the shame of his treatment at the hands of a "mob." Bertram Wyatt-Brown, an authority on the role of honor in Southern culture during Howard's time, explains, "The chief aim of the notion of honor was to protect the individual, family, group, or race from the greatest dread that its adherents could imagine. That fear was not death, for dying with honor would bring glory. Neither was it the prospect of damnation in the hereafter. Judgments of that kind were in the hands of God. Rather, the fear was of public humiliation."[30]

Over the course of three unforgettable days, armed men had bound and dragged Charles Howard from Ysleta through Socorro to San Elizario, yelling, hooting, and jeering at him all the way, thrown him tied hand and foot into a guarded room, weakened his resolve by talking over his murder within his earshot, and forced him to sign away his father-in-law's property rights and to admit that his own actions had given the people "just cause" to treat him so foully. When finally released in front of an angry crowd demanding his head, Howard had been saved, not by his own courage, but by that of the priest who shielded him.

As Wyatt Brown and other scholars on the subject of honor among Southerners and politicians (and Howard was both), have pointed out, "honor is reputation." He had certainly lost face among Paseños. However "ignorant and brutal" he considered them, they were still the majority voting block, a necessary tool of political power and influence until a white immigrant tide could ride in on the railroads. Worse, his friends and rivals in El Paso knew he had folded. More damaging still, he knew word would spread. Before long everyone would know, from his family in East Texas and the comrades he had ridden with in the 8th Texas Cavalry, to the moneymen and power brokers who held the keys to his future. Among these was Zimpelman, who had entrusted his capital venture in West Texas to his son-in-law. From high to low, they would all know he had been arrested and "jailed" like a criminal, bound like a slave, and humiliated and intimidated into what would have been regarded as the cowardly act of signing his name to save his skin.[31]

Stripped of honor, Howard concentrated on its restoration. If his intentions were not immediately clear, they soon would be, for, as Wyatt Brown explains, "Dread of being ashamed is a powerful incentive for revenge and violence." As his future statements and actions would make clear, Charles Howard had been given all the incentive he needed.[32]

CHAPTER 11

"That never satisfied greed in man."

News of the riotous overthrow of county government spread beyond Texas and New Mexico as far as New York and San Francisco. Politicians, generals, and the press argued about its meaning. Was it purely a local matter, something for the sheriff and district court to handle? Was the army needed to put down insurrection? Was the nation headed for a second Mexican-American war?

Governor Hubbard was bombarded with cries for help. Adding to the urgent pleas from El Paso, A. J. Fountain wired, "The situation is critical, most prominent citizens have fled. Authorities are powerless to assist; can only protect fugitives, large force will be recruited to restore order and enforce law." What others called a "mob," Judge Blacker, then at Fort Davis, described in military terms. "Four hundred Mexicans are reported in the organization and have scouts, sentinels, and pickets. . . . The insurrectionists claim they can augment their force largely by volunteers from the other side." Blacker warned Hubbard that the few "Americans" were "considered to be in imminent peril of their lives and property." The pressure on the governor to act was magnified by press publication of private and official correspondence under such headlines as "The Mexican Abduction Outrage at El Paso."[1]

Instead of sending troops from Fort Davis, Ord telegraphed Pope in Leavenworth. That general again passed the question of jurisdiction to Sheridan. Viewing the tempest as a matter for diplomats ("There has been no invasion, nothing of the kind; simply a row about some salt."), Sheridan suggested that Howard could file his complaints with the state department.[2]

The sheriff finally reported the tumultuous events to Adjutant General Steele. Kerber cautioned that the arrival of troops would trigger an invasion by hundreds of Mexican nationals. With some bitterness he added, "I telegraphed to everybody I thought could help me, but all for nothing. It seems that we have no claims on our government and we have to give up all our property for the sake of a lot of greasers." In the middle of writing his report, Kerber received two messages from Father Bourgade, warning that "the mob are worse than ever, and will kill all the Americans."[3]

On October 5, five Anglos in San Elizario—Atkinson, Ellis, Dr. John K. Ball, J. S. McCarty, and John Clark—petitioned the sheriff for help. "Can nothing be done for the Americans who are unfortunate enough to live in this part of El Paso Co." They worried that the insurgents remained under arms and were "furious at having been duped in the Howard matter & openly state that they will massacre every man who worked for his release, as soon as they hear that troops are coming." Armed men not only shouted "death to the Gringos" and "kill the Shivas," but also fired shots into houses. The five feared that their "lives + property depend upon the caprice of an ignorant + furious armed mob."[4]

While Anglos in San Elizario fretted for their lives, those safe in Franklin were more hopeful. Cautiously optimistic, Magoffin assured the governor, "Everything is now quiet, and I do not believe the people will do anything further." Less confidently, he added that unless troops were sent, "we may expect worse times than ever."[5]

A few citizens attempted to put the county's broken pieces back together. A committee of Cardis, Magoffin, Blanchard, Slade, Stein, and Joseph Schutz invited Paseños to form their own committee and meet for discussions at Ysleta. The Anglo committee expressed their "good intentions and the desire of all to harmonize." According to Cardis, the invitation was well received, and "the people to meet us and harmonize."[6]

Kerber saw things differently. He complained to Steele, "I am . . . without any power to enforce law, bringing these men before court. I would not dare to arrest any of these rebels if not backed by State troops: neither can I collect any taxes, and the only way I see is to put the county under martial law. If we allow such rising against law and order pass unpunished, then we had better all leave our homes and property." To the governor, Kerber grumbled, "None of the American citizens are safe so long as we have no troops here to enforce law and order."[7]

Troops were indeed on their way from Mesilla to El Paso, but only to guard U.S. property and assist Lieutenant Rucker in ascertaining the cause of trouble. The soldiers were expressly forbidden from interfering in "civil affairs." Slade took advantage of their arrival to intercept a party of salt gatherers from Mexico. One night the soldiers and Customs Inspector Magoffin spotted the gatherers camped with three wagons, "laden with salt." He halted his party about two hundred yards away. The next morning, the customs man approached the wagons until six or eight *salineros* aimed their guns to warn him off. During the night, the Paseños tossed the salt out of the wagons, hitched up fresh oxen, and escaped across the river. "It was well that they escaped," Mexican friends told Magoffin, "for every one of us would have been killed."[8]

The unlucky salt gatherers blamed Howard for their loss, according to Miguel Garcia, the Texas Ranger who had complained in 1876 of Telesforo Montes's handling of patrols. One day, Garcia, fresh from the Mexican side, stopped at Dowell's and alerted the saloon-keeper "that 40 men had come up the night before from Guadalupe, and crossed the river two or three miles below here. . . . they wanted to kill Howard because he had sent Magoffin down and caused them to throw out their salt."[9]

Howard had returned to El Paso, riding from Mesilla with the army detachment as far as Franklin. With wisdom seldom displayed in this drama, Lieutenant Rucker refused to allow Zimpelman's agent to accompany him to San Elizario. Howard had indiscreetly expressed his consuming hatred for Cardis to Capt. Miles L. Courtney, who warned Cardis of Howard's "desperate threats" on the legislator's life.[10]

Rucker reached San Elizario on the evening of October 8. The "more timid" fled upon the cavalry's approach, but many armed men remained in town, ready to fight. In two conferences with insurgent leaders on October 9, Rucker thought he succeeded "in quieting their fears of arrest" and convincing them to disperse.[11] With misplaced confidence, he advised Colonel Hatch, "They will allow the law to take its course."[12]

On October 8, Cardis's committee met with their Ysleta counterparts. The committee urged the people of San Elizario and Socorro to reinstall "at once every legal civil officer of the county, and assist them in the proper execution of their several duties." Cardis hoped that the insurgents had given Rucker "the pure and simple truth," but wondered, "if he has not a good interpreter, and has not consulted with those that are not afraid to speak the truth, who knows how he will understand the case?"[13]

On Wednesday, October 10, Rucker was invited to attend a junta. There, "the leading spirits of the mob" informed him that "the people would let the law take its course; that they had reinstated the deposed county officials and that all trouble had ceased." The lieutenant took them at their word, but anticipated trouble if the sheriff attempted any arrests without the backing of troops. Rucker was heartened by the lack of bloodshed. He credited "the untiring efforts of the Rev. Bourgard [*sic*]" with the peaceful resolution of the crisis. Unfortunately, the padre reported that "his influence over the mob was exhausted."[14]

After "careful investigation," the lieutenant determined the riot was "the result of a misunderstanding on the part of the ignorant Mexicans regarding the location last July of the salt lakes in this county, by a firm in Austin, Tex. These lakes have for generations been considered public property, and when their crops failed it has been customary for the people to go to the lakes, get a load of salt, take it to Chihuahua, and trade it off for provisions. . . . Being very ignorant, they cannot understand why they are deprived of this means of subsistence."[15]

So much for the root of the problem. The "immediate cause" was that Paseños regarded as persecution the prosecution of two men "for merely signifying their intention of going to the lakes and getting salt." Rucker blamed "several persons of intelligence" (presumably including Cardis) for advising Paseños "that they had a right to take all the salt they required and that they (the parties who instilled their belief into the Mexicans' minds) would stand by them and see that the authorities did not interfere, in fact assume all responsibility in case of trouble." But when trouble arrived, wrote Rucker, "these parties failed to fulfill their promises and the Mexicans, finding they were doing wrong did not know how to get out of the scrape. The facts of the matter is, the Mexicans had been lied to, to such an extent they did not know who or what to believe and the most ignorant of them thought the best way of settling the matter was to drive the Americans out of the country."[16]

Who was to blame? Howard? Cardis? Howard *and* Cardis? The sheriff harbored no doubts: "Hon. Louis Cardis is considered here the Commander of the mob. . . . He tells them in his speeches that they have a right to organize into armed bands, if they think that they will not find justice." Indentured clerk Ernst Kohlberg laid the blame on "a couple of sharp lawyers, with a great deal of influence. [They] have taken away the means of livelihood of the . . . three villages. Imagine how embittered these people were at this snide trick."[17]

Customs Collector Slade reckoned that the triggering event, the arrest of Juarez and Gandara, had been the work of a county judge [i.e., G. N. Garcia] "considerably indebted for his election" to Howard. The riot ensued only because "[t]his high-handed and unjust proceeding so exasperated the Mexicans." Slade was unsure whether Cardis pulled the insurgents' strings: "Many of the Americans here believe that Mr. Cardis, who is to the Mexicans what Montezuma was to the Aztecs, as they say themselves, was the instigator of the mob, and advised them to this step in order to get the salt lake agent killed, for it will be remembered that they were bitter enemies. It is also believed by many that the Mexicans needed no urging in a matter so vital to their own interests."[18]

One hundred miles to the west, Silver City's *Grant County Herald* identified

Howard as the villain: "The trouble in El Paso county, Tex. all comes from that never satisfied greed in man, which urges him on to the acquisition of wealth even at the expense of the entire community surrounding him. . . . Judge Howard knew that he was getting into a hornet's nest when entering the land embracing the Salt Lakes."[19]

Howard himself asserted that adversaries "were trying to have him assassinated because they knew they could not use him any longer as a tool and they he knew too much about their iniquities."[20] The *Mesilla Valley Independent* gave him a forum to excoriate his foe. "The present condition of the county," said Howard, "is the work of one man, whose evil counsel for years has hung over the Mexican population of El Paso, like a pall. He is a liar, a coward, a mischief-maker, a meddler; such a thing as could only spring from the decaying carcass of an effete people." Cardis was "a vampyre." Feeding on the people's "ignorance and prejudices," he taught them that, as "owners" of the region, "they have a right to murder or assassinate Americans, and should do so, when they interfere with them."[21]

The Mexican people, said Howard, are "naturally quiet, simple and law-abiding," but "filled with the most deadly and insidious hatred against the American Government and people" by Cardis. The army's abandonment of Fort Bliss in January 1877 enabled Cardis to convince many "that the United States and the State of Texas had given up the county, and would not trouble themselves any more about it; that the Mexicans were strongest, and it would be there own fault if they did not drive out the Americans and take their property." The salt lakes dispute, Howard averred, was merely the pretext for revolution. "But what ought to be the punishment of that infamous monster who has brought these poor, ignorant, deluded and credulous people into this trouble for his own cowardly and selfish ends. I can think of nothing in Earth or Hell bad enough for him."[22]

Meanwhile, a protest to Zimpelman's claim to the salt beds had been filed with the state land office by the Texas and Pacific Railway Company. "The certificate on which the location was made, came from the old Southern Pacific Company, and this opens an interesting field for litigation." The *Galveston News* presciently commented, "Whether these lakes are subject to location by any or all of the parties thus claiming the right is perhaps a question that . . . will have to be settled by the courts or the Legislature."[23]

Other voices were less interested in blame or legal ownership than in whether the "riot" justified war with Mexico. News of the San Elizario "emeute" reached Austin, Washington, and New York at precisely the moment when Texas seemed to be preparing for war in response to disturbances along the lower Rio Grande. In a speech given at the Williamson County Fair, Governor Hubbard confirmed that "it would not be the province of Texas to solicit a collision, nor to precipitate a war with this unfriendly nation; but that when all laws and treaties were set at defiance, and totally disregarded, Texas could only appeal to the national government, whose duty it is to protect our people." That same day, the *Galveston News* reported, "The people living in the section of the country between the Nueces and Rio Grande rivers are much excited over the threatening aspect of affairs, and are very anxious for Governor Hubbard's policy of preparing to resist Mexican invasion to be carried out." "Rip" Ford was reported touring the border towns on "some special and confidential mission" for the governor, presumably to identify appropriate safety measures.[24]

Ignoring Rucker's report of Mexican involvement, Phil Sheridan viewed the situation in El Paso was strictly a civil affair: "The trouble is with our own population. . . . I have no reason to believe that any citizens of Mexico have taken part in the troubles, but the population of El Paso County is mostly composed of citizens of Mexican birth."[25]

Questions persisted. Was Mexico somehow to blame? Had her officials supported the insurrection by permitting armed men to cross the border? Did Mexico's failure to control her population provide fresh evidence that the U.S. Army needed to occupy her neighbor's northern states? The question of invading Mexico did not escape the notice of the New York papers. One warmongering *Galveston News* article picked up by the *New York World* caused the *New York Times* to weigh in: "It would appear that we must have a war with Mexico. Precisely why we should have a war with anybody . . . may not be obvious to the average reader. But in two or three Texas newspaper offices, and in the editorial rooms of at least one New York journal, it is understood that we must fight. . . . The infuriated rabble, as the [*Galveston*] *News* account calls the invaders, clapped the county officials into jail. Why they should thus harass a peaceable county clerk, an innocuous recorder of deeds, and a senile justice of the peace and the quorum, is beyond the comprehension of the American mind. These Latins are such queer fellows." The *Times* observed that "Lieut. Gen. Sheridan . . . flouts the whole story of outrage, flagrant and otherwise. . . . [As] the general ruthlessly puts the case, 'it was simply a row among the Mexicans about some salt.'"[26]

While the *Times* accurately reflected Sheridan's attitude, the *Army-Navy Journal*, the self-appointed organ of military thinking, disagreed. In an October 13 editorial, publisher William Conant Church predicted, "How long it will be before the example of the mongrel inhabitants of El Paso County is followed by other towns of Texas where the Mexican blood predominates is hard to say, but it is probable that if a check is not soon put to this lawlessness, it will spread very rapidly." Either Mexico must properly police its northern border, or she must witness again "the spectacle of an American Army dictating terms of peace in her capital."[27]

The *Journal* had every reason to fear guerrilla war. Irregulars had done much to keep Mexican independence alive during the regime of the puppet Emperor Maximilian. Kohlberg feared that Americans were underestimating the potential Mexican adversary. "It is very difficult and dangerous to invade a country that has no railroads, no roads, no bridges and very little water. On the other hand, it is an ideal country for guerrilla warfare. The Mexican has lots of endurance. He can take long trips and hard marches with only corn and sugar for food. If the American soldier does not have his rations of bacon, coffee and sugar for a couple of days he easily becomes exhausted."[28]

War would not be triggered in El Paso County, if Cardis could help it. He telegraphed Hubbard on the ninth day of the month, "The late disturbance in this Co. is over and everything seems to be quiet. It is mainly due to my efforts, that peace has been established, though reports may come to you to the contrary. Full particulars by mail."[29]

CHAPTER 12

"I feel like killing until I am killed."

In the early afternoon of Wednesday, October 10, Cardis entered the store of S. Schutz & Brother, to have bookkeeper Adolph Krakauer prepare a letter. He seated himself in a rocking chair near the back of the store, with his back to the front entrance. The letter, addressed to "Friend Cipriano [Alderete]," contained the legislator's plan to correct the "erroneous and exaggerated reports" sent by Howard and Kerber to the governor. Cardis reassured his friend, "Tranquility and peace and the truth will manifest itself in time. . . . [The] wish of the good citizens will prevail, and we have justice and peace in the future." In an ironic postscript, Cardis added, "Do not pay any attention to the slanders that you hear against me, and my life. Let the people remain tranquil, and we will get justice, and this is what we wish and need no more." Cardis, at least, saw a way out of the trouble in El Paso.[1]

Charles Howard had other ideas. He had returned from Mesilla armed with a deep sense of injustice. As he told Wesley Owens, "I feel very restless and must have my revenge." After a quiet mid-day meal, he picked up his shotgun, walked into Joseph Schutz's store, and unloaded both barrels into his nemesis.

According to Howard's version, he was preparing to leave the county and "await further developments" in Mesilla when he walked into Schutz's around two in the afternoon to buy wads and bird shot for his shotgun. "While sitting on a counter stool waiting to be served by the clerk, Howard observed that the clerk and others present were greatly perturbed, looking around to ascertain the cause he observed Louis Cardis with a pistol on his person pass behind a high desk some ten or twelve yards distant from where he (Howard) was sitting; from his position Howard could see Cardis from his waist down, the upper portion of his body being concealed behind the desk (which stood on a high table), his right hand held a pistol which was also exposed below the desk. Howard raised to his feet and cocked his gun." Seeing this, Schutz cried, "[D]on't shoot in the store." Howard answered, "[L]let him come out and I will not shoot." Howard spotted Cardis, still behind the desk, raise his pistol.

Howard immediately "dropped on his knee and fired underneath the desk. Cardis staggered and exposed his left shoulder and a portion of his breast, when Howard again fired the charge taking effect to the upper portion of Cardis' person." Cardis died within fifteen minutes.[2]

Other witnesses saw a murder take place. Joseph Schutz had called out in a loud voice intended as a warning, "How do you do, Judge Howard?" Alerted, Cardis jumped out of the chair and hid behind the high desk that exposed his body from his navel down. After Howard snarled, "Here you are," Schutz begged him, "Don't shoot here, judge; respect my house and my family." Howard replied, "I won't, if he gets away from there," but shot Cardis almost instantaneously.[3]

Edmund Stine was home when he heard the shots. A neighbor brought Stine to Schutz's store on the run. He found Cardis lying on his left side, face down, with his knees drawn up. H. H. Harvey arrived a few moments later. As the pool of Cardis's blood inched beyond his body, Harvey moved to take his pulse. Unaware Harvey was a deputy sheriff, someone barked, "Don't touch him! No one must touch him until an officer comes." The legislator breathed for another ten or fifteen minutes, but never spoke a word. When his body was turned over, to place it on the plank pallet, Stine noticed a large revolver on half-cock still in its scabbard belted around Cardis's waist. Harvey, one of Kerber's men and so perhaps no friend of Cardis, testified to seeing two scabbards, one empty, "the other had a pistol in it, partly drawn out, the handle turned outward from the body, at full cock."[4]

Cardis's true role in sparking the arrest and captivity of Howard, and his ultimate responsibility for the tragic events that followed his death, remain a matter of conjecture. Early El Paso chronicler Owen P. White absolves Cardis of plotting to assassinate Howard. He was "too clever to implicate himself directly in the killing of Howard when he felt morally certain that that killing, which he probably desired, would occur without any action on his part." But did he stir the people to rebellion? Accusations that he did so generally came from his enemies, but they may have been correct. As early as 1872, Howard—then his ally—told their mutual friend Mills that Cardis was even then "inciting people to riot and murder." Whatever Cardis' responsibility in 1877, the Paseños did not need anyone to tell them to resist Anglo claim to the salt lakes.[5]

After the shooting, Owens retrieved Howard's rifle, loaded with sixteen rounds, and exchanged it for the empty shotgun. Howard attempted to place himself under the protection of the U.S. customs agent, expressing a desire to give himself up unless it meant getting killed. Magoffin persuaded him "to go and not to compromise the citizens here." His grasp on reality restored, Howard hitched up his wagon, and with two servants rode out of town, leaving behind a county in chaos. The next evening, A. J. Fountain was having dinner when a troubled Howard appeared at his door. "I am a fugitive," he announced. Fountain invited him to share a meal. "I killed Cardis yesterday." Howard admitted. "I think his friends are in pursuit of me and I have come to you as an honorable enemy to solicit your assistance to defeat any effort they might make to kidnap or mob me." Fountain assured his support and took in Howard as his guest.[6]

Meanwhile, Franklin was gripped by fear. Although there were soldiers with a cannon in town, they were "only interested in guarding the customhouse." According to Kohlberg: "For two nights we stood guard armed with pistols and guns with 100 rounds of ammunition. We were on the roofs of the houses. We lost our sleep for nothing as the Mexicans did not attack. They raided the outlying districts, drove off stock and stayed out of range of our Winchester rifles."[7]

Within a few days, panic loosened its hold on some. Two days after the shooting, the self-styled "committee" of Magoffin, Slade, Stine, and Joseph Schutz wired the governor, "All is quiet in El Paso Co." In a separate letter, Magoffin elaborated, "The people in fact (as well as the leaders of the late faction) are all willing to sustain order. There was no uprising on account of Hon. L. Cardis such as expected but his Mexican friends demand that justice must be done him and that Judge Howard must be dealt with according to law."[8]

As usual, Sheriff Kerber was anxious. "It is feared," the *Independent* reported, that several "leading citizens of San Elezario [*sic*], both American and Mexican, will fall victims to the fury of the mob." Howard, of course, was under sentence of death, and the sheriff had been warned that he would be killed unless he brought Howard back to justice in El Paso.

The "insurgents," Kerber reported, were camped only two miles from Franklin and would "fight until the last man is killed before one of them shall be arrested. If they are not arrested and punished, we (the Americans) must leave this part of the country, that is if they will allow us to do so." The *Independent* added that "all the better class of citizens are opposed to the mob, which, with the exception of a few leaders, is composed of the worst element to be found on either side of the river."[9]

Kerber delayed until October 18 before formally advising Steele of the Cardis killing and suggesting an answer to the county's problems: "I made arrangement with the [Dona Ana County] sheriff that he would protect Mr. Howard till I go after him again. The Mexicans intend to revenge the death of Cardis, but a temporary company of your State police [i.e., Texas Rangers] would quiet them very much." The sheriff also advised "that certain Jews from El Paso send statements to Austin, contrary to the real condition of our rebellion. I beg to assure you that, if it is so, they only do it out of fear that they would have to pay a little more taxes." He admonished Steele to recall "the treacherous character of the Mexicans and Indian population. You too well remember . . . the cowardly attack on some of your troops [in 1862] at La Mesa [la Mesilla?]." He followed this canard with a renewed plea for Texas Rangers: "[R]elease us from the great danger of being assassinated because we are American citizens and white men."[10]

Kerber became angry when one Jewish merchant, Joseph Schutz, attempted to retrieve two Remington carbines previously loaned to Kerber but now required back for a paying customer in Chihuahua, Mexico. Kerber, said Schutz, "told me that undoubtedly these arms were to fight us with. . . . He said . . . he had lived among them in New Mexico, and he would give almost anything to have a war with them, so that he could have a chance to fight them, the sons of bitches."[11]

Meanwhile, Steele scolded Kerber, "[I]t would appear that you should have held Howard for the killing, unless there has been some judicial examination which exonerates him. You do not mention any of the circumstances, nor whether or not there was any inquest or what was the result. All that is known here is the fact that Howard killed Cardis." Bridling at the criticism, Kerber explained the suddenness of events, adding, "There will yet come a time when Judge Howard will be the biggest man in the county, and if it was in any other county but this, a monument would be erected to his memory for delivering us from a tyrannical, unscrupulous scoundrel."[12]

The "quiet" observed by Magoffin and others existed only so long as county officials accepted the complete collapse of legal authority. No one dared file complaints against the "rioters" who had falsely imprisoned Howard and others. No legal authority dared issue or execute any process. When Sheriff Kerber tried to collect state and local taxes that month, he was told the people "had stopped being d———d fools [and] the government could maintain itself."[13]

Leaders of the *insurrecto* were not eliminating "law and order," merely redefining it through their own government "of the people, by the people, for the people." In place of collapsed Anglo rule, on October 23 the people convened a junta at Ysleta, attended by commissioners from all the towns, including Franklin. The commissioners resolved to again force Judge G. N. Garcia's resignation under threat of death and replace him with a judge who would "not prosecute any persons taking part in the late insurrection, and to prosecute C.H. Howard to the utmost." (Edmund Stine refused the offer.) When Martin Alderete stood up and asked the assembly to "recollect that we are citizens of the United States," the whole gathering reportedly arose and with one voice cried, "No, we are citizens of Mexico; *hasta la muerte, hasta la muerte; viva Mexico.*" In the eyes of John Atkinson, "The revolution in our town has . . . gained new strength." In the view of one "Old and Reliable Citizen of El Paso, . . . I tell you candidly, if Americans ('Gringos' as they style all who refuse to join the mob) are not murdered, they are in danger of being stripped as bare as the day they were born. . . . [If] Uncle Sam does not come to our relief, who will? Should there be a war between the United States and Mexico, another Mountain Meadows massacre in this county will be inevitable." While riding to San Elizario that day, Kerber met San Elizario commissioners Salcido, Granillo, and Juarez, the man bound over for trial back in September. Kerber found them "armed to the teeth" on their way to the junta. That assembly was still underway when he returned and joined it. The crowd went "wild" when he declared that Howard had every right to return. Jesus Telles of Socorro, apparently wanted for rustling but now feeling safe from arrest, taunted the sheriff, asking what the punishment would be for killing Howard.[14]

In a letter reporting the meeting to the governor, Maximo Aranda, Atkinson, Cobos, Ellis, Judge G. N. Garcia, and others provided the most detailed look at the insurgents' purpose, organization, and leadership. The signatories advised that county government had ground to a complete halt in the face of "an armed and organized body, whose avowed object is to resist the laws of the United States and of the State of Texas; this force is composed of 450 Mexicans, the majority of whom are residents of this county, but who are in their manners, habits, customs and feeling Mexican to the backbone, and who hold the laws of the United States in the most supreme contempt. They are commanded by one Sisto Salcido, a citizen of the Republic of Mexico; their second in command is one Leon Granillo, a citizen of the Republic of Mexico; their third command is one Jose Ma. Juarez, also a citizen of our sister republic."[15]

The letter identified by name twenty insurgent "captains, lieutenants, sergeants, and corporals," including "Franco [i.e., Francisco] Barela." The "mob" was said to include nearly one hundred Mexican citizens, "backed by hundreds of their relatives and friends in Mexico who are ready to come to the assistance of their compatriots whenever called upon."[16]

The citizens reported that "neither law nor order [existed] in the county; the county authorities are powerless and every man does what seems good in his own eyes. The lives and property of the Americans and their few friends in this county are in imminent danger. They are liable to be sacrificed at any moment. They hang upon the caprice of an ignorant, prejudiced, and bloodthirsty Mexican mob."[17]

After this feverish prose, the letter writers submitted the facts "without comment." The message provides insight into the worried minds of El Paso's Anglos. Hubbard's correspondents were making the case that an "invasion" was underway. The allegation that the three chief popular leaders were Mexican citizens is of doubtful accuracy. The charge that the people of *la Isla* could call upon "hundreds of their relatives and friends in Mexico" was understandable. The fear of violence was real.[18]

The letter's signatories described the popular movement as "an organized mob." But what did they seek? Did they intend, as some feared, revolution against the United States and reversion of their land to Mexico? Did they intend the massacre of all "Americans" in the valley? Or were their aims limited to some sort of legal or political redress within the confines of continuing American citizenship? The direct evidence on this issue is scarce. In the absence of junta minutes or the published account of any Paseño leader, we are left with circumstantial evidence and the views of enemies.

Even if Paseños overwhelmingly thought of themselves as Mexicans, it appears unlikely that nationalism drove the uprising. The little we know about the junta of October 23 indicates that some, perhaps most, attendees expressed a desire to politically rejoin their relations on the other side of the river. Others, such as Martin Alderete, opposed the idea. Certainly, the presence of Lieutenant Rucker's small detachment provided evidence that the U.S. Army's earlier abandonment of Fort Bliss had not signaled the federal government's relinquishment of the Pass. Subsequent Paseño efforts to avoid open conflict with U.S. authorities and later the Texas Rangers appear to signify that their leaders ultimately agreed upon the achievement of more limited goals. Chicana sociologist Mary Romero considered the question of Paseño aims and concluded that their "purpose and goals were aimed at eliminating Howard, and thus alleviating economic and political oppression. At no time were border Mexican-Americans and Mexicans attacking the United States government."[19]

Undoubtedly, the Paseño population did not feel any particular bond with their governments in Austin or Washington. The violent death of Cardis, their only trusted representative in the state capital, loosened whatever tenuous ties existed; but it cannot be assumed that Paseños looked to Mexico City for support. The civil wars of the 1870s between *Juaristas, Lerdistas,* and *Porfiristas,* while of great moment to the *caciques* who governed Chihuahua and Paso del Norte, meant little to the general population beyond double and triple taxation. Historian Mark Wassermann views Chihuahua as "literally and figuratively a land apart." He asserts that, "The region's isolation meant that settlers confronted the Indian terror and hardships with little help from central authorities (Spanish or Mexican). The people of the *haciendas,* mining camps, and villages learned to survive on their own. . . . Not surprisingly, the *norteños* guarded their independence, regularly resisting the attempts of governments in Mexico City to impose centralized authority." And although Chihuahua (including the Pass of the North) was a victim of "invasion, military defeats, loss of territory and humiliation," the United States was a market for products. "*Norteños* admired the enterprise and technology of Americans."[20]

Borderlands historian Oscar J. Martinez has pointed out that, "Until the advent of modern transportation, the North was practically cut off from the Mexican interior, and population centers in the region itself were isolated from one another. . . . Remoteness from the central government in Mexico City has bred regionalism and independence, and *norteños* have demonstrated in a strong way their dissatisfaction with national policies at various points in Mexican history." If anything, *norteños* had a predilection for self-rule.[21]

Despite some strong sentiment for reunification with Mexico, the evidence supports the view that the Paseños ultimately decided upon the more limited objective of "government by the people," whichever nation's flag flew over *la Isla.* Father Bourgade testified to the Paseños' view that the people, and not their government, were the ultimate arbiter of law. As Agatón Porras put it, it did not matter whether they were "wrong" (i.e., whether they had acted outside the confines of Anglo law). In arresting Howard they were in the right, "because they were the people and the people were the law; that for the same reason they were right in forcing Howard to sign his relinquishment to claim to the salt lakes, and because they had no hope in collecting the forfeited bond in the courts, they had a right to take measures to collect it themselves by force."[22]

Several months later, Judge Blacker further explained the Paseño principle of self-government: "The Mexican people of El Paso County are, by association, practice, and sympathy, opposed to the American system and method of government. They believe it to be an inherent, inalienable right that whenever the people become dissatisfied with the officers whom they have elected to administer their laws, they instanter, without waiting for the timely arrival of the elective system, they have the right 'to rise,' overthrow the government, and put in other officers. They . . . conceive that there are grievances so intolerable and wrongs so monstrous, that it can or will be overlooked by those who have the administration of the government in hands." Blacker overlooked the fact that persistent electoral corruption had persistently robbed Paseños of elected officers. Cardis was an exception, and he had been gunned down.[23]

The judge might have added that Paseños on both sides of the border were equally opposed to any government antithetical to their economic security. The people backed whoever left them alone and opposed those who oppressed them. When in May 1877, the hapless *Lerdista* Colonel Machorro levied one *prestamo* too many on Mexican citizens, they quickly drove him out and killed his men. El Paso historian C. L. Sonnichsen drew a parallel between the independent-minded Paseños and the Sons of Liberty. The people of the valley, he noted, "had no claim which would have stood up in an American court—but they had a claim. From their point of view, they had about as much on their side as our forefathers did in resenting the Stamp Act."[24]

If Paseño aims were limited to the restoration (or confirmation) of their fundamental political and traditional property rights within the American system, there still remains the question of means. Though warning that a "mob" had taken control of El Paso, the citizens who wrote to Governor Hubbard contradicted this notion by clearly describing the resistance to Anglo legal authority as "organized." What they called an "organized mob" is perhaps best defined by what modern international law would term an "insurgency." Such an uprising is marked by the semblance of a government, control over a certain portion of territory in which the civil conflict takes place, and an organized military force. Depending on the geographical situation of the country, insurgencies are also often marked by a tendency to internationalize a conflict.[25]

This closely describes what took place in El Paso. For a period of eighty-four days, the Paseños effectively overthrew—or at least rendered impotent—local Anglo government. Insurgent juntas represented the shell of a home-grown government. Father Bourgade would later state that the insurgency's leaders "did not very well agree with each other, some being moderate, others rash." Differences among the general population were probably even more pronounced, as some opposed any form of insurgency. These divisions may account for the fact that the people were unable in the short term to replace constitutional government with any effective replacement. Even the long-standing system of enforcing the maintenance and repair of ditches apparently came to a halt during the period of unrest, a strong indication that even this traditional form of Paseño self-government suffered at this time.

It is in the military sphere that the uprising most closely resembles a popular insurrection. While some skepticism must be applied to the Anglos' description of a highly structured insurgent military organization, the movement certainly relied upon a rudimentary chain of command. The people of El Paso were quite capable of organizing an effective military force grouped into ad hoc units tasked with carrying out discreet military missions. The Paseños of 1877 were the inheritors of a long military tradition. Those who took to the streets now included citizen-soldiers who had fought Sibley's and Steele's Confederates in 1862, threatened a school war in 1872, and campaigned as citizen militia or Rangers against Apaches. This last group included Salcido, a two-time Ranger, and Barela, a corporal under Montes. Interestingly, Barela rated only an "also named" status from Hubbard's correspondents. Perhaps he was still growing in status among the people. Before long, all would identify this gifted leader as captain of the uprising.

Anglo fears went beyond a local massacre to general war with Mexico. Fearing the worst, they attacked both the state and federal governments for manifesting "a singular apathy and indifference to our situation." Seeing their salvation in the army, one Anglo reminded Washington City that, with war on the horizon, the Pass of the North was "a strong strategic position . . . the only good opening for hundreds of miles to the North Western States of Mexico."[26]

Fearing that war was only one spark away, Customs Agent Schutz alerted Secretary of State Evarts that "[a] most intense & bitter feeling has manifested itself . . . among the Mexicans on both sides of the river against the Americans, which is likely to break forth at any moment, in which event it is my opinion not a single American would be spared." Should war with Mexico break out, all Anglos at the Pass would have to flee or lose their lives. Even without war, "the few Americans on this frontier are living on a mine which may explode at any time & bury every one of them in its ruins."[27]

Determined that his soldiers do nothing to trigger war, General Pope warned Colonel Hatch, "Inform Lieut. Rucker at once by telegraph and courier that his functions at San Elizario are limited to preventing encroachments upon American soil of armed bodies of Mexicans. He has nothing to do with local disturbances of the peace by citizens of Texas, whether they be Mexican or Americans."[28]

Rucker's orders rendered his detachment worse than useless, Howard charged, since the soldiers' impotence only emboldened the more aggressive leaders of the insurgency. On October 25, Howard vented his rising frustration at Steele. He listed the usual complaints: the false impression of peace in El Paso, the collapse of civil authority, the restraints on Rucker, and his fear of being murdered. Howard then got to the point. Texas Rangers were "absolutely necessary to restore order. . . . If the State will not help us, we want to know it, and will try and help ourselves. I do not wish any general punishment inflicted on the rioters, because most of them are as ignorant as mules, and men misled by others; but the leaders ought to be punished. . . . If the governor don't help us, I am going to bushwhacking."[29]

Howard also poured out his anger at government inaction to a friend in Austin (probably Zimpelman). "All law and authority are still trampled under the feet of the mob," he wrote. "Should anything happen to displease them, they will rise again. . . . There is no use to mince matters—the Mexican population of El Paso county can not be controlled except by fear. We cannot have the lawful authorities reestablished until there is a sufficient force at their disposition to overawe the mob and enforce the orders of the courts. The longer the matter is being handled the harder it will be to deal with it. If the American citizens of the county find that the Governor is not going to help them they will abandon the county altogether." But Howard was thinking along other lines. "Sometimes I feel like arming myself to the teeth, going to El Paso and killing until I am killed."[30]

Howard framed the struggle as a battle for the soul of Texas. The values all Americans cherished were under attack, he wrote, from Mexicans who were "ignorant and brutal to an inconceivable extent. . . ."

> Think of it, will you, that in the United States, and in the State of Texas it has become a danger and a disgrace to be a citizen; that to speak English renders a man an object of suspicion; and to say that one is a Texan puts his life in peril from a Mexican mob. Will the State of Texas permit this? I tell you that every man in El Paso, who favors law and order, or who sympathizes with Americans and the American government, as distinguished from Mexico and Mexicans, is in hourly danger of his life. . . . . I ask for myself that the government of the State afford me what the Constitution guarantees to every one—a fair and impartial trial; that when disarmed and submitting to the law I be protected by the law.

Howard's vent then turned to the personal matter of his recent debasement:

> When a native-born citizen of the United States and for more than twenty years a citizen of Texas is forcibly taken by a Mexican mob, dragged about through the county, hooted and jeered at, called a d———d Gringo by two or three hundred Mexicans, imprisoned and guarded for several days and nights by *Pelados,* the matter discussed within his hearing as to whether he should be murdered or not, his death finally determined on and the hour fixed for it, and finally released on condition that he would abandon the county forever, and only twenty-four hours to leave it. I think he ought to be allowed to grumble a little about it. . . . If the governor will do nothing, we must give up the ship.

Just weeks before, Howard had forgiven the "armed men" who dragged him, "hooting and jeering at me." Then, his anger was aimed solely at Cardis. Now, Cardis was dead, but Howard was still getting "warmed up" over the incident. His rival's blood had not washed away the bile still rising as he continued to live with the dishonor he had endured.[31]

The unrest in El Paso continued in an atmosphere of war fever. Thanks to continuing disturbances along the lower Rio Grande, San Antonio's Congressman Schleicher pressured President Hayes for military action. The hawkish *Army-Navy Journal* advised that General Ord's annual report showed "conclusively that without a war with Mexico we shall probably never have a quiet Texas border." The Austin *Statesman* countered that the United States had only to "Americanize" Mexico. "Railways alone can effect a desirable conquest of Mexico, and the United States government, when equipping an army for the Rio Grande frontier, should supply picks and shovels instead of muskets."[32]

In this atmosphere, General Pope cautioned Sheridan about El Paso, "The controversies between the Mexican and American Citizens of Texas are beyond the reach of any jurisdiction by the military forces and it will be necessary therefore and very difficult to discriminate in a riot, or other disturbance between those Mexicans who are citizens of Texas and those who are foreigners." Pope had just 779 enlisted men in Hatch's District of New Mexico, only 450 of them belonging to the 9th Cavalry, including those detailed as "cooks, clerks, teamsters, etc., etc." Hatch's command was already stretched keeping an eye on several Indian nations. "Indeed at this time nearly the whole disposable force in New Mexico is in the field and has been for some time hunting up the band of these Indians who broke away from the San Carlos Agency in Arizona." It was "impracticable" to furnish the cavalry force recommended by Lieutenant Rucker. Pope warned, "If troops are sent there they will of course be kept there as the demand for them and the reasons therefore, actual and imaginary, or manufactured will increase with every day of occupation."[33]

While the U.S. government continued to dither, Texas finally decided to send in the Texas Rangers, or rather one Texas Ranger.

CHAPTER 13

"Twenty good men"

The state's chief executive had decided to act. Governor Hubbard and Adjutant General Steele had discussed sending a company of Texas Rangers to El Paso as early as October 8, two days before the death of Cardis. Steele, whose regiment had been bloodied by Paseños in 1862, should have pressed for the dispatch of state troops to the scene at once. Since logistical requirements would have delayed their arrival for at least one month, Hubbard deferred making any decision. Now he could wait no longer. On October 24, Steele ordered Maj. John B. Jones, the thirty-eight-year-old commander of the Frontier Battalion of Texas Rangers, to "proceed to El Paso County Texas and investigate the recent troubles in that County. You will also place a company of your Battalion in position to march at once in case of the necessity for the use of troops." Jones departed for El Paso at once, riding the rails north to Topeka, Kansas, then southwest to Santa Fe. From there he took the stagecoach south, stopping at Mesilla on Saturday, November 3. He stayed throughout Sunday to discuss the situation with Fountain and Howard before moving on to Franklin on Monday.[1]

Jones was the only leader the battalion had had since its inception in 1874. He owed his appointment to Gov. Richard Coke, who served with this former Confederate captain during the Civil War. As Walter Prescott Webb says, "It is quite certain that Governor Coke cold not have found in all Texas a man more competent for the difficult job ahead." Just three weeks before this assignment, one gushing newspaper correspondent described Jones "as one fitted in every way to turn like a thunderbolt upon the 'greasers' and drive them back into their holes. Major Jones is just in the prime of life, with a frame and will of iron. To the dash of Murat and the courage of Ney, he combines caution and prudence in a degree rarely found in the 'make-up' of a man." Following this effusion, "Mervyn" (the correspondent's name) added that Jones was, "in short, a commander, a detective, a quartermaster, a clerk, a soldier, a scholar and a gentleman. . . . Major Jones is an excellent judge of horse flesh and will receive into his command nothing but sound, swift animals. He is also very select in recruiting men, and has gathered into

Lt. John B. Tays. From the author's collection

the battalion quite a number of scholars whose attainments would grace any position in life. Perfect discipline is maintained, strict obedience to orders is exacted."[2]

In the meantime, the insurgency continued. Some Paseños wanted to drive out all "Americans," while others would be satisfied with securing Howard's twelve thousand dollar bond. The trigger for action came on November 6, when a rumor swept through the towns that Jones had come to make arrests. That night, Captain Garcia learned that bondsmen Atkinson, Cobos, Ellis, and Tomas Garcia were in imminent danger. Several hundred people gathered at Granillo's home in San Elizario "for the purpose of getting the money, or imprisoning the bondsmen and afterward killing them perhaps." Captain Garcia, Dr. Ball, the bondsmen, and four to six others armed themselves and barricaded the doors of Tomas's house to resist. They dispatched a message to Lieutenant Rucker, pleading for the army's help. Meanwhile, Paseños friendly to the bondsmen idled over to Granillo's house, where threatening voices from the crowd were overheard. According to one account, "Granillo could rally no more than forty-five or fifty men to his standard, and he gave up in disgust." Some hoped the dawn would bring courage. Father Bourgade was informed that the insurgents planned to arrest the bondsmen in the morning.[3]

Overnight, Sheriff Kerber handed Jones an urgent note from Howard's bondsmen. Salcido, they said, aided by José María Juárez, Ramon Zambrano (one of Captain Garcia's sons-in-law), Luciano Fresquez, and two hundred well-armed men, was demanding twelve thousand dollars or the bondsmen's lives. Lieutenant Rucker told Jones he wanted to relieve the bondsmen but was constrained by orders. Jones wired Steele, "The San Elizario mob has not disbanded. Have Howard's bondsmen surrounded now, threatening to kill and plunder them. Can get no assistance here. United States troops ordered not to interfere. Mexicans are unreliable. Not enough Americans for a posse. I go there this moment to see." Jones then set off for the lion's den with a guide, the ageless Price Cooper. The sheriff warned Jones, "Be on the lookout Major. *Greasers* are very treacherous."[4]

In San Elizario, Jones found the bondsmen and their allies still holed up, safe but "very much alarmed." The major's first action was to authorize "Captain Garcia, a respectable law-abiding citizen, to call together as many as he could of the better class of Mexicans, to get them together and arm them, and hold them under arms, subject to [Jones'] orders, to preserve the peace." In all, "twenty-two other native citizens" volunteered to join Garcia. According to the *Independent*, "This was the first serious check the mob had received. The rioters and their leaders knew Captain Garcia; they knew him to be a sagacious, brave determined man, and they knew that when he proffered his services to the State Government he meant business." The captain's appointment had what Fountain called "a salutary effect" on the situation, at least momentarily.[5]

Before the insurgents could consider their next move, Jones asked to see their leaders. The people "sent a commission composed of Guillermo Gandara, Ramon Zambrano, Higinio Loya, Ambrosia Olguin, Vidal Garcia, Agatón Porras, and others." José María Juárez, the man whose defiance of Howard and Judge Garcia six weeks before had first incited the people to riot, was also present. Salcido, Barela, and Granillo were notably absent from the conclave.[6]

The discussion was wide-ranging, the mood confrontational. Jones tried an even-handed approach. "I told them . . . it was my duty to arrest Howard and have him tried for killing Cardis, and if any of them violated the law and complaints were made against them, that I would have to arrest them." He wanted the people "to disband, to hold no more unlawful assemblies, and submit hereafter quietly to the enforcement of the laws by the courts." The major sought assurances that the bondsmen would not be harmed. Tomas Sanchez replied that "the bondsmen had freely volunteered; that they [the people] had no hope of collecting the bond in any court, consequently they had to see how they could collect it themselves." As for whether any Paseños should be subject to arrest, Porras undoubtedly spoke for many: "[He] did not care whether they had been wrong before or not, but now they wanted to settle it. They said that in making the arrest of Howard and the judges they were right, because they were the people and the people were the law; that for the same reason they were right in forcing Howard to sign his relinquishment of claim to the salt lakes, and because they had no hope of collecting the forfeited bond in the courts, they had the right to take measures to collect it themselves by force." The commissioners understood the basic law of the land. "They produced a book containing the Constitution, to show that they had a right to assemble and to bear arms." Nevertheless, said Father Bourgade, who witnessed the meeting, the commissioners "promised to Major Jones to keep quiet, disband, and go to work, and not to give cause for any more complaints."[7]

The insurgent activists reassembled that night and invited the major to meet once more. When he arrived, Jones found "60 or 70 in the room, and some outside." The Paseños objected to the enlistment of their own people by Jones. Instead, they "wished to raise their own company and elect their own officers, and if they could not do this they preferred having Americans enlisted." Jones took note. He again asked them to promise to have "no more juntas, and to keep quiet in the future, and to abide by the decision of the courts and to obey the laws." Once more, the leaders gave their pledge, but Bourgade "doubted the sincerity of some of them, from hearing side remarks and private talk in the assembly." The major's own view of affairs was revealed in a telegram he sent Steele: "Serious trouble here. Think can control mob few days; in mean time will organize a three-months' minute company."[8]

Captain Garcia's company gave Jones a two-day window to devise a solution. If the governor had authorized the transfer of a Ranger company when that idea was initially floated, the unit would be in place. Now Jones needed to organize a local Frontier Battalion unit. Lacking authority to create a new company, he formed the "Detachment of Company C," to be commanded by a lieutenant reporting directly to him, not to the Company C commander. Only a few vacancies in the Frontier Battalion existed at the moment, not enough to fill the twenty-man detachment he had in mind. Accordingly, he issued Special Order No. 109, instructing Companies B, C, and E to reduce their establishments by four men each. Their places would be taken by El Paso recruits.[9]

Jones discussed the critical matter of the detachment's leader with Joseph Magoffin. The latter urged him to commission the stalwart Captain Garcia, "a good man to put in charge." Jones demurred, because "he wanted American citizens or people who had not mixed themselves up in the trouble if they could be got." Garcia did not fit the major's profile. Jones countered with a name that "had been recommended to him very highly," John B. Tays. Asked to assess the candidate, Magoffin replied, "as far as I knew he was a very good man." Jones made the offer and Tays accepted. On November 10, he was sworn in as the lieutenant commanding the new detachment.[10]

Tays has been relegated to the trash heap of Texas history. Very little reliable information has been published about him. Too often the "facts" available have ranged from gossip to gibberish. It is said that he "had a rather poor background for command of a Ranger company." C. L. Sonnichsen describes him as "a sort of handy man [who] made furniture, dug wells, and puttered around." Walter Prescott Webb criticizes him for lacking "resourcefulness, initiative, and that combination of dash and judgment so essential in a crisis." Texas Ranger Sgt. James B. Gillett damns him with the judgment of his own men, "most of [whom] believed Lieutenant Tays had a streak of yellow in him." Beyond the usual biographical gaps and errors, historians have consistently confused Lt. J. B. Tays with his brother and successor in El Paso, Lt. J. A. Tays. Incidents from James's life have been dropped unwittingly into John's, creating a thoroughly muddled account of the latter's Ranger career, to the detriment of his reputation.[11]

But the life of J. B. Tays provides ample evidence of an adventuresome spirit, considerable intelligence, and some scientific knowledge, perhaps self-taught. He may have volunteered to help around his brother's church, but he was also a capable engineer, farmer, and businessman, always on the lookout for new entrepreneurial frontiers. That he married the most eligible Anglo widow in El Paso County is evidence that he was more than a "sort of handy man."

John, born in Nova Scotia in 1842, was the seventh of John and Jane (Ellis) Tays's eight children. He was six feet tall, with brown eyes and a fair complexion weathered by fifteen years in the Canadian and American west. John left the family home in 1862, traveling west to the gold fields of British Columbia with older brother James (born circa 1836). After several years of "indifferent success" in the rough camps, the pair moved on in 1869 to the mining camps of Montana. The success of their brother Joseph convinced James and John to join the "Parson" at the Pass. They arrived by early 1872, when James enrolled in the state police. James and John supplemented their incomes smuggling Mexican cattle into Texas, either for sale or building up a Tays herd. Over the years, John also invested actively in El Paso properties.[12]

The appointment of John B. Tays has been roundly criticized, but, as previously noted, his candidacy was supported within El Paso's Anglo community. Howard thought Tays a good man, "but . . . very slow." The major weighed their views, studied his candidate, and made his own assessment of Tays as "cool and determined." Initially pessimistic about Jones's ability to put a lid on the county's troubles, A. J. Fountain considered the new Texas Ranger outfit and expressed hope: "The command is assigned to Mr. John Tays, a good selection. Major Jones is confident that this force will be sufficient to enable the county officers to effectually enforce the laws, the company is composed of good material, the men are well armed, and their Commander is a man who will stand no trifling."[13]

Identifying one suitable officer was an easier task than finding twenty men fit to be Texas Rangers. By largely limiting himself to "Americans," Jones handed himself a recruiting problem. Not only was the population almost entirely "Mexican," so too was virtually every Texas Ranger veteran in the county. At least eighty-one men had served in the Garcia and Montes companies, seventy-nine of them Paseños. In truth, perhaps very few of El Paso's ex-Rangers would willingly have served the unpopular cause of defending discredited Anglo laws and institutions. Jones hoped to avoid the dangers posed by conflicting loyalties.

The county's population of Anglo males amounted to only a few score in 1877. Civil War veterans such as Howard, Atkinson, and Ellis were now the very professional men and entrepreneurs who needed defending. They were not about to close up their businesses to draw forty dollars a month in Ranger pay. Given the shortage of suitable recruits *willing* to serve, the major and his new lieutenant accepted men of any age and looked as well to New Mexico. Jones announced in the *Mesilla Valley Independent* his intention of "recruiting a company to put down the El Paso County mob. Reliable men with horse and Winchester rifle will be paid $40 per month, and board. Here is a chance to serve your county!"[14]

On November 20, Jones reported to General Steele that he had "twenty good men" under arms. Historians have balked at that description. Sonnichsen remarked, "All that can be said for them is that where the sheriff despaired of raising a posse, these men were assembled out of the holes and corners. Not a one of them would have been a Ranger under normal circumstances." Texas Charles M. Robinson III echoed the view, while Robert M. Utley judged the new troops as "mediocrities or worse." These comments, accurate in some cases, are dead wrong in others.[15]

Enrollment of the detachment began in Franklin on November 11. Possibly the first name on the roster was that of thirty-two-year-old Canadian John E. McBride. The appointment of Howard's "enforcer" as First Sergeant sent a clear and unfortunate signal to Paseños. The other noncommissioned officers were both twenty-two years old. Second Sgt. C. E. Mortimer was another Canadian. Beyond being a "mechanic," nothing is known of him. Irish immigrant Cpl. Henry Oliver Mathews, known to his companions as "Rocky," was a keen observer who proved a fine war correspondent.

On the surface, the eight privates enrolled on the first day illustrate the challenge facing Jones and Tays. Only one was under thirty-five years old; the eldest was in his sixties. The "boy Ranger" was nineteen-year-old William Marsh. The son of D. C. Marsh, Fountain's man in the Customs House back in 1869, Billy was no stranger to El Paso's political quarrels. His father was dead now. The son's career choice would soon give his widowed mother new cause for grief. Tays was probably lucky to enlist two stagecoach drivers, Ohioan Frank Thrall and Frank Kent from Mississippi. Anyone who plied that trade across the dangerous roads in and out of El Paso was probably a steady man, unlikely to lose his nerve in a perilous situation. Frontier physician John K. Ball, a Confederate veteran from Tennessee, could kill an enemy and patch together a comrade, if it came to that. One who would soon need his services was North Carolinian John J. Eldridge.

Jones and Tays also enlisted John Boyle and California Column veteran John Evans, both forty-eight. The next was even older, though no one was sure just how old. Price Cooper said he was sixty-one, but he may have been older. He was still healthy, his courage was unquestioned, and his ability to survive tough scrapes must have made him a local legend. Cooper's presence must have given his comrades comfort.

The officers then moved their recruiting operations to the heart of public discontent, San Elizario. Miguel Garcia, Lieutenant Montes's grumbling private, signed up. So did Price Cooper's son Santiago, revealing some of his father's adventurous spirit. Two solid Paseños enlisted on November 19, Gregorio N. Garcia II and Pablo Mejia. The captain's son was intelligent, educated, and one of the unit's married men. Mejia was in many ways the direct opposite. Illiterate, tough as nails, somewhat brutal, a man feared by Apaches, Mejia was a coup for Tays.

While Jones and Tays scoured El Paso County for recruits, Charles Howard was in Mesilla, attempting to stack the Ranger deck in his favor. He telegraphed Jones that he could "bring down fifty good fighting men." Smelling trouble, Jones wired back, "Can you raise ten good fighting men, sober, of the right kind?" Howard replied positively, but Jones was dubious. "Can you raise ten sober men of the right kind?" Then, "Can you raise five good fighting, sober men?" Howard: "I can't raise five of that sort." Not yet despairing, Jones wired, "Can you raise two sober, good men?" Howard replied, "I have found two good quiet men, who will start down tomorrow to El Paso to enlist. There are a good many men here who would go, but all noisy and drunken." Just one day later he wired, "After agreeing to go men refused. You cannot get one good man." Jones explained to the *Galveston Daily News,* "New Mexico was made up of the loose rocks and ends and ragged remnants of creation, after the world was completed, and . . . Mesilla is a fit capital for it. Now it is a land of refuge for numerous outlaws, driven from Texas and other States."[16]

Howard may or may not have talked to rustler Jim McDaniels. Much had happened since the two men first met in August 1875, when McDaniels persuaded Judge Howard to hand him the Maese gang extradition papers. That conversation resulted in the demise of the gang and a little embarrassment to Howard. By now, Jim had no small reputation as a killer. In addition to the mysterious Maese killings and the murder of three men in Las Cruces, McDaniels nearly added another victim in a dispute over a horse and rifle. By this time, McDaniels was a full-fledged member of the Jesse Evans gang. Lincoln County historian Frederick Nolan wrote that this tough bunch "went where they pleased, stole what they wanted, and terrorized or killed anyone who got in their way." On the night of November 16, a "large band" broke Evans out of jail in Lincoln. The gang headed east and then south. No mention is made of Jim McDaniels, but on November 19 he showed up in El Paso County, taking the oath of a Texas Ranger, making his mark with an X. The horse he rode in on was appraised at eighty dollars, not the most valuable among the detachment's mounts. Perhaps this man who knew and appreciated good horseflesh did not have time to choose a more valuable animal before he left for Texas.[17]

Another troublemaker joined on November 21. Like McDaniels, Sam Frazer listed his occupation as "farmer," but "big-mouthed back-shooter" would be a more apt description. He would pay for his flawed character, though not soon enough for the people of *la Isla*. The first round of recruiting came to an end on November 26, with the enlistments of brothers Tom and Andy Zickefoose. These West Virginians rode in on "two fine horses" (valued at three hundred and two hundred dollars, respectively), provided by their employer, John Campbell, the man who had tried to gun down John Atkinson the year before. Campbell also supplied the Zickefoose brothers with weapons.[18]

Tays's detachment was a mixed bag of young and old, of Anglos and "law-and-order" Paseños, of community pillars and man-killers. Jones and Tays made some obvious mistakes in their selections, but very few. At least half were local men, with a stake in the community. Thrall, Kent, Mejia, the Garcias, and the Coopers were all solid choices. Civil War veterans Evans and Ball and the Zickefoose brothers could also be relied upon. A few of the men are unknowns at this date, and only three can, with hindsight, be said to be poor choices. McBride sent the wrong signal to the population, and loudly. McDaniels's reputation should have preceded him, and Frazier's behavior proved so bad that it is likely his character should have given advance notice. Like their new officer, Tays's men were game, if unprepared for what was to follow.

Not every recruit came to his unit possessing Ranger-worthy firearms. Even before recruitment, Steele wired Richard Coke, now a U.S. senator, "Get if possible an order on New Mexico for arms and ammunition for Major Jones." On November 14, Jones asked Steele, "Send by Stage to Lieut Tays San Elizario ten Winchesters, twelve pistols, belts and scabbards also cartridges

Texas Rangers Andy and Tom Zickefoose (with unidentified man in the middle). Courtesy of Larry Wayne Zickefoose

for both." The state's tight military budget was an issue raised in the AG's telegraphed reply, "Can you not buy arms and ammunition at less rate than it will cost by stage, which is not less than ten cents per mile?" Steele soon bowed to necessity and shipped the requested weapons and ammunition.[19]

The major had one other task to perform before he departed. On November 15, he, Tays, and several Rangers rode up to Joseph Magoffin's magnificent adobe home. As Magoffin was entertaining friends, Jones asked him to take a private stroll outside. Jones now confided what he had told no one else, not even Tays. Howard was arriving that night. Later that evening, Jones and Magoffin rode out of town to the home Carabajal, JP of Guadalupe. The jurist's command of English was doubtful, but he was intelligent and quite literate in his native Spanish. Magoffin translated as Jones asked the justice whether Howard could waive examination and be admitted to bail. First, said Carabajal, a complaint against Howard would have to be filed. The major asked Magoffin to do it, but the justice had no paper to write on hand. Magoffin went home and returned with a complaint charging Howard with having shot and killed Cardis. The question of bail remained, and there was no government prosecutor to give advice on the matter. "The justice of the peace looked over his law-books a little while and finally agreed with Major Jones that he would put Howard under bail, and . . . let him leave until the district court met . . . in March, and by which time Major Jones promised that he would have more help here to assist to enforce the laws."[20]

Early the following morning, Jones, Magoffin, Tays, "six or eight rangers," and Charles Howard rode up to Carabajal's house. There the complaint was filed, Howard waived examination, and the JP bound him over for an appearance before the district court. Howard paid the four thousand dollar bond, and Tays and Magoffin countersigned as sureties. The paperwork finished, Magoffin warned Howard, "for Lord's sake . . . stay away until the court met." Howard agreed, expressing a desire to visit Chihuahua. The wisdom of letting Howard loose on bail was later questioned in light of ensuing tragedy. Since Jones was the moving force behind the decision to release Howard, was he, and the state of Texas, culpable for the deaths and destruction that began less than four weeks later?[21]

Howard remained in El Paso for two days before departing for Mesilla. He used the time trying to mend fences. Among those he approached was Benigno Alderete. Before October 10, Cardis had been godfather to Alderete's four children. Now he administered the dead man's estate. Alderete had been shaken by the loss of his friend and *patron:* "When I found that Cardis had been killed, I remarked that it was a misfortune I was not at Franklin, either to kill or to be killed by Howard, for I regarded Cardis as my second father, and on the spur of the moment felt as though I ought to avenge his death." Through Price Cooper, Howard asked Alderete to meet him in his hotel room. Alderete hesitated, but on Cooper's advice took Jones with him to see Howard. The men talked over Cardis's death and then, said Alderete, "made friends. . . . Howard said if left unmolested he would not molest any others, and we agreed that we would meet, where people would see us, in public, and let them know that we had made friends, and we met sometimes after that, in the streets, and spoke." If either man thought their personal treaty signaled a new beginning for El Paso, they were sadly mistaken.[22]

On November 19, the major wired Steele, "Much excitement and many threats in consequence, but do not apprehend collision." As Jones later reported, "It was known all over the country that [Howard] was here, and no demonstration was made against him." Jones may have understated the danger. According to Kerber, when Howard's reappearance became known, "the 'rebels' assembled together, and at ten o'clock at night I received notification that they had 200 men and intended taking Howard prisoner. Judge Howard thought it best to return to Mesilla under such circumstances."[23]

Although Fountain's *Independent* warmly approved of the major's activities, the paper expressed pessimism regarding the lasting results of his visit. The insurgents were quiet only because they were unmolested. "But let an effort be made to arrest any of them, let the officers of the law attempt to enforce the collection of taxes, in short let anything be done or attempted to be done that is distasteful to the mob, and see how promptly the last one of them will be in arms prepared to resist, unless the officer is backed by a sufficient force to overcome them." The government of El Paso County would remain usurped until sufficient Rangers or soldiers moved in to suppress the rebellion.[24]

Fountain reported the rumor that the government was contemplating reopening Fort Bliss. He called the rumored action "wise and necessary," since war with Mexico was "by no means an improbable event." Though talk of war had actually cooled by November, the conflation of El Paso's troubles with continuing *Lerdista* filibustering threatened to raise temperatures again. President Diaz was reported dispatching "several regiments" to the border to counter the *Lerdistas*. According to the *Galveston Weekly News*, "It is evident from inside intelligence that there are officers of the Mexican army in Lerdo's interest inciting to mischief, hoping to make trouble; some of them, it is believed, in the El Paso mob. Gov. Hubbard . . . indicated to Major Jones that it was his duty, if there was a conflict and the Mexicans retired to the south of the river, to pursue and capture them."[25]

On November 20, Jones departed El Paso for Austin. The major had done all he could and anticipated no trouble before the district court reconvened in March. He departed El Paso, placing his trust in Lieutenant Tays and the makeshift detachment of Company C.[26]

Before leaving, Jones issued Special Order No. 110, which set forth the new lieutenant's operational and disciplinary duties. He was ordered to "maintain strict discipline and attention to duty" and was given discretion "to discharge them from the service for drunkenness, riotous and disorderly conduct, willful disobedience of orders or neglect of duty, or for incapacity or unfitness for the service in which he is engaged." Tays was directed to establish his post at San Elizario, the center of disturbance, but to patrol El Paso and, if necessary, adjoining counties. He was also to "assist the sheriff in making arrests and escorting prisoners when called on by him to do so. Will make arrests of offenders against the law upon proper process of the courts directed to him and will also do all in his power to give protection to the lives and property of citizens against unlawful attacks or depredations from any source whatsoever."[27] Tays would ultimately be criticized for his actions in defense of Charles Howard, but they were taken in full compliance with these orders.

CHAPTER 14

"We do not care what happens after we even the score."

The new Texas Rangers went to work immediately. On November 14, Miguel Garcia carried a dispatch from San Elizario to Franklin. Riding along the Mexican side, he overtook another traveler, Juan Montez. From him, Garcia learned that men were at the dam, watching for Howard, "on the lookout for to kill him." Thinking Garcia was with the insurgents, Montez confided that Mexicans planned to ignore the injunction against crossing the river. Montez added that the *jefe politico* advised that if men insisted on crossing, they should do so "only a few at a time."[1]

An American soldier now arrived in El Paso. He was Capt. Thomas Blair of the 15th Infantry. Blair was either thirty-seven or thirty-nine in 1877, a native of Scotland, where he either worked as a journeyman carpenter or "never did a day's work in any trade." He immigrated to the United States, either in 1861 or 1865. What is certain is that on April 7, 1865, just two days before Lee's surrender, he enlisted in the U.S. Army's Engineer Battalion. Blair transferred to the Infantry in 1867 as a Second Lieutenant in the 15th Regiment. Just five months later, this Johnny-come-lately was promoted to First Lieutenant and assigned to act as regimental quartermaster.[2]

In the summer of 1869, he returned to Scotland on extended leave. Blair's urgent need was "to marry a lady there, to whom he had been for a long time engaged and who was being forced by relatives or guardians to marry against her inclination someone else." He returned to duty in the summer of 1870, without a wife, forlornly explaining "that his betrothed had married & gone to Australia before his arrival in Scotland."[3]

The 15th Infantry received a new colonel in December 1870, Brevet Major Gen. Gordon Granger, a Union hero at Chickamauga. On February 1, 1873, he appointed Blair regimental adjutant at the Santa Fe headquarters. Blair also made the acquaintance of the colonel's wife, Maria Letcher Granger, the daughter of a Lexington, Kentucky, physician. In late 1875, Gordon Granger was paralyzed by a stroke. A second stroke in January 1876 killed him. Lieutenant Blair accompanied the colonel's remains to Maria Granger's family home in Lexington. Whatever bond they shared at this time would grow into something deeper.[4]

A severe fall from his horse in February 1877 left Blair with fever, violent headaches, dizziness, and loss of sleep. He applied for four months' leave on March 20, after the regimental surgeon certified he would need four months' rest *and* a change of climate to recover. Granted "permission to go beyond the sea" to Scotland, Blair left Santa Fe on May 3. Shortly after arriving in Scotland, he applied for an extension, citing a doctor's report that "a return in less than *10 or 12 months* would endanger his ultimate recovery." The doctor reportedly prescribed "residence in South of England or in France." Tragically for El Paso, this "treatment" was denied. Blair was ordered home, where he was promoted to captain and given a company to command.[5]

Blair arrived in El Paso on November 18, replacing Lieutenant Rucker, who returned to New Mexico. Four days later, he rode to San Elizario with Joseph Magoffin, serving once more as interpreter. The captain found "a good deal of apprehension" amongst the people. He was advised of "a strong organization of disaffected Mexicans, numbering not less than 400, probably more," whose dual object was to kill Howard and resist arrest. "Strenuous efforts are being made to force the peaceably inclined to join the organization." Blair himself witnessed a meeting held for that purpose in the Socorro home of Juan Armendáriz. Cipriano Alderete warned Blair that San Elizario alone would furnish three hundred armed men to resist the arrest of any who had participated in "the Howard affair." Their primary object was to kill Howard at any cost. "We have marked everything down," Alderete warned, "from the time when the rooster trod the hen until she laid her last egg. Everything is going to be paid for now, and we do not care what happens here after we have evened the score. You can take all these old houses. What are they worth? We will go across the river or somewhere else." Alderete's declaration spoke volumes about the insurgents' focus and determination, as did one of the captain's observations: "The Mexican citizens of this State in the vicinity of San Elizario are, as a rule, not planting or preparing to plant, a circumstance in itself showing apprehension." Though Blair did not report it, he and Magoffin also attempted to interview Father Borrajo. That priest allegedly cut short the meeting, explaining he was "tired of reporting to United States officers."[6]

Blair had no illusions that trouble would involve only American citizens. The residents of *la Isla* were being "encouraged, incited, and aided by a large number of idle and more excitable of the citizens of Mexico along the river." One meddling Mexican was a man named Lerma, reputedly an ex-officer of the Mexican army. He had been seen "helping them to perfect the organization" on the American side only nine days before. Another prospect was far more worrisome. Blair reported it likely that "any resistance to the State authorities, any disturbance or commotion of any kind on this side, will be made the occasion for swarms of thieves and plunderers coming from the other side, ostensibly to assist those on this side to resist the authorities, but really to seize everything valuable they can lay hands on, of Americans first, and next of American sympathizers." Mexican citizens would be hard to detect because they were likely to cross the dry Rio Grande riverbed "singly and in twos and threes."[7]

Sheriff Kerber shared his apprehensions with Judge Blacker. He observed that "the 'rebels' are not making the least preparation yet to plant the fall wheat crop, but keep up their meetings and have all their arms together. They have even introduced sixty stands of arms from the other side of the Rio Grande. Our rich Mexicans (who are for peace, which is in their interest) are even more scared than we are now." He complained that Paso del Norte's *jefe politico* had ostensibly forbidden "armed bodies" to cross, but in reality permitted "small squads" to do so. Even with twenty Texas Rangers to back him up, he had too few guns to attempt any arrests.[8]

In Austin, Judge Blacker met with Jones, who suggested that he should return to El Paso in early February and "look the ground over." If Blacker thought Tays's company insufficient to handle court and prisoner security, Jones would increase the force and, if necessary, personally take charge in El Paso. The major was optimistic that "no further disturbance would occur" prior to the session.[9]

The notion that everyone would patiently wait for Blacker's court to sort everything out in March was not universally shared. Vidal Garcia, a forty-five-year-old farmer on friendly terms with some of the insurrectionists, described the dread permeating the valley. "All the people of this town were afraid of Howard, and feared that he would come back." Even with Howard away, "the trouble here again commenced by some drunkards, friends of Howard, and his party getting drunk and insulting the other Mexicans. We resolved not to respond, but to suffer these insults which were given by Mexicans of this town at all times, whether drunk or sober."[10]

As November slipped into December, the festering crisis competed for attention with more prosaic concerns, like the weather. A cold snap brought ice to the valley during the last ten days of the month. As the *Independent* put it, "The oldest inhabitant has never witnessed colder weather in the Mesilla valley than we have had for the last week. The thermometer crawls down

18 degrees below the freezing point, and ice so coveted during the summer months has become a burden." With the high not much above freezing on December 29, and the following overnight low in the low teens, Lino Granillo and scores of like-minded Paseños prepared for a journey that would jerk everyone's thoughts back to the question of Zimpelman's salt. They set out on December 1 for the Guadalupe salt deposits, determined to force the issue. As the cold snap continued through the first week of December, the *salineros,* clad in their rough wool coats, drove their animals on the two-hundred-mile roundtrip trek.[11]

Though equally set on defending Zimpelman's claim to the salt, Howard tried not to let such thoughts occupy his every moment. Late November was hunting season, "a paradise for hunters just now," trumpeted the *Independent.* With plenty of targets he could shoot legally, the exiled marksman answered the call. One day, "while retrieving a wounded duck, [Howard] dropped into about ten feet of mud and water from which he emerged after a hard struggle in a rather sorry plight. The Judge says he regrets that no one was present but himself to enjoy the fun, and he did not enjoy it worth a cent."[12]

But as the days passed, Howard became increasingly restive. His eventual return to El Paso has been portrayed as an act of supreme arrogance, one that risked every citizen caught between that arrogance and the community's rage. Evidence, however, suggests a tragic figure propelled by fate and an overwhelming sense of honor. Howard and Fountain spoke daily in Mesilla, passing the evenings together in conversation. Howard, said Fountain, "spoke fully and frankly regarding his personal and business affairs, he was subject to fits of depression during which times he would talk of his wife and her early death, he would at such times express the belief that he had not long to live." On Howard's last evening in Mesilla, Fountain "earnestly advised him not to go. . . . that his return to El Paso County would not only jeopardize the life of himself, but the lives of Atkinson and Ellis, my warm personal friends." Howard felt he had no choice. "[He] was bound by every principle of honor and gratitude to protect Col. Zimpelman's interests at the risk of his own life." His father-in-law had bailed him out of a "pecuniary embarrassment" resulting from the Murphy estate case, that Zimpelman had continued to advance him money "to enable him to extricate himself from troubles that were weighing heavily upon him, that he had received several thousand dollars from Zimpelman, to pay which he had conveyed to Zimpelman certain property, that he was still indebted to Zimpelman. [He] said that he realized the fact that his life would be endangered by his return to El Paso County, but that his, Howard's, quarrel with Cardis had imperiled

Zimpelman's interests in El Paso County, and that it was his duty to protect those interests and that he would not allow himself to be driven from his duty by threats of personal violence." For Howard, personal honor trumped every other consideration.[13]

Howard returned to El Paso County on some business in late November, contrary to his promise to Jones. On his way south, while crossing a ranch at Canutillo, Texas, fifteen miles north of Franklin, Howard's driver shot a ranch dog. Reportedly, "fifteen armed men rushed out on Howard with the intention of killing him but for some nice talk and over $100 he got off but badly scared." With bigger fights over weightier issues before him, Howard bought his way out of this useless scrape. The one hundred dollars was a hefty price for a cur, but perhaps the fatalistic Howard knew he would not need that money again. Howard stayed in Franklin only briefly before returning to Mesilla.[14]

Howard's inability to stay away from El Paso now did what John Campbell and a chair-hopping doctor had not been able to do, convince John Atkinson that his life was forfeit if he remained in the county. At this time he sold all his merchandise to Charles Ellis. A few loose ends remained in the business, debts to collect and to pay, but these he would take care of and then leave the Pass of the North.[15]

In early December, Charles H. Howard entered the cauldron of El Paso County one last time. He had heard about the latest salt train and was determined to have any stolen salt legally sequestered. Tays agreed to provide him with a Texas Ranger escort to San Elizario. The lieutenant has been roundly criticized for this. There was no doubt in anyone's mind that trouble would ensue if Howard returned to *la Isla,* trouble possibly greater than the little Ranger detachment could handle. But Tays, if he followed orders, had no choice.[16]

The key man in this decision was Sheriff Charles Kerber. He was Howard's man, with an admiration bordering on the absurd. The two were kindred spirits, preferring confrontation over diplomacy. Kerber intended to seize the salt now en route, like Howard, to San Elizario; but he possessed too few deputies and could not raise a big enough posse. His only means of enforcement was Tays's detachment. On December 10, he received Governor Hubbard's authorization to call on Tays for help. The lieutenant's standing orders instructed him to "make arrests of offenders against the law upon proper process of the courts directed to him and will also do all in his power to give protection to the lives and property of citizens against unlawful attacks or depredations from any source whatsoever." Tays had no choice in the matter. Should he, in violation of orders, have refused to arrest the *salineros* or protect Howard? In light of events predictable in their violence, if not their exact course, the easy answer is yes, but Tays acted according to orders. Neither Jones, Steele, nor Hubbard ever publicly questioned his decision.[17]

John B. Tays was not the slow-witted handyman so often portrayed. On the eleventh of the month, he dispatched the intrepid Pvt. Pablo Mejia to discover and report on the progress of the *salinero* train. With both Howard and the *salineros* converging on San Elizario, his small command would need support. He had already made a bargain with Captain Blair to secure it. Around December 8, the two officers spoke at some length about the anticipated meeting of the salt gatherers and the salt owner. Tays alerted Blair, "Howard intends going down to serve a writ of sequestration, and more than likely there will be some trouble, as they have made threats to that effect."[18]

As Blair listened to the Ranger, he undoubtedly made a mental check of his own orders from above. As recently as November 24, General Pope had expanded and clarified his orders to Hatch regarding the limits of the army's authority. The troops in El Paso, he instructed, had authority only to "prevent incursions of armed bodies of men" crossing the international boundary from either side to the other "or any interference" by Mexican nationals with the internal affairs of Texas. The troops were also to protect U.S. government property and assist federal officers in carrying out federal court orders. "They have nothing to do with preserving the peace between citizens of Texas, whether Mexican or American, nor in assisting officials of that State in executing process of State Courts."[19]

Pope broadly expanded Blair's responsibilities in a succinct message communicated to Hatch on December 6. "Instruct Blair that his duties are to prevent citizens of Mexico from interfering with the civil officers or laws of Texas, by force if necessary." The first clause clarified Blair's role in assisting Texan officials. With the second clause, Blair possessed, for the first time, an explicit authorization to use force, an instruction that previously had been implicit, if obvious. Blair later asserted that he never received this final order with its specific authorization to interfere with force.[20]

Understanding that Blair could intervene only if Mexican citizens interfered with civil authorities, Tays agreed to immediately inform Blair if that happened. Expecting trouble, Blair asked, "How much ammunition have you?" "Fifty rounds to the man," replied Tays. The army captain had only twenty rounds per man, but expressed confidence: "I think we will be able to stand them off with that much." Blair ordered his men to put three days' provisions in their saddle-bags. Finally, he assigned one trooper to stay with Tays as a courier.[21]

On Tuesday, December 11, Blair reported, "All quiet here. Reports from San Elizario are twenty-four wagons with eighty men gone to Salt Lakes will be joined by one hundred and fifty armed men before returning to town. All are citizens of Texas none from Mexico. . . . Trouble anticipated when they return three days. Hence I am going down Thursday [i.e., December 13] to see myself what takes place." Blair's claim that the *salineros* in Granillo's train were all "citizens of Texas" was probably accurate. Of greater consequence was its convenience to Blair. The officer's later insistence on no Mexican involvement was a major cause for the tragedy soon to engulf El Paso County.[22]

Kerber had already alerted the governor of the impending crisis. In reply, Hubbard instructed Kerber to deliver this message to three Paseño leaders: "I am advised there is trouble and prospect of armed collisions in your county. I call on you to obey the laws and to respect civil authorities and the State military force in El Paso. You can control your people, if you will, I'm informed. Do so."[23]

On December 12, Kerber stopped first at the door of his Ysleta neighbor, Chico Barela. Unfazed by the crowd of twenty armed men, "three of them from the other side of the river," Kerber delivered the governor's telegram. He then went down to San Elizario, to hand the same message to Sisto Salcido and Leon Granillo. Neither man was home, he was told. Granillo had gone to get salt, while Salcido was said to be one of four commissioners lobbying the governor of Chihuahua. The sheriff was unwilling to leave Hubbard's message in English for absent men who read only Spanish. He went to the home of Mauro Lujan, who supported the insurgency. There he found, lying sick, James P. Hague. In Hague's presence, Lujan translated the governor's wire and attached the two documents together. His mission accomplished, the sheriff headed back to Ysleta, where he expected Howard at any moment.[24]

Upon his arrival home, Kerber found Tays waiting with important news. After dispatching the four-man escort to accompany Howard, he learned from Father Bourgade that the road was watched by armed men. Fearing an ambush, he took another twelve Rangers to meet Howard and bring him in safely. Along the way, the Rangers met eighteen men led by Barela, evidently unimpressed by the governor's admonishment. Some of the Rangers recognized Mexican citizens within Barela's band. Tays composed a note regarding these men from "the other side," and dispatched Blair's courier on ahead to Ysleta, where the captain had left a second trooper to carry any note back.

According to Tays, Blair had assured him, "In four hours from the time I get word, I will be in San Elizario without fail."[25]

Howard arrived in Ysleta around 4 P.M. There he filed papers against Granillo, Salcido, Juárez, and eight others, alleging that they were in the process of "forcibly and with violence" stealing eight hundred bushels of salt. He also prepared a writ of sequestration, commanding the sheriff to seize the salt and hold it until a court ordered it "replevied" (restored to the owner). He also sought ten thousand dollars in damages. Kerber agreed to execute the writs. Howard and his Ranger escort then rode to San Elizario, reaching town around 6 P.M., just after dark. The citizens watched as the men rode by, but "everything was quiet."[26]

CHAPTER 15
"Now is the time."

A visitor to San Elizario standing in the gazebo in the center of the town's serene *placita* can look in all directions and see signs of progress and decay, of spiritual needs and community ambition. Looking past the edges of the park, our viewer sees in turn a church, a museum, a community center, and a playground. Between and beyond these stand a number of old adobe structures, some in sad disrepair. Just a few feet away, a "Salt War" plaque, one of six describing key moments in the town's history, states that a "riotous mob" killed Charles Howard and several state troops. The words speak of a drama that neither the peaceful plaza nor the scenery surrounding it convey. But if our visitor looks again in all directions, this time breaking the restraints of time to imagine the sight of armed and determined citizens, the sound of crackling gunfire and reciprocating taunts in Spanish and English, and the acrid smell of black powder mixed with the stench from open wounds through the days and nights of mid-December 1877, one can be in the midst of it all.

East of the gazebo stands San Elizario's pretty white chapel. Possibly the fourth church on this site, it was not built until the 1880s. But if this structure was not a place of comfort and refuge when the guns went off, it occupies the spot where its predecessor played that very role.[1]

To the south stands Los Portales. What began as the home of Gregorio Garcia is now a museum and information center. In 1877, it was the county courthouse, until local law collapsed. Doubtlessly, it was taken over by insurgents, but whether used as a makeshift barracks, armory, or hospital for dying and wounded men the record does not reveal. Turning to the west, our visitor faces the Adobe Horseshoe Theater, built in 1971. This large building constructed in the territorial style obliterated the crumbling adobe evidence of homes, corrals, and garden walls.

Spreading out to the northeast is a large playground. In 1877, the hacienda of Juan Mauro Lujan stood here, an elegant house, well furnished, undoubtedly one of the finest in the county. Lujan's home was then the scene of discussions among the insurgents about strategy and tactics. Its reputation as an insurgent hangout forced the owner to flee precipitously. Today the only

Wednesday, December 12: *After Howard arrives in town, Lt. Tays positions his Texas Rangers in the barracks and El Molino. After dark, the insurgents halt Captain Blair 300 yards from the center of town and intimidate him into turning his column around.*

Thursday, December 13: *The first shot in the morning mortally wounds Sgt. Mortimer as he walks from El Molino to the Ranger barracks. The battle continues throughout the day.*

Friday, December 14: *The Rangers repel eight assaults against their positions.*

Saturday, December 15: *The insurgents capture El Molino after dark, taking Captain Garcia and five men prisoner. This success renders the Ranger barracks untenable.*

Sunday, December 16: *After an insurgent mine fails to breach the barrack defenses, Barela offers a truce. Tays agrees.*

Monday, December 17: *Howard agrees to surrender. Through a stratagem, Barela tricks the Rangers into surrendering. The insurgents execute Howard and two others, reportedly near the Borrego building.*

Tuesday, December 18: *The Rangers are allowed to ride back to Franklin (modern El Paso), without their arms or their pride.*

Birdseye View of San Elizario, December 1877. Courtesy of the artist, Bob Boze Bell

voices raised on the hacienda grounds are those of school children and those of teachers trying to bring order.

If our visitor at the gazebo looks first to the southeast and then to the northwest, gazing to beyond where the old Spanish presidio walls once stood, two structures can be seen, both of them once belonging to merchant Charles Ellis, each one the scene of key moments in the five-day battle of San Elizario. The building to the southeast behind the church is vacant, unoccupied for nearly half a century. Decades of erosion and neglect have left the ruins too dangerous for tourists to enter, but not too dangerous for vandals to steal into at night and do further damage. The place is decrepit now, but in 1877 it was one wing of an impressive hacienda, whose walls were graced by beautiful murals, and whose contents were emptied despite the pleas of the newly widowed Teodora Ellis.

At the other end of the historic district, to the northwest across the playground, is a two-story building once used as a mill and granary. It is for that reason still called *El Molino,* but has since been used at various times as an apartment building, gas station, gift shop, post office, and church. The most imposing nineteenth-century edifice in town, apart from the church, it was in 1877 a strategic military post, whose capture by insurgent forces proved the battle's turning point. It was the place where Charles Ellis was last seen by his friends.

Behind *El Molino* is the Borrego building, formerly the Madrid residence, with its three-sided façade, a structure that reportedly provided the wall against which executions took place. In the other direction, across Main Street on the cleared ground between the former granary and the present elementary school, is the site of a building demolished around 1927 to allow the right of way for the widening of the road to Socorro and El Paso. In December 1877, it was the center of the storm that lashed the tranquility of San Elizario. This long absent, long forgotten structure was the "war barracks," the Ranger headquarters.

Texas Ranger records, available newspaper accounts, and military commission testimony are all silent as to the barracks' owner, prior use, and exact location, but other evidence, though less trustworthy, allows us to make an educated guess. Elceario (or Elizario) Martin Montes, relating a story told to him by his uncle Agapito de Jesus Montes (a son of Lt. Telesforo Montes), stated that the Rangers "quartered themselves at the house of G.N. Garcia the First. (He was not a relative of the Captain.) Right in the center of town[.] [T]hey bored port holes, (I remembered those) in the front wall of the house and mounted their guns there. At the back of the house was a corral where they kept their horses and sheep. They had a water well and facilities to withstand a long siege."[2]

Montes's account must be treated with caution. Contemporary evidence indicates that the Rangers' quarters was not "right in the center of town," unless that term encompasses an area just outside the old presidio walls. In fact, evidence places the barracks just within or just outside the presidio's northeast corner. The account of Corp. H. O. "Rocky" Mathews makes clear that firing positions in the Ellis store, the *El Molino* building, controlled the ground between it and the Rangers' quarters. This could only be true if there were no intervening structures, which places the barracks relatively close to *El Molino*. The barracks, Mathews also stated, was in line of sight with Mauro Lujan's home. This places the quarters to the west of *El Molino*.[3]

The "G.N. Garcia the First" Montes refers to might be Judge Gregorio N. Garcia. Given his standing among the insurgents, he certainly had reason to want the Texas Rangers to guard his home and his person. But despite the parenthetical remark in the Montes account, circumstantial evidence actually points to the Captain, who was called "Gregorio N. Garcia the First." The captain is the leading candidate, because of his relative wealth, his prior donation of Los Portales to the county, his Ranger history, his readiness to lead an ad hoc unit at Major Jones's request just weeks before, and his actions of the coming week. Captain Garcia had long since donated his home on the south side of town. Did he now live near *El Molino*? I have not located a deed that nails down this point, but a branch of the Garcia family did own this plot of ground in the 1920s.

The evidence that this location marked the Ranger quarters is inconclusive, but not contradicted. In the absence of conflicting reports in either the uncovered written record or oral tradition, the narrative that follows is based on this supposition.

WEDNESDAY, DECEMBER 12

Tensions ran high in San Elizario. The salt train was approaching town from the lakes, Howard from Ysleta. Only the day before, John Atkinson approached several Paseños on the street, including Vidal Garcia. The salt question would soon be settled, Atkinson sneered, "if they were men." Feeling even cockier on Wednesday, Atkinson chortled that, "the heads of those who went to the salt lakes would have to dance on the sands." Stopping Cipriano Alderete, he instructed, "Look here, I have but a few cartridges (showing a handful) for use when Howard gets here." Unimpressed, Alderete replied, "Very well, we've got plenty."[4]

Instead of lying low following his arrival, Howard challenged those who had imprisoned him. He walked openly on the streets, moving from the Ranger quarters to Father Bourgade's home and on to Charley Ellis's store. Jesus Cobos met him at the priest's house. Howard's bondsman was agitated. Already that evening he had twice bumped into bands of armed men on the outskirts of town. Now he urged Howard "to remain quiet in order to prevent a fight, knowing there were but few of us and many opponents." Howard would not be budged. He would "insist upon his rights if it cost him his life." He finally went to Ellis's store to be among the friends he was endangering. According to Corporal Mathews, Howard "conversed freely with them in a friendly manner for several hours."[5]

While Howard held court, the people of San Elizario "took possession of the streets." Rattled by Howard's belligerence, Cobos went home, grabbed his arms, and headed for the safety of the Rangers' quarters. On his way, a messenger alerted him to the crowd forming at Desidario Apodaca's house. Cobas returned home but was too agitated to sit still. Hearing "considerable noise" at Apodaca's he went outside and learned that soldiers were on their way from Franklin. Cobas watched as "many ran up the road, both [on] horseback and on foot, to meet the soldiers."[6]

Their departure may have created a false quietude. Some Rangers felt easy enough to roam around town, visiting private homes or joining in the conversation at Ellis's store. Not certain the calm would continue, Lieutenant Tays placed double guards on the quarters. Sometime after 8 P.M., Atkinson, Sergeant McBride, and Jim McDaniels climbed the roof of the post office to ask the crowd below what they wanted. The reply was the expected, "We want Howard!" Atkinson shot back, "If you want him, come and get him." The crowd grumbled but took no action. The three men climbed down and returned to the Ranger quarters.[7]

Tays retired at 9 P.M. A Ranger roused him around 10 P.M. reporting noises coming from the direction of Lino Granillo's house. Tays instructed Mathews to find and retrieve the First Sergeant. At 10:05, the corporal located McBride at a Paseño's home. McBride said he would be along in a few minutes. Returning alone, Mathews spotted Chico Barela and another man entering Mauro Lujan's home. The corporal sneaked up to eavesdrop through a window. He did not see anything but guessed from the noise that "there were a number of people in the room." He "heard some one reading in Spanish. They stopped reading, and a voice in English said, 'Will that do?' The answer in English was, 'Yes.' Mathews thought the speakers sounded like Lujan and Hague." His eavesdropping finished, Mathews returned to quarters.[8]

The same noises that roused Tays also prompted Ellis to step outside his store. According to Juan N. Garcia, Ellis placed a pistol in his boot and walked off in the direction of Granillo's house to determine what was happening and possibly to talk some reason into the crowd. "What does this mean, *muchachos?*" he asked. "Don't act foolishly. Let me advise you for your own good." Then, before he could utter that advice, Ramon Zambrano (a son-in-law to Captain Garcia) cried out, "*Ahora es tiempo!*" ("Now is the time!") "Eutemio Chaves rode up on horseback and threw a lasso over Ellis and started on a run, dragging the unfortunate man; after he had dragged him some distance, he then got down and cut his throat, and the body was thrown to the coyotes."[9]

In testimony given March 1, 1878, Juan N. Garcia admitted that this story was hearsay. A very different account was passed down through the family of Cipriano Jesus Parra, the husband of Leonides Bustillos de Parra, a niece of Teodora Ellis. Cipriano maintained throughout his life that Ellis did attempt to calm the people down, but when they accused him of being a spy, he ran and hid in a wine cellar. He laid low, undetected for the moment.[10]

The ubiquitous Mathews "heard a shot, and then someone shouted (in Mexican) 'Stop him! Stop him!'" A "loud yell" came next, followed by silence for five minutes, and then cries of "Death to the *gringos!*" and "*Bravo! Bravo!*" Howard had left Ellis's store at the first sign of trouble, heading for the barracks. Alerted that Ellis was missing and that a "mob" intended to storm the Ranger quarters, Tays put several men on the barracks roof, a few in the Ranger corral, and some more in the home of John Clark, just opposite the quarters. Others cut portholes through the barracks' adobe walls. The lieutenant also dispatched Miguel Garcia to Ellis's store, with orders to keep it safe until Ellis returned. There Garcia joined his cousin, Captain Garcia, who was already standing guard with a few friends. Tays gave strict orders not to fire unless fired upon. According to Mathews, the Paseños came within seventy-five to one hundred yards of the quarters. Then, "finding us prepared to receive them [they] retired without firing a shot. The leaders in vain tried to make them turn and charge us, they fled, and we saw no more of them until day light."[11]

It is difficult to determine the peripatetic ramblings of Rocky Mathews on the evening of March 12. He seems to have been everywhere, and his travels are not always easy to square with other accounts. His next departure from quarters would seem unlikely, because of the danger, and yet his companion's mission renders this trek extremely plausible. At Second Sergeant Mortimer's request, Corporal Mathews went for a walk. The rangers "went up the Ysleta road about 800 yards, turned to the left, and went across a wheat-field, crossed the Allquio [*acequia?*], and came back to the quarters on what is called the lower road." The venture was both a reconnaissance mission and an opportunity for Mortimer to visit a lady friend. (According to Mathews, at least four other rangers were "visiting Mexican houses" that evening.) The two Rangers saw only a handful of townspeople on their trek. Once Mortimer was ready, he and Mathews went back to the center of town, reaching Ellis's store sometime around 11 P.M. The shopkeeper was not present. The Rangers ate some sardines and by 11:15 were back at their quarters in bed.[12]

With his men controlling the approaches to the barracks, the horses safely corralled, a cistern providing water, and plenty of food and ammunition, Tays was fairly secure for the moment. What the lieutenant needed was Captain Blair's expected relief column.

Early that day, Blair had telegraphed Colonel Hatch in Santa Fe: "Four rangers have just arrived here to escort Howard to San Elizario." Blair failed to advise his superior that he anticipated the possibility of action. Only later did he acknowledge, "It was anticipated that [Howard's] appearance in San Elizario would be the signal for action on the part of the Mexicans, and Lieutenant Tays, of the rangers, had promised to send me word the very first indications he saw of the appearance of armed [bodies of] men from the other side of the river." At 4:15 P.M., Blair got that word from his couriers. Blair had promised, "In four hours from the time I get word, I will be in San Elizario without fail." The captain ordered Lt. Samuel Pague to organize the men, and at 5:20 Blair, Pague, and fourteen mounted soldiers, trailed by three more with a pack mule loaded with three-days' rations, set off down the road. Blair "traveled fast." All except the lagging soldiers with the pack mule should have arrived in San Elizario as promised by 8:15. Kerber says the captain only reached Ysleta at 10 P.M., but the sheriff's timing may be off. In any case, night had fallen and the moon was up as the captain approached San Elizario.[13]

Along the way, Blair "caused the men to load their pieces, as I anticipated they might try to surprise us, but with strict orders that no one should fire until ordered." He was challenged by an armed man two miles short of town but kept going. At a distance of three hundred yards from the town plaza, at a point where the road was hemmed by a low wall on one side and a "brush-hedge fence" on the other, Blair's rescue effort ground to a halt. The soldiers were "challenged and halted by not less than a dozen voices at the same time and from various directions." Riding forward, Blair asked "who they were and by what authority they dared to halt me; they replied that it was none of my business, and that they had been posted there by their captain with orders not to permit anyone to enter the town." Blair insisted on talking to their leader.

Spooked. *Courtesy of the artist, Bob Boze Bell.*

After some delay, he threatened to advance his command in five minutes. Chico Barela then appeared. In a vain effort to impress, Blair demanded "by what authority he had presumed to halt me, an officer of the United States Army."[14]

The target of this question has been described as "a quiet man, fair for a Mexican, with brown hair and blue eyes." His enemies made a great deal of his illiteracy, but he nonetheless was "a leader of great fire and talent." He certainly possessed an uncanny ability to lure his enemies into the easy way out, and he used that to full effect now. Whatever was going on in San Elizario, Barela said, was "no business" of Blair's. The Paseño was forthright: The people wanted Howard, and they would resist any effort by the soldiers to interfere. Attempting to sound firm, the captain said Barela would regret stopping his column. But Blair was outmatched, and not just in troop strength. Barela simply "shrugged his shoulders; said they wanted Howard and would have him, and if I tried to enter, he would fire on us."[15]

Blair now asked if there were Mexicans in the armed force facing him. No, said Barela, all were residents of the county. Blair later acknowledged that he disbelieved Barela's claim: "I did not deem it my duty to interfere although I had reason to believe there are men from Mexico in the mob."[16]

At this point, the captain ordered the lieutenant to withdraw the men covering him from behind an adobe wall. Then he asked to be taken prisoner! Blair "wanted to see what they were doing." What he would have done with any information once he was a captive he didn't explain. The irregular request would have had the advantage of separating Blair from any continuing responsibility of to rescue Tays. Barela, desiring no such entanglement with an army officer, led him back to his troops. Blair turned his detachment around and withdrew to Ysleta. The soldiers arrived about midnight and remained for the night.[17]

THURSDAY, DECEMBER 13

The insurgents have been described as a mob with officers and noncoms. All witnesses agree they possessed leadership and organization. One account asserted that the insurgents "have an excellent military organization, they have their commander-in-chief, are divided into companies with captains, lieutenants, sergeants and corporals; they maintain a respectable degree of military discipline, understand military tactics, are not badly armed, and when together present as respectable a military appearance as any Mexican soldiers." There is no evidence that active Mexican army officers were involved, as some bellicose Texans claimed, but they had received some drilling from the ex-officer Lerma. More important was the military schooling, however limited, and military ability, quite considerable, of the insurgents' acknowledged general, Francisco "Chico" Barela. He, Sisto Salcido, and others had been Texas Rangers in the Garcia and Montes companies. Paseño campaigns against Apaches had taught these men the value of organization and discipline when facing a deadly foe.

Despite their obvious efforts to operate in an effective military fashion, it would be a mistake to exaggerate the insurgents' cohesiveness. They included men of all stripes, from outraged community leaders and usually quiet farmers to notorious Mexican rustlers, poor men from the Mexican side bent on plunder, and other young males with testosterone to burn. Some insurgents, like Barela, aimed only to kill Howard and ensure communal ownership of the salt lakes. Others were more bloody-minded. For them, Howard's death was but a step along the path to cleansing the county of *gringos*. Before the insurgents could win this war, they would have to decide among themselves what constituted victory.[18]

The available evidence makes no mention of *soldaderas,* women soldiers, within the insurgent army, but there can be no doubt that some took part in the coming siege. Female participation in Mexican warfare was a tradition dating back centuries, one that received considerable publicity only a generation later during the Mexican Revolution. Historically, women commonly served in support capacities and notably if less commonly as armed warriors on the battlefield. Since Paseño women shared an interest in the outcome of the struggle equal to that of their men, and since the forthcoming battle took place within sight and sound of many of their homes, it is hard to believe that San Elizario, and perhaps other towns, did not produce *soldaderas* who cooked for the insurgents, treated their wounds, and, in some cases, picked up a weapon and fought.[19]

Insurgent numbers swelled with each passing day. For one who had lived in El Paso for five years, Howard had demonstrated egregiously bad judgment in choosing this moment to exercise his legal rights. His return to San Elizario coincided with the opening of what Hatch later described as "an annual fair" at Paso del Norte, one that always attracted a large gathering from Chihuahua and even Sonora. This was the Feast of Our Lady of Guadalupe, which, according to tradition, marked the appearance of the Virgin Mary to an Aztec convert to Christianity in 1531. *Nuestra Señora de Guadalupe* was an important symbol of Spanish nationalism, and that role was transplanted to Mexico. Long before and after the Salt War, the rebel forces of Miguel Hidalgo y Costilla and Emiliano Zapata carried banners displaying the Virgin of Guadalupe. Devotion to her was particularly strong in Mexico's northern borderlands. Whatever the limited political aims of the insurgents, this uprising occurred at a moment when feelings of Mexican nationalism were running high at the Pass of the North. Lino Granillo and the other insurgent leaders may even have chosen this moment to challenge Howard in order to swell their army's ranks and its fervor. If religion and national identity were not enough to motivate some who joined the Paseño struggle, other, more prosaic lures existed. Many were attracted to the festival by the opportunity to buy or barter; others by the concentrated wealth waiting to be stolen; still others by the music, dancing and adventure of it all. Travelers across the Mexican borderlands naturally carried firearms for protection, from Indians and from each other. Those who sought adventure, plunder, or a chance to redress grievances against Americans now saw their opportunity in San Elizario. They crossed to the American side singly, in twos and threes, and in larger groups, ultimately numbering in the small hundreds and swelling the agitated crowd to as many as six hundred men.[20]

When the Rangers awoke on Thursday morning, they found that the insurgents had been reinforced and "were raising a fuss." The Paseños had ringed the Rangers with three lines of pickets and with horsemen blocking roads in and out. Barela's men, said Tays, "had stretched rawhide ropes across the openings, and cut holes to command them [i.e., the openings in the street] in such a manner that it was impossible for us to go out and charge them." Tays repositioned his men. After first light, Captain Garcia sent for more men to defend the two-story *El Molino*. Tays answered with privates Gregorio N. Garcia II and Frank Kent. Besides these, Garcia could count on Tomas Garcia and perhaps other sons, Ranger Miguel Garcia, and Frank Campbell, a miller in Ellis's employ. Campbell brought his wife and small child to the Ellis store for protection.[21]

Tays still had fifteen rangers (Private Mejia had not yet returned) and five citizens to dispose. The latter included Howard, Wesley Owens, Pecos County Deputy Sheriff Andy Loomis (there by accident), an unidentified man (possibly Cobos), and John Atkinson. The latter brought with him to the Ranger quarters a trunk containing at least seven hundred dollars in specie. Corporal Mathews claimed that Atkinson also packed at least ten thousand dollars in "greenbacks and drafts" in his pockets. With morning, Tays placed Atkinson, Sergeant McBride, Private McDaniels, and another Ranger on the roof of John Clark's house, which doubled as the post office, across the street from the barracks, leaving twelve Rangers and four citizens to defend the main compound. It may have been at this point that Atkinson parleyed with the insurgent leaders. Predictably, where this merchant was concerned, "angry words passed between them."[22]

That morning, Sisto Salcido arrived at the Texas Ranger barracks with a message. Little is known about Salcido, one of only two Paseños to serve in the Texas Ranger companies of both Captain Garcia and Lieutenant Montes. He must have been intelligent, articulate, and forceful. He repeatedly appears in the Salt War story as the junta leader entrusted with communicating insurgent positions or negotiating with their enemies. Eleven weeks before, he negotiated with Bourgade to extract a settlement with Howard. In early December, he was one of four commissioners seeking the help of Chihuahua's state government. Now he delivered an ultimatum to Tays: Deliver Howard within three hours or face an assault. If it came to that, said the junta leader, "We will burn you also, as you are interfering in our business." Tays refused, but at 9 A.M. took the opportunity to again shift his forces. The men were ordered off Clark's roof, while those on the Ranger roof were increased to seven. All was quiet as Sergeant Mortimer walked across to Ellis's store, checking Captain Garcia's position there, and then to the corner of Atkinson's house. As he walked back to the quarters, a single shot from the window of Nicholas Kohlhaus's store, reportedly fired by a man named Zuñiga, struck him in the back. "I am shot!" he cried, as gunfire became general. Mortimer ran a short distance before faltering in front of the grist mill. In an instant, John Tays dashed out and, braving the Mexicans' fire, grabbed hold of his man, pulling his body to safety. Inside, Dr. Ball examined the sergeant. The bullet had entered Mortimer's back and came out below the right nipple. The sergeant lingered until sundown before dying.[23]

"At the sound of the shot, the mob commenced to fire from all directions." The Rangers returned fire, except the men on the barracks roof. These quickly discovered that the low parapet afforded no real protection against sharpshooters taking aim from the roof of Mauro Lujan's house 150 yards away. One shot cut the rim of Atkinson's hat as he hugged the roof. Another passed close to Rocky Mathews's head, "cutting the hair." Said Mathews, "To attempt to escape over the side of the house was certain death." Tays ordered a hole axed through the ceiling, a move which quickly resulted in collateral damage, two chopped fingers belonging to Private Eldridge. Otherwise unharmed, the men extricated themselves. Tays redistributed them to loopholes around the quarters and to positions within the corral. The men in Ellis's store maintained their rooftop position.[24]

The battle continued throughout the day. The Rangers had difficulty spotting their targets, for the insurgents made good use of adobe walls and buildings. Mathews "never saw a man after the first half hour's firing." Nevertheless, the state troops "kept up a steady fire through the port holes." Several Mexican charges against the Ranger positions were beaten back, but they succeeded in drawing the noose tighter. Up on the Ellis store roof, one Ranger nudged his companion, young Gregorio Garcia II, and pointed to a sombrero visible behind an adobe wall some seventy-five yards away. Garcia, "lying prone on his stomach, shifted his position slightly and moved the butt of his rifle to a more comfortable position against his shoulder." Soon, the sombrero moved higher, exposing the head and shoulders of the wearer. The two Rangers fired, and their target grabbed his throat and fell forward over the wall. Some sixty years later, Garcia recalled, "I don't know whether I killed him, or if it was my friend's bullet. If my aim was true—then that was the only man I ever killed."[25]

Around 4 P.M., the insurgents sent a JP in their custody to request a truce. Tays agreed, and took the opportunity to send Pvt. Santiago Cooper back on the roof to retrieve some blankets left there that morning. No sooner did Cooper show himself than one shot tore a hole in his hat while another ripped through his coat. He gamely returned fire and came back down through the hole in the roof. Tays barricaded the doors and ensured that all loopholes and doors were well covered. Firing continued sporadically throughout the moonlit night, enough to keep the Rangers awake.[26]

Captain Blair's whereabouts must have been topic number one among the Rangers. If the Alamo's fate did not pop into the head of Canadian transplant Tays, it must have preyed upon the minds of Texans like Howard. Lacking a James Butler Bonham to pass through the Paseño cordon around their positions, the defenders had no way of discovering if help was on the way.

Blair had an excuse for failing to force an entrance the night before: "It would have been worse than useless had I done so. We were surrounded by no less than from 120 to 150 men, all well armed, and both sides of the street in front of us lined with them, and ropes stretched across it at intervals." But if he can be excused for withdrawing from an impossible tactical position on Wednesday, he deserves censure for failing to regroup Thursday morning and make another attempt to take the pressure off Tays. By Thursday, he had justification beyond his orders for moving back to San Elizario. The three soldiers with Blair's pack mule were now missing. After losing sight of Blair's main column, the sergeant, two privates, and their mule had apparently taken the wrong turn and approached San Elizario from a different road than the captain used to advance and retreat. Having a firmer grasp of duty than their captain, the baggage detail "attempted to force their way in and were fired at." In the exchange, the soldier leading the mule was thrown from his horse, and both the horse and the mule ran off. The two mounted soldiers remained in the area all night. Unable to link with their comrade, they rode back to Ysleta to make their report.[27]

By Thursday morning, Blair understood that his own command had come under attack, and that one man was missing and possibly a prisoner, along with a horse and mule loaded with supplies. War was underway, and Blair was under orders to assist state authorities, *by force if necessary*. He claimed, of course, not to have received this instruction. He could, therefore, hold to his previous orders instructing him to take no action if only American citizens were involved. Barela had told him that was the case. But Tays had told him otherwise, and his own reports show he knew Mexican citizens would inevitably become involved. For Blair to turn his back on Tays, knowing that Mexicans were involved and that his own men had been fired upon and possibly killed or taken prisoner was a dereliction of duty. Unless, of course, he could make a good case for inaction.

Leaving Lieutenant Pague and the detachment back at Ysleta, Blair headed back to El Paso and away from the battle, to telegraph a report that would cover his actions. In his wire, he acknowledged receiving Tays's call for help. He found San Elizario "in the hands of an armed organized mob all claiming to belong to El Paso County. I did not deem it my duty to interfere, although I had reason to believe there are men from Mexico in the mob. There is no possibility of identifying them. My force would be entirely useless should I attempt to do anything. They claim fifteen hundred (1500) men in town. I don't believe there are so many. Lieut Pague and the men are now at Ysleta."[28]

There, in a nutshell, was Blair's rationale for nonintervention. He would not act since he could not identify exactly which men in the crowd would justify action. But this interpretation went far beyond the strict limits already placed on him by Pope. The instructions of November 26—orders he did not deny receiving—specified it was his duty to "prevent any interference with internal affairs of Texas by Mexicans from Mexico." Were there Mexicans from Mexico in the insurgent crowd interfering with the Texas Rangers and Texas courts? Tays said so; Blair believed so. Did Blair need to know exactly who in San Elizario were Mexican citizens? No. He simply needed to support the Texas officers as long as the insurgents contained Mexicans in their midst. It is worthwhile to note that Blair did not mention the missing soldier with the pack mule.

Blair asserted that his force "would be entirely useless should I attempt to do *anything*," a claim that begs the question, "Why?" The apparent, if obliquely formulated, reason was the number of insurgents. Instead of stating *his* estimate based on observation and other information, including that given in his November 23 report, he provides a grossly exaggerated number, a claim appearing nowhere else in the records of the Salt War. He then discredits that high count after inserting it in the record. Blair was right. His detachment would have been useless, so long as he was in command.

Some time later, Blair received an update from Lieutenant Pague. Father Bourgade, having been prevented from entering the center of San Elizario by one of Barela's sentinels, had withdrawn to Ysleta. He had heard "much rapid firing." Pague also reported "there was undoubtedly many there from the Mexican side—mob getting drunk serious results feared." The captain now had to figure out what to do about a drunken mob with guns. Fortunately, nothing. "On application of the Special Deputy Collector at El Paso," S. C. Slade, Blair pulled Pague and the detachment out of Ysleta and brought them to protect the Custom House. He wired these fresh developments to Hatch. Within hours, the colonel directed Blair to defend the Custom House. The captain's defensive posture was now sanctioned.[29]

News of the siege in San Elizario quickly traveled outside the county. Magoffin telegraphed the governor, "San Elizario holds immense mob. Howard and State troops in imminent danger. Federal commander told me he would be resisted if he entered." H. Harvey wired Silver City, New Mexico, that, "[t]he mob commenced firing on the rangers at 12 o'clock. Three Mexicans left to cut the telegraph line—Howard is gone up, I think."[30]

Pablo Mejia was still out on the road. He returned from his reconnaissance to the Guadalupe Lakes on Thursday, to discover the path to his surrounded comrades blocked. A resident of San Elizario, he found a place to hide until nightfall. With a fresh mount, he continued downriver and crossed over to Mexico a few miles below San Elizario. There he found "a great many camps of families in the *bosque.*" Already the war was creating refugees.[31]

FRIDAY, DECEMBER 14

General Pope responded quickly to Blair's telegrams, directing Hatch "to take every man [you] can without leaving the posts in the absolute powers of the Indians." On Friday evening, the general wired Sheridan, cautioning for the third time in five weeks that Hatch had barely enough men for "ordinary duties" in New Mexico, and could actually bring little strength to El Paso. The general supported Blair's reading of the problem. The dispute appeared to be "almost if not entirely confined to citizens of Texas, Mexican or American." While some Mexican citizens might be mixed up in the trouble, "it is difficult if not impossible to identify them." Furthermore, the salt troubles had "no relation to the troubles on the [lower] Rio Grande." Having undercut the rationale for or feasibility of federal military action, the general observed it was doubtful "that State officials can send a sufficient force for the present to execute the State laws or processes." In other words, this was not really a matter for the army, which was stretched too thin anyway, but there is no other solution available.[32]

Meanwhile, the "dispute" was renewed in San Elizario on Friday morning. Though few details have surfaced, the next twenty-four hours may have been the most harrowing for both the small band under siege and their brave adversaries. From one dawn until the next, the insurgents made no fewer than eight attempts to cross open ground and force admittance to the tightly shut Ranger bastions. Barela's men first charged the Rangers' corral, but were beaten off. They next attacked the rear of Ellis's store, but were repulsed by Captain Garcia and his six men. At noon, an hour-long assault was launched against the front of Ellis's place with the same results. The failure of a fourth assault, this one against the Ranger quarters, appeared to leave the revolutionaries in some disorder. Lieutenant Tays determined that this was a "sham retreat" designed to draw his men out into the open. When the stratagem failed in its purpose, the besiegers attacked a fifth time, and were again beaten off. The insurgent guns fell silent in the evening, prompting the Rangers to believe the cavalry had finally arrived. To be certain, they "dressed out a log of wood and stuck it up. It received a general salute from all directions," that the Rangers returned "with interest." The attacks continued throughout the night; Mathews calculated three night assaults were made before daylight Saturday. At one point, the besiegers threw "rockets" in an unsuccessful attempt to fire the haystack in the Rangers' corral.[33]

It is likely that the insurgents suffered heavily in these vain attempts to overcome their enemies. The repeated courageous assaults on the Ranger strongholds were not the work of a mindless mob. The men who organized and led the charges, and not a few who took part in them, were themselves ex-Rangers, veterans of skirmishes with tough Apaches. Barela, Salcido, and others were old enough to have fought Sibley's Confederates in 1862, and that experience had taught them not to give up, whatever their casualties.

On the other side of the adobe walls that sheltered them, the Texas Rangers and their allies counted veterans of equally perilous situations. Only five years before, young Jim McDaniels and his Chisum cowboy companions had withstood the assaults of over fifty Indians at Horsehead Crossing. Old Price Cooper could reassure the men with tales of his own battle with a hundred Comanche behind the crumbling walls of the Butterfield Head of Concho station. John Ball could contrast the hail of Paseño bullets with the withering Yankee musketry and cannonballs that had not found their target in him. Captain Garcia, leading the defense at the Ellis store, had brought Lieutenant Lazelle's ambushed column to safety at Dog Cañon, while Miguel Garcia had survived another battle in that gorge. John Atkinson, armed only with an unusable revolver, had "walked down" his opponent until the man ran out of bullets and away. If the Detachment of Company C was a motley outfit, hastily thrown together, it at least had a strong leavening of gutsy frontiersmen and ex-soldiers who could be counted upon to stiffen the spine of any nervous comrade. The man they defended, Charles Howard, was second to no one in his self-confidence, battlefield experience, and disregard of danger.

In Paso del Norte, *Jefe Politico* Jesus Padilla issued an "Aviso al Publico" (Public Notice), officially noting the fight for salt underway in San Elizario. He instructed Mexican citizens not "to interfere in this question, either by crossing with arms or by taking any part in the disturbance, and whoever violates this policy will be severely punished."[34]

While Blair hid in the Customs House, Sheriff Kerber focused his energy on saving the trapped men. At risk to his own life, he lingered on *la Isla* well into Thursday evening in an effort to determine what was going on in San Elizario. Upon reaching Franklin at midnight, Kerber reported that the Rangers could hold out for five or six days. The sheriff surrendered any notion of raising a posse of "Americans" to face several hundred armed Paseños. He needed the army's help. Kerber thought he obtained Captain Blair's agreement "to go with me to San Elizario, but after having saddled my horse, he refused to go."[35]

Kerber sent telegrams in all promising directions, to Hatch, Hubbard, and Texas' U.S. senators in Washington, Richard Coke and Samuel Bell Maxey. Typical was this telegram, which stressed the federal, even international flavor of the crisis: "Chas. H Howard with twenty men of Texas forces surrounded at San Elizario by Mexican mob from both sides of River over three hundred strong fighting all day yesterday. Communication cut off. United States troops not allowed to enter town. One colored soldier taken prisoner. Unable to raise posse among our population being Mexicans."[36]

Rumor competed with news throughout Friday. Someone brought word that "[s]everal rangers have been killed." In Austin, Adjutant General Steele—more confused than he realized about the situation—ordered Lieutenant Tays, via a wire to Magoffin, to "[r]eport condition of affairs by telegram." That same day, someone telegraphed much more positive news to the *Galveston News*. The Texas Rangers, said the wire, "are intrenched [*sic*] and are well armed; and it is presumed will hold out until national troops reach San Elizario from the nearest camps. There are only twenty Texas troops when the trouble began, but it is almost certain other Americans in the county will have joined them."[37]

It was Kerber who dispatched possibly the most significant telegram that day, this one directed to the governor: "Ellis and five of State force killed. Only twenty United States soldiers here and cannot help. Allow me to raise 100 in New Mexico." With this extraordinary request, the El Paso Salt War took an unexpected turn.[38]

CHAPTER 16

"Howard they wanted; nothing else, nothing less."

SATURDAY, DECEMBER 15

On Friday, Gov. Richard Hubbard formally requested federal intervention: "The Mexican force being too strong to be expelled by the Texan troops, and it being impossible to raise a civil posse from the citizens, who are nearly all of Mexican blood and sympathy, and having no reinforcement within 700 miles, I ask the aid of such United States troops as may be nearest to the scene of action to repel this invasion of our territory."[1]

Hubbard elaborated upon his request in a letter to the president. El Paso County's "ignorant Mexicans," he explained, had been "led by designing persons to believe that they (the people) are the law." Hubbard warned that the insurgents would, "receive speedy re-enforcement" from Chihuahua. With the nearest Texas Rangers more than five hundred miles away, Hubbard needed federal troops "to repel this invasion and suppress insurrection."[2]

Hubbard separately begged Congressman Roger Q. Mills, secretary of the Texas congressional delegation, for immediate federal help. The Hayes administration understood the importance of keeping this delegation satisfied as to army affairs. As Sherman himself acknowledged, "The Texas members [of Congress] claim that we of the army owe them a debt of gratitude for saving the army Bill this extra session, which is true, for the democrats had the power and were resolved to cut us down [from 25,000] to 20,000 men this session, and to 17,000 in the regular term. There is some force in this claim, and unless we can reconcile the Texas democrats in the House, we will be slaughtered this winter." Reinforcing the army in Texas was no easy matter. Troops were needed in the northern plains to keep an eye on the recently defeated Sioux. As Gen. Winfield Scott Hancock, commander of the Division of the Atlantic, reminded Sherman, "last summer's experience [putting down railroad strikes] demonstrates the absolute necessity of keeping the force, mostly artillery, where it is, nominally guarding the lakes . . . and . . .

coasts, but really for use in case of an organized resistance to the Revenue Laws, and labor mobs."[3]

Representatives Mills and Schleicher buttonholed the president as he visited Capitol Hill on Saturday. The Texans asked Hayes "to adopt measures to preserve peace and protect citizens." The commander-in-chief assured them he had already ordered federal troops to El Paso "from the nearest point, [but] he feared they would arrive too late to do much good." Secretary of War McCrary let it be known to the press that he wanted the force sufficiently large to serve as a buffer, "to prevent conflicts between the Mexican population and the Texas militia, who are very much aroused and prejudiced."[4]

Sherman's specific instructions for Hatch in New Mexico were to "do all he can concerning the demand from the Governor of Texas as that required by the Constitution in case of insurrection and invasion." Hatch must not cross into Mexico, however. Sherman also directed Sheridan to order a company of the 10th Cavalry Regiment to march from Fort Davis on El Paso. "It is important to prevent the mob getting the better of state officials. We are fully conscious of the great distances and only want you to do what is possible." In his own orders to Pope, Sheridan noted, "I am willing to assume the responsibility of withdrawing the troops from other posts to meet this emergency. When trouble exists, we must send the troops to where they are at once needed, instead of keeping them at places where trouble is only to be anticipated." Pope assured the Lieutenant General that Hatch would "take his whole force if necessary at once."[5]

The president and his chief generals had spoken. The time for the army's ambivalence was over. Pope repeated to Hatch his orders from Sheridan. Before the day was out, the first troops were on the march, though these were in response to Blair's earliest telegrams, rather than the president's order. The 9th Cavalry's L Company moved out from Fort Bayard, and the 15th Infantry's A Company from Fort Craig. At least one report had four companies of the 9th departing Bayard that morning.[6]

One more telegraph was sent that day, a communication intended to save El Paso County from hordes of lawless Mexicans. In response to Kerber's Friday plea for authorization to raise a posse in New Mexico, the governor wired: "Raise 100 men at once, to resist invasion, put down insurrection, and re-enforce State troops. Have telegraphed President Hayes, for aid of United States troops. Call on all good citizens to aid you at once, by my order."[7]

For three days the junta soldiers had used their superiority in numbers and tactical options. They had corralled the blustering, blundering Howard, the valiant but inexperienced Tays, and the veteran but outnumbered Garcia.

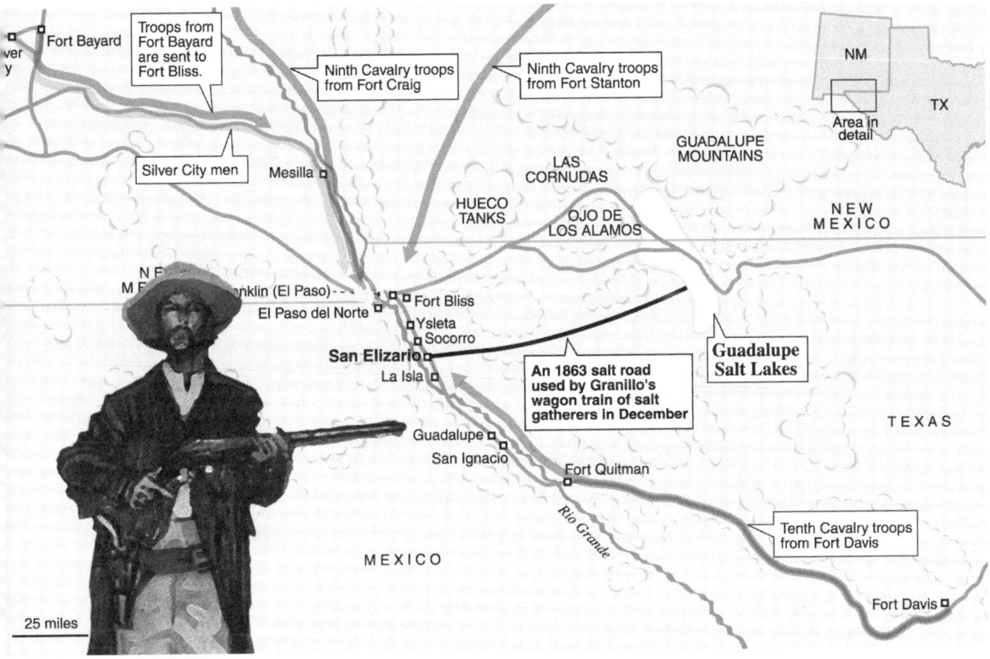

"Theater of War"–the Rio Grande Valley and surrounding country.
Map by Gus Walker, courtesy of Bob Boze Bell

Having fixed their adversaries, the insurgents had only to overpower them. Time was their enemy. They must conquer before help reached the Rangers. Chico Barela had reason to feel confident. He gained time the moment he imposed his will over the tough-talking but timorous Blair. By cowing the captain into inaction, he had, it appeared, permanently taken the U.S. Army out of the fight. Now he could focus on gaining a quick decision and presenting the distant Texas government with a *fait accompli.*

The morale of the junta army at this point is an unknown. The available record does not speak to it. The insurgents appeared to be winning, and their numbers were certainly growing. But the little army had been bloodied in three days of assaults. The new recruits included outsiders set on plundering. Many a Paseño, and not just those forted up with Garcia or taking shelter in their own homes, must have been anxious about the rebellion. As much as they hated Howard and wanted his death, as the siege wore on, they had time to think and to worry. Who among the family providers would die? Would looters leave anything behind? Would Texas retaliate? And if all went well, there was still the matter of those untended crops. The men and women of the besieging force had much to consider between assaults on the Ranger barracks.

On this Saturday, though he did not know it, Barela lost the strategic initiative. The shift was made possible in 1844 by Henry Morse's perfection of the electromagnetic telegraph, or more precisely to its military coming-of-age in the 1860s. If that was the means of federal action, some might have seen the motive in the overarching political need of Hayes and Sherman to satisfy the Texas congressional delegation. So in mid-December 1877, in an age before email, text messaging, or even transcontinental telephone service, Governor Hubbard's plea to Washington was, within twenty-four hours, transformed into orders for action from the nation's commander-in-chief, through the secretary of war, the commanding general of the army, and the subordinate commanding generals in Chicago and Fort Leavenworth, to the commanding officer in Santa Fe, New Mexico Territory. Together, the call for help and the orders to act had traveled some four thousand miles by electromagnetic wire. The U.S. Army was now committed to the war. On top of that, Sheriff Charles Kerber had carte blanche to raise an army of mercenaries. Technology, politics, and racial enmity had done their part. It remained for men and animals to cover the miles separating Charles Howard and the Texas Rangers of Detachment Company C from deliverance.

The telegraph carried news of the San Elizario "riot" north, east and west. The first accounts of the dramatic events were published as early as December 15 in various borderlands newspapers. Fountain's *Independent* trumpeted the story with nearly a full column of breathless headlines ("The Rangers have a fight with the Mob! The Hair begins to Fly! Three Leaders of the Mob Killed by the Rangers!"). The most significant news, beyond the fact that fighting had broken out, was Kerber's telegram: "I have authority . . . to raise 100 rangers. Send at once well armed and mounted Americans." Editor Fountain smirked, "By which we judge that a scrimmage is imminent. Who'll go? Don't all speak at once." Among the other "facts" reported by the *Independent* were the deaths of "Ellis, Price, Cooper, Kent, and Eldridge of the State troops." (Fountain, who knew Price Cooper, must not have reviewed his typesetter's work.) Readers of Silver City's *Herald* learned that, "The mob are going to try to wash out [i.e., flood], the rangers' quarters." The newspaper commented, "It is impossible to say what the result will be, as a mob which has once scented blood, knows no bounds to its fury."[8]

Interest in border mayhem was shared by readers as far away as Brooklyn, where the *Daily Eagle* carried the news of "A Battle on American Soil." In the nation's midsection, the Chicago *Tribune* relayed news of the "considerable excitement" in Washington and advised, "The Administration would do all possible in the premises." In Washington City, the commanding general downplayed the crisis. Sherman, said one press report, "takes no stock in the sensational telegrams concerning the troubles now alleged to be existing in Texas, and says it arises out of the desire of a few speculators to demand payment for the produce of a salt well which has been free for years alike to Texans and Mexicans."[9]

Word of mouth also carried news of the battle to the citizens of Mesilla and Las Cruces. Mary M. Phelan and her husband, newly arrived from Michigan, heard "persistent rumors [of] a very hot and bitter fight over the salt beds going on in and around San Elizario." As Mary later told her daughter Edith, then just an infant, "Texas was then the end of the earth to me and neither of us gave it much heed, when one day there came stamping into the house a body of well-armed, vicious looking Mexicans and ordered Mr. Phelan to 'get ready.' Thinking the peremptory words were advice for Mr. Phelan to prepare to meet his Maker, the couple was relieved, and confused, when their baker said, no, Phelan needed 'to go with them and fight in the 'salt war.' [Mary] wept, plead and implored, trying to explain that there was plenty of salt back in 'the States,'" and that she would die if left alone with her baby, "all to no avail. One woman or baby more or less was of no importance and men to fight for those salt beds were important." The men "generously offered $30.00 a month" for Mr. Phelan's services. And so Phelan "got ready" and was marched off. That night, the disconsolate Mrs. Phelan barricaded herself and little Edith behind their front door, while "a bunch of very drunken natives danced and whooped in front for hours, at their joy of seeing the Gringo carried off to war."[10]

Hunkered down in their adobe bastion, John Tays and his men knew nothing of the world beyond the rings of insurgents surrounding their positions. Firing remained heavy. The Paseños' reported attempt to flood the Ranger quarters failed, probably because the water level in the *acequias* was too low. It was then that the besiegers began tunneling. During a morning lull in the fighting, Mrs. Frank Campbell and her two children crossed the open space from El Molino to the Ranger quarters. Though fired on twice, they made it safely across. Mrs. Campbell carried the news that the insurgents had succeeded in entering John Atkinson's store and were busy breaking through the wall that separated it from Ellis's. She figured her children were safer with

El Molino, San Elizario 1923: Captain Garcia's bastion in December 1877. From the author's collection

Tays than in with Captain Garcia's endangered men. She also conveyed the news that her husband and Ranger Frank Kent wanted to make a dash for the barracks. Tays would need to cover them and keep the door open.[11]

Campbell and Kent then successfully ran the gauntlet to the barracks. Kent confirmed that the insurgents had smashed through the Atkinson-Ellis shared wall and were threatening to overrun Garcia's outpost. The captain had urged all his men to escape, but at least three—young Billy Marsh, veteran Ranger Miguel Garcia, and son Gregorio II—remained at their posts throughout the day, holding back insurgent efforts to come through the shared wall. After dark, the defenders ran out of ammunition and were forced to surrender. Captain Garcia had received head and leg wounds. Miguel was apparently so seriously wounded that virtually all accounts misidentified him as one of the dead. The victors fortified the store's windows with sacks of flour, quietly preparing for the next assault on the Ranger quarters.[12]

Sometime that Saturday, while Tays held out and Garcia's men expended their last cartridges, Charles Ellis was discovered. His hiding place in the wine cellar was at first known only to Cipriano Jesus Parra, whose wife Leonides was Teodora Ellis's niece. Knowing it was too dangerous for a well-known Ellis relative to be seen mysteriously carrying food around town, Parra enlisted the help of another man. The courier was crippled and so usually ignored as he hobbled around town. According to the Parra family, the man brought Ellis a meal and then revealed his hiding place to the insurgents.[13]

At this point, the story matches some of the hearsay evidence provided by Juan N. Garcia to the *Independent.* A band of insurgents pulled Ellis from the wine cellar, tied him to a horse's saddle and dragged him out of town past the Parra home on what is now de la Rosa street. Parra followed the party at a safe distance. From behind sand dunes and scrub, he watched while Ellis's throat was slit and his body left lying. Parra returned home. Retrieving a large burlap sack, he went back for the corpse, stuffed it into the sack, and carried it in his wheelbarrow to his home. That night, Parra loaded the body into the wheelbarrow and as unobtrusively as possible, though bravely enough, wheeled it to an orchard next to the Ellis home. He dug a grave and by morning Ellis was buried.[14]

In the meantime, Captain Blair finally made his way back to San Elizario, this time without his troops. After receiving what he called "numerous very exaggerated reports," he set out, with only Dr. E. Alexander as translator, to determine the facts. The pair were stopped outside of town by "a guard of 25 well-armed mounted men" who took them to the insurgents' "headquarters." Blair estimated that the Rangers faced "350 sober, well-organized, well-armed, determined men, with a definite purpose. Howard they wanted; nothing else, nothing less." The captain told them he "cared nothing" for Howard but "would be very sorry if anything should happen to Lieutenant Tays."[15]

Tays was a good man, the Paseño leaders acknowledged, and they had nothing against him, but why was he defending Howard? Blair explained that Tays was a state officer obeying orders. "This seemed to be a new idea to them; they had imagined he was a friend of or hired by Howard to defend him." Some of Barela's men wanted to hang the captain on the spot, but their leader refused them that pleasure. The captain turned the conversation to why the insurgents had fired on his men and stolen the horse and mule. "They shrugged their shoulders, and said their sentinels had orders to fire on any one who attempted to come in." Blair was promised the return of his horse and mule. Then he and Dr. Alexander were led back via the Mexican side of the river because, as they learned later, a party at Socorro was determined to shoot them on the return trip.[16]

Once he was safe in Franklin, Blair telegraphed Pope. Of the "Mexican" force, he declared there was "not one I believe from the other side. Mexicans say rangers have not had one hurt. Town completely in their hands & seem to be terribly in earnest & think they are right. They ought to get well thrashed & soon. Mexican authorities have forbidden interference by parties from Mexico & also forbidden their crossing the river singly or in parties with arms."[17]

The captain had again abandoned his duty to collect accurate military intelligence in favor of regurgitating Barela's misinformation: There were no Mexican citizens besieging the Rangers. No Rangers had been hurt. The first statement flatly contradicted Lieutenant Pague's earlier report of Mexican involvement. Blair was sticking to his story, despite Tays, despite Pague, despite any evidence to the contrary. He supported this by indicating the orders of Mexican authorities, although he failed to mention whether the *jefe politico* made any effort to back up his orders with force. The captain simply implied that Padilla's *pronunciamentos* were effective. As to whether or not any Rangers had been hurt, Blair had accurately characterized this information as coming from the Mexican side, but he offered no independent assessment of this intelligence. What was worse, he spoke of no effort to collect information independently. Having downplayed the crisis as much as the facts would permit, from the safety of the Custom House, Blair came out swinging. Like some toothless old hound tethered for his own safety behind a strong fence, he barked, "They ought to get well thrashed and soon." All in all, the telegram was a bravura Blair performance.

SUNDAY, DECEMBER 16

Charles Kerber conveyed no mixed messages. "Howard and State troops are doomed," he wired Hubbard. "Will be all murdered unless deliver Howard to the mob to be beheaded. No help in reach to save him." The governor's reply assured Kerber that the president had ordered troops from New Mexico to San Elizario. Hubbard instructed the sheriff to "Send messengers to them to make forced marches to our aid." His message to Congressman Mills conveyed Kerber's gloom: "[O]ur State troops are surrounded and will all be murdered unless relief can be had at once. Ask Secretary of War to please order his troops in New Mexico to make forced marches for their relief. They are American citizens."[18]

While Texas attempted to spur the federal government into action, the nation's newspapers were already arguing over what President Hayes should or should not do. At least one Texas newspaper was already pointing to previous *inaction* by the Hayes administration as an ultimate cause for the crisis in El Paso. The "wanton and unprovoked" attack on the Rangers was "invited by the recent action of the Washington government in countermanding the orders to cross the Rio Grande in pursuit of raiders just at the time the whole Mexican population was aflame with indignation and hatred against Los Gringos."[19]

B. C. Murray, editor of the *Denison News,* expressed the different opinion of some Texans living far from Mexico. Murray traced the larger border problem to the "stolen election" of 1876. Echoing Senator Blaine, he also blamed "ambitious men in our own State, dazzled with a view of the Senatorial toga, in their mind's eye, see in a war with Mexico their own selfish ends obtained. Territorial acquisition in the South can be seized and used with great effect in restoring the waning fortunes of their party by rousing sectionalism anew."[20]

Washington's inaction was just fine, said the *Brooklyn Daily Eagle,* which warned the *"de facto* President" against building the border troubles into a case for war with Mexico. "There will not, we venture to say, be any war with Mexico, and Mr. Hayes will not try to force one, while there is a Congress in Washington that would not be pained if a real good chance to impeach him were presented. The American people don't want any war and won't have any. . . . even to oblige the speculators and adventurers of the Rio Grande."[21]

John Pope, meanwhile, at last had acquired a sense of urgency. On Sunday he directed Hatch, "that if you have not already done so, you are to order all troops at Forts Stanton and Bayard, except one company at each post, to move with the utmost dispatch to El Paso. Order also such troops as may be necessary from other points. . . . you are to assemble at El Paso, at the earliest moment a force sufficient to insure the execution of all laws, *both State and National,* and to act in support of *either U.S. or State* officials as the emergency may demand." [Italics added.] Pope again warned his colonel: "You are under no circumstances to cross the river into Mexico, unless you receive orders to that effect from here." Hatch could take the gloves off but must not lead the nation into an unwanted war. By this time, most of the 9th Cavalry was en route from Forts Bayard (140 miles distant) and Stanton (200 miles away). Two companies of the 15th Infantry (Blair's regiment) were also reportedly on the way, from Stanton and Fort Craig (170 miles). Hatch expressed confidence that the troops would arrive "in time to prevent any serious trouble."[22]

It was only after the sun rose over San Elizario on Sunday morning that the Rangers realized how precarious their situation was. The usual exchange of gunfire, "a lively deal," Corporal Mathews called it, began about the time the U.S. mail coach reached the outskirts of town. The insurgents warned off the driver, ex-marshal William Treyer, who detoured his team around the excitement. Moments later, the insurgents opened fire from the roof and windows of the Ellis store, as well as the roof of Mauro Lujan's home. Only then did the Rangers discover the loss of Captain Garcia's bastion with its firepower and

Rooftop sniper. *Courtesy of the artist, Bob Boze Bell*

controlling fields of fire. The besiegers were now able to run right up to the outer wall of the quarters. Once there, they fired through the Rangers' own portholes, compelling the defenders to abandon the rooms on that side of the building. According to one account, the besiegers "sprung the mine" they had dug, "but it was not successful, only slightly injuring the quarters."[23]

Late in the morning, Pecos County Deputy Sheriff Andrew Loomis decided he was needed elsewhere. Tays agreed to let him go and in the hour before noon put out a white flag to negotiate his exit. The besiegers were happy to permit Loomis's departure. The probable reaction of the Rangers to this desertion is not hard to imagine. Barela then sent word that he wanted a parley with Tays. The lieutenant stepped outside and met with an emissary, who offered a ceasefire until morning, with negotiations to follow. Tays agreed.[24]

During the lull, two women came to the door of the ranger quarters. Ellen Marsh and Mrs. [Joseph?] Campbell were determined to see Barela about getting their family members out. It had been eight years since the Marsh family had come to El Paso, when Dr. Dwight C. Marsh took up the post of customs collector. Since then, Ellen Marsh had lost her husband to insanity, separation, and death. Now a widow at forty, she had the comfort that came with a bit of property, but the insanity of this war over salt was threatening to take the life of her eighteen-year-old son. Taken first to Barela's headquarters, Mrs. Marsh pleaded with the insurgents for the release of Billy Marsh. Barela agreed, since putting the boy in his mother's care kept him out of action just as effectively. The second supplicant, Mrs. Campbell, was the mother-in-law of the woman who, with her two children, was now in the Ranger quarters. The insurgent leader agreed to their release also.[25]

Inside the Ranger position, the women delivered some bad news. The besieged learned of what Corporal Mathews sarcastically called "the gallant conduct of Captain Blair." Until this point, they could imagine that Blair was attempting some stratagem on their behalf. Now they understood his absence. Howard commented, "There is very little hope for me, I should have known better than to depend on [Blair]." With their numbers reduced, with the loss of Garcia's supporting position and strength, and with the news that help was not on the way, the defenders realized the noose was tightening. By now, according to Wesley Owens, some of the Rangers "were growling and grumbling at being kept prisoner on [Howard's] account." Hearing the talk, the bold salt agent became resigned to his fate, telling Owens, "I'm going to give up and let them kill me. It won't do to rely on these fellows [i.e. the Rangers]. I'm not going to stay here and have them all killed for me."[26]

The Campbell and Marsh families proceeded out of San Elizario. On the road, they were stopped by a patrol of citizens who refused to honor Barela's agreement to the young Ranger's release. Mrs. Marsh was permitted to proceed, but Billy was again put under guard.[27]

The junta leaders put the ceasefire to good use. According to Blair, the besiegers sent men to Paso del Norte to obtain a small cannon located there. *Jefe politico* Padilla refused to let them have it. Men "belonging to the mob" also swarmed over Ysleta, taking "all the ammunition they could find." In San Elizario, the night was quiet, except for the sound of digging outside the ranger quarters.[28]

CHAPTER 17
"I heard my men had surrendered."

MONDAY, DECEMBER 17

Despite the wonders of the telegraph, no one outside San Elizario had an accurate picture of what was taking place on the firing lines. The county lawman and army officer whose job it was to inform their state and federal superiors transmitted daily reports at such variance with one another as to describe two different crises. Interested citizens who sent private messages added to the confusion. Newspapers reported and editorialized in ways that reflected both their politics and a need to sell product. As a result, the politicians in Washington and Austin who made decisions and the generals who translated orders into action lacked a shared understanding of just what was going on. As the siege dragged on, distance, delay, and wildly differing perspectives only deepened misunderstandings of the conflict newspapers began calling the "Salt War" or "El Paso Emeute."

The sun was still low on San Elizario's eastern horizon when Representative Mills visited the war department on the fourth morning of the siege. Secretary McCrary had good news the congressman could convey to Hubbard: "Troops are moving with all possible speed from New Mexico to El Paso. Four companies of cavalry left Fort Bayard Saturday morning; should be at El Paso to-night. Others on the march from Stanton, Davis, and Santa Fe." General Steele immediately advised Kerber that relief was imminent, instructing the lawman, "Send word to Tays if possible." To the press, Steele predicted that, should help not arrive, "the Rangers will not agree to surrender Howard, but will fight it out to the death of all."

The Davis contingent was the last to march. After some delay, Capt. Louis H. Carpenter's Company H, 10th Cavalry, received orders to "leave at once and proceed as rapidly as possible to San Elizario." Post commander Col. George Andrews directed the fort's quartermaster department to furnish the seventy men and seventy-six horses of Company H—a large formation compared to those of Hatch's Ninth—"with 31 days' rations and 200 rounds of ammunition per man and 8 day[s] of grain forage."[1]

Not everyone agreed with Hubbard that Texas was under invasion. In an editorial sympathetic to Paseños, the *New York Herald* warned its readers of "Strong Efforts to Breed Trouble with Mexico." The crisis was of "purely local origin" and not fomented by Mexico. "For more than two hundred years the people of that region have had free access to the salt pond or springs in their neighborhood. . . . it is no wonder that they resent being shut off from their usual source of supply. There is not a community in the world that would not act in the same manner under similar circumstances." The *Herald* blamed Hubbard for failing to send state troops when trouble first erupted in October. "Governor Hubbard is an imbecile who has not done his duty."[2]

Despite the army's commitment to action, its leaders were not about to get caught up in any war fever. Sherman let it be known, "The officers of the army thoroughly understand the predicament that would inevitably wait an indiscreet act, and as the regular army officers are more anxious for peace than for glory there need be no apprehension of trouble." Phil Sheridan's headquarters released "Official information" downplaying the potential for a second war with Mexico. "[N]o Mexicans have crossed the river to take part in the trouble, and . . . Mexican authorities have given positive orders that none of their citizens should cross. . . . All officers of the army in a position to know anything about existing troubles in Texas, or the views of either Government, are satisfied that there is no possible ground for warlike preparations." Sheridan also warned, "This withdrawal of troops leaves the Indian frontiers exposed."[3]

Sheriff Kerber remained the best-informed and least sanguine reporter on the situation. From Ellen Marsh and the two Mrs. Campbells, he had first hand intelligence regarding the Rangers' situation. On the plus side, they had plenty of food and cistern water on hand. The Texas troops began the fight with at least fifty rounds per man, though much of that had been spent. Kerber also knew that Tays had lost more than a third of his strength (Mortimer, Marsh, Loomis, and the seven men in Garcia's band). Of the eighteen men remaining in the ranger quarters, Eldridge was nursing a mangled hand, and Price Cooper had been wounded in the head. Worse, the fall of the Ellis store had compromised the barracks' defense. The defenders were tired, having been subjected to gunfire day and night until the Sunday lull. They had been gut punched with news of Blair's cowardice. How much of this the women were able to convey to Kerber we do not know, but enough to give him heightened concern. Also, from some source, possibly the women, the sheriff was aware of the insurgents' tunneling. Worse still, army reinforcements were still far away. With these factors in mind, Kerber restated the gloomy situation to the governor: "The first help expected in forty-eight hours. I fear it will be too late."[4]

Meanwhile, Captain Blair telegraphed his superiors, "Mrs. Marsh went down to San Elizario yesterday and got out her son who was with the Rangers. He reports one Ranger and Ellis killed. Nothing new has come to my notice." The captain's message suggests advancing torpidity, certainly a passive approach to intelligence gathering. The one scrap of news he offers, the release of Billy Marsh, suggests that he neither bothered to interrogate Mrs. Marsh nor to listen closely to even a secondhand report of her mission. As the officer on the scene, he should also have reported whether or not the companies of the 9th Cavalry were drawing near. He did not, again indicating passivity. Blair was safe in the Custom House; that much he knew.[5]

Mary Phelan was visited by a neighbor. From her, Phelan learned that "my husband had been shot on the way to San Elizario and I had best return east, but return with what or how, she did not say." Her next visitor was the baker, "drunk as a lord and brandishing a hideous looking knife, yelling in Spanish, 'Death to the Gringos.' I was almost insane with fear that he would kill me and do some awful thing to the baby." Fortunately, the baker passed out, and Mrs. Phelan made her way with Edith to the home of an elderly *Nuevo Mexicana* who had befriended the family. "Dear old soul, she took us in, fed me hot coffee and tortillas, warmed and fed my baby and put us to bed." Exhausted, Mary slept until close to sundown on the following day. All the while the old woman had cared for the infant. Declining the invitation to stay, Mary returned home and once again barricaded her home for the night. At 11 P.M. someone began "a most terrific pounding on the door. I was dumb with fear and knew that my time had come." Terror-stricken, Mary hid with Edith in the bakery's large flour bin, "nearly choking with its dust."

The following morning, the third since Phelan had watched her husband disappear, she peeked out of her window to see Las Cruces photographer W. L. Jerrell rushing up to her door. "Thank God you're alive!" he exclaimed. "Where's the baby?" Through her tears, she told him of their ordeal, Jerrell remonstrated, "You poor little idiot, that was me" pounding on her door. She cried even more. Jerrell bundled up a few of her things, "and our odd little procession straggled up the street to safety."[6]

Two hundred miles to the west and north, ex-Sergeant John Kinney received a telegram from another old sergeant. Kinney was in Silver City at this time because on November 2, 1877, in Mesilla, he had shot Ysabel Barela in the jaw, mortally wounding him. The *Independent* said Barela was unarmed. In Kinney's version, Barela precipitated his own death with a surreal conversation: "He said he was a man and liked bad men. I said, 'you can't like me, I am a good man.' He said that he was a man, and I said that I was not a woman. He said, 'I am a man, I will show you.' I said, 'how?' He said 'look! look! look!' and put his hand on his Winchester rifle as if to pull from the scabbard, when I pulled my pistol and shot."[7]

Kinney fled west to Silver City. On Monday morning, December 17, he "received a dispatch from Sheriff Kerber . . . conferring authority to raise a force for the relief of the rangers besieged at San Elizario. Kinney went to work at once and by evening had twenty-five names on his muster roll. The party started for El Paso the same evening." By the time they left the lower Mimbres river (the neighborhood of Fort Bayard), "Kinney . . . had recruited his command up to about thirty men." Kerber had chosen well. Kinney was not a man likely to deviate from the sheriff's "gloves-off" approach to the deteriorating situation in El Paso.[8]

When Mrs. Marsh and the two Mrs. Campbells walked out of the Ranger quarters on Sunday afternoon, they again closed the entrapped men's door to the outside world. Tays and Howard probably now knew that Kerber had telegraphed to New Mexico for help, but they knew nothing of New Mexico's response. They knew the distances to Forts Bayard, Stanton, and Davis, but not how fast the army was marching nor whether Colonel Hatch saw his duty in the same light as Captain Blair. Beyond their portholes, they could spot no slackening of determination among their foes. Instead, when they awoke on Monday morning, the Rangers discovered that the insurgents had "erected fortifications and dug rifle pits."[9]

The junta leaders had their own set of problems. Though direct evidence is lacking, the insurgent force of citizen-soldiers and (probably) *soldaderas*, farmers and festival goers, community leaders and cattle rustlers, may have shown early signs of fracture. With the capture of the Anglo merchants' stores, some looting of San Elizario was taking place. Homes and shops in Ysleta and undoubtedly Socorro also had been ransacked in the insurgents' search for ammunition. Whatever the passion for their cause, ordinary Paseños cannot all have viewed the pillaging in their towns as a good thing. They understood that victory would be followed by more looting. Their leaders could promise victory but could not guarantee order.

Also hanging over the junta army was the knowledge that the U.S. Cavalry was approaching. We do not know what the insurgent spy network looked like, how far it operated, how fast its word traveled. We can be certain Barela valued military intelligence and would have used it. As a Texas Ranger, he had campaigned against Apaches. Knowledge of that enemy's whereabouts and intentions was all the more precious for its rarity. In this struggle, on more than one occasion he used misinformation to mislead, divide, and weaken his opponents. The successful effort to divert Blair's column from its path is evidence that Barela knew the importance of keeping the U.S. Army out of the fight. This imperative had not changed. Indirect evidence suggests Barela would have kept himself advised of the army's movements as it approached El Paso from the north and Quadrilla from the east.

Perhaps sensing the balance was shifting, Barela dispatched an envoy to treat with the Rangers. The insurgents volunteered Vidal Garcia for the job. He and his wife Refugia were safely tucked inside their nearby home until a delegation appeared to inform the reluctant emissary of his "election." Persuaded, Garcia took Barela's message to the Ranger quarters, where he was met by Lieutenant Tays.[10]

Despite misgivings, the beleaguered officer and salt agent thought it best to listen to Barela's offer. Stepping into open air, Tays, with Atkinson and McDaniels, accompanied Garcia to the Gandara home. The men would have presented a bold face to the armed insurgents, townspeople, and others who watched, masking a mixture of anger, anxiety, and curiosity. Atkinson came along to translate. McDaniels was there to watch their backs while the two men haggled with the insurgent leaders. Tays may have had several volunteers from a crew of Rangers tired of being cooped up for four days, but McDaniels had several qualities to recommend him. From working with vaqueros on the trail, he probably understood some Spanish, enough to listen for trouble. The instigator and victor of several gun or knife fights, he had something of a reputation. If the negotiations suddenly turned heated, an insurgent might think twice before making a wrong move with Jim in the room. At the same time, he was generally affable, and so unlikely to instigate trouble at the wrong moment.[11]

Half way to the Gandara home, Barela met the delegation and Vidal Garcia was allowed to return home. Inside the house that now served as headquarters, Barela spelled out the situation. If Tays "did not give Howard up they would blow the quarters, as they had gunpowder under it." But it need not come to that, for "if Howard would go down to them of his own will and relinquish all claim to the salt lakes, they would not hurt him." All Howard had to do was sign the necessary prepared documents, and all would go free.[12]

Upon his return, Tays explained the conditions to Howard, who disbelieved Barela's guarantee. "I will go, as it is the only chance to save your lives, but they will kill me." Tays remonstrated. The Rangers would defend him and themselves to the last. Howard demurred. "It is useless for us to attempt to stand them off any longer; this is our only chance to escape." He bade goodbye and, says Mathews, "walked out apparently unconcerned."[13]

Gossip branded John Tays a coward, but at this moment he did not remain behind the thick walls of the Ranger quarters. Duty and honor impelled the officer to accompany Howard back to Barela's headquarters, hardly a cravenly act. Unfortunately, Tays's personal courage would be the Rangers' undoing, for his departure left the men without either clear direction or steadfast leadership.[14]

The meeting had difficulty getting started. Barela did little talking, while neither Howard nor Tays knew enough Spanish to negotiate terms. The lieutenant sent for Atkinson to translate. There are conflicting stories of what happened next, but it seems that Tays and Howard were held in one room while their interpreter, John Atkinson, was immediately directed to another. There, according to Juan N. Garcia, the JP held by the insurgents against his will, Atkinson was persuaded to hand over eleven thousand dollars in specie, currency and drafts, "on the condition that they would permit him, Howard, McBride and all the rangers to depart without molestation. Chico Barela swore by the Holy Cross that he would faithfully keep his part of the agreement." Barela apparently set one more condition on the deal: Atkinson had to write out an agreement for the Rangers' surrender. Whether acting in trust, or grasping at the only remaining chance for life, Atkinson agreed, telling Barela, "Well you have received a better price for us than we would bring if sold at public auction."[15]

"Atkinson came back in a short time," according to Mathews, "and said that Tays had ordered us to go down with our arms as we were wanted as witnesses, and that all had been peaceably settled." At that moment, according to Dr. Ball, the Rangers were well stocked with water and provisions. "There was no disposition on the part of the lieutenant or the men to surrender, and no disaffection among the men. We would not have surrendered, but Atkinson . . . told us that is was the lieutenant's order to surrender. We hesitated even then, but he assured us that everything was arranged; we would not lose our horses and guns, and the men who accompanied him said that it was all settled, and a guard would be placed over our things." Perhaps to prevent embarrassment, Ball fails to mention any dissension within the Ranger ranks. "They induced us to surrender," he avers. "Up to that time there had been no talk of surrender."[16]

Did Tays authorize the surrender? The lieutenant said no, "The [insurgent] leaders then came to me and told me I could go with my men. I objected, and said that I had brought Howard at their request and would not leave him till all was arranged. Then twelve men filed in and showed me to another room, where the sheriff of Pecos County, whom they still held, was imprisoned. Then, for the first time, I heard my men had surrendered, and that John Atkinson, to save his own life, had betrayed them."[17]

Atkinson did not live to contradict the lieutenant's version. Still, from the start Barela's strategy had been to *divide and conquer* through deception. Virtually every Anglo who listened to a Barela promise was taken in. Captain Blair had been the first but not the last. The continuing presence in San Elizario of the Pecos County deputy sheriff also testified to the success of such tactics. Andrew Loomis was last seen by any of the Rangers on Sunday, as he was given safe escort out of the quarters. "[U]pon arriving at the first rifle pits," he later testified, "I was immediately placed on a strong guard, searched and robbed of all I had . . . , and then marched to the general headquarters." There he was kept in the small room where Tays found him. It is not hard to credit that Atkinson honestly believed he had bought his comrades' freedom and lives. The false order tactic would have been just one more sham in a long line. The evidence points to Atkinson and away from Tays.[18]

Ball's account hints that the Rangers discussed whether to trust Atkinson's message. They should have waited for word from Tays, but by this time they were tired. They had probably gotten some sleep the night before, but it had been their first real rest—probably fitful at that—since the siege began. They assuredly began the siege with faith in their arms, their ammunition, and their manhood. Since Wednesday evening they had been in constant danger, their defensive positions had steadily eroded, no help had arrived, there was talk of gunpowder waiting to be lit just a few feet beneath where they stood, and their commander had apparently left no instructions covering orders delivered by third parties. Finally, by this time not every Ranger was of a mind to die for Charles Howard's salt. After four days of battle, Atkinson's message sounded good enough to overcome any misgivings.[19]

The men marched out, two of them carrying Atkinson's trunk. They were directed into a corral, where they were surrounded and disarmed, and then led into the home of Ramon Sambrano. Mathews reported, "We were then marched in a dirty room and the door closed upon us. We were so exhausted with our three days' siege that many of us fell asleep."[20]

Meanwhile, Barela and the other junta leaders held a meeting to determine their captives' fates. The deaths of Howard, bondsman Atkinson, and McBride were quickly decided upon. Howard was taken out first, his hands bound behind him. Surrounded by hundreds of insurgents and onlookers, he was taken to a wall nearly a hundred yards from the Gandaro house and stood against it. Desidario Apodaca, Barela's "second-in-command," gave the order to a firing squad. According to the most lurid "eyewitness" account, Howard

> walked erectly with his hands bound behind him; when he reached the place of execution, the command was given, halt! Howard instantly stopped, and turning, faced the mob; they were drawn out in a line. All was silent as death. Desidario Apodaca, with a firing party of eight men came up and took positions about ten feet from the doomed man, who stood quietly watching the proceedings. When all was ready Howard spoke. He could not speak Spanish very fluently, but enough to make himself understood; he said, "You are now about to execute 300 men," then, baring his breast, he gave the word, "Fire!" then the firing party fired, and Howard fell and kicked and squirmed on the ground; then Jesus Telles ran up to the body, and raising a large *machete* (large knife) in both hands, struck at the body; the blow fell, but Howard turned and the *machete* fell on Telles' feet, cutting off two of his toes. The body was then hacked and mutilated, after which it was dragged to an old well and thrown in.[21]

The "eyewitness," Juan N. Garcia, later admitted under oath that the details were hearsay. Maria Cooper, who was there, testified only that she saw [a different] "Juan Garcia, the Sonoranian, shoot Howard in the face after he was down," thus confirming the need to administer a coup-de-gras to the wounded man. She also witnessed "Jesus Telles cut Howard across the breast with a machete," though she makes no mention of Telles's self-inflicted wound. Did Howard actually tear open his shirt and give the command to "Fire!"? We do not know. But the hearsay account's depiction of Howard is consistent with the sang-froid others said he displayed during his last moments before exiting the Ranger quarters. If honor still mattered to him, and surely it did, it is easy to believe he walked erectly to his place of death, tore open his shirt, bared his breast, and gave the command that ended the shame he carried within. It was a fitting end for a classic tragic protagonist, one who had no choice but to propel himself and his world toward catastrophe. Arrogance and bullying marked his first steps toward chaos. But it was duty, loyalty, and most of all honor—otherwise noble qualities—that ultimately prevented Charles Howard from seeing any other course of action, from backing away from the cliff over which he consciously drove himself and the people of El Paso County.[22]

Atkinson and McBride were brought to the place of execution and shot in turn. McBride, who had been sleeping until awakened to meet his doom, apparently faced death with a look of sadness, perhaps for the orphan daughter he left behind. According to the same hearsay account, McBride was killed quickly.[23]

Not so his companion. Speaking in "excellent Spanish," Atkinson asked the insurgents whether they "intended to violate their solemn pledges; the crowd shouted, 'Acabenlos!' (finish them). 'Then,' said Atkinson, 'there is no remedy!' 'No! No!' shouted the crowd. 'Then let me die with honor,' he said, 'I will give the word.'" The melodrama continued as the condemned man "took off his coat and vest, opened his shirt as to uncover his breast, looked at the party of eight men who stood with their guns ready to fire, and said in a cool manner, 'when I give the word, fire at my heart—Fire!' As he gave the words, five bullets struck him in the belly; he staggered, but recovered himself and shouted, 'Mas arriba, cabrones! (Higher up, you [bastards]' Two shots were then fired, and he fell, but was still not dead; he motioned towards his head, and Desidario Apodaca, the commander of the firing party, put a pistol to his head and finished him."[24]

All three corpses were mutilated and dumped into a well. The mob—many in the crowd definitely played the part now—was not satiated. What happened next was etched into the mind of Maria Cooper. The wife of one captive Texas Ranger and the mother of a second had come down from Ysleta to be close her unfortunate men. She heard one bloody-minded rioter ask, "Who shall we kill next?" With horror, she listened as others replied, "Let them all be killed, both Americans and Mexicans." Fortunately, "Don Chico Barela then stepped forward, opened his shirt and said, 'No, no other man shall be killed, and if you propose to kill anybody else, I will take out my people and fight you all.'" Tays also testified to Barela's live-saving display of courage and humanity.[25]

While the throng of people focused their attention on the executions or on looting anything not tied down, Captain Garcia and the other captives from *El Molino,* held separately from the other Rangers, saw their chance. Slipping through the guard, they made their way to a nearby farmhouse where they found horses. They rode across the Rio Grande to the town of Guadalupe, where they asked for the protection of the municipal president. "Cura Borrajo advised the authorities to drive him away and turn him over to the mob to be murdered; but Don Francisco Escajeda stood up like a gallant gentleman he is, and said he would protect Don Gregorio with his life; and so he was saved, although most of the *Guadaloupenos* were of the mob."[26]

Tays was still furious at the treachery and turn of events. Satisfied with his victory, Barela gave him permission to send two men back to the quarters, one to prepare meals for the imprisoned rangers, the other to look after the horses. Tays himself was given permission to return to the quarters under a strong guard. (He did not report whether he accepted the offer.) The day ended with the Rangers waiting anxiously to learn their ultimate fate.[27]

So ended the siege of San Elizario. The Texas Rangers and their companions reportedly killed between five and fifteen Paseños and wounded as many as forty to fifty more. They themselves had lost two dead (Sergeants Mortimer and McBride) and at least five wounded (Price Cooper, John Eldridge, Miguel Garcia, the last probably seriously, Corporal Mathews, slightly, and Captain Garcia). The rangers had also lost their arms, the man they were protecting, another who had sought their protection. Ellis, who died alone, can be added to the toll. Beyond that, they had achieved a lasting reputation as the only Ranger unit ever to surrender to a "mob." Some would label their commander a coward. Charles Howard, like his nemesis Louis Cardis, was dead. Their personal quarrel was over, but the Furies were not finished with El Paso County.[28]

TUESDAY, DECEMBER 18

Tays reported that on Tuesday morning, "There was a review of the mob, and after maneuvering the troops the leaders made speeches." Two Mexican nationals urged the slaughter of "all Americans in the county." But for Barela, said Tays, "The mob . . . would have rushed in and killed us." Instead, the Rangers were given their horses and saddles and set free.[29]

Price Cooper may have been instrumental in obtaining the release of the Rangers' horses. At the urging of Tom Zickefoose, the oldest Ranger approached Barela, an old friend and Ysleta neighbor, with the idea. Barela agreed and convinced other junta leaders to so speed the Rangers on their way. Tays demanded the return of their firearms "in the name of the State of Texas" but was refused. The insurgents also seized eighteen hundred rounds of Ranger ammunition. The unarmed Rangers mounted and rode out, reaching Franklin at 6 P.M.[30]

Sheriff Kerber took stock of the situation. Nothing now stood between the "mob" and the lives of the remaining "Americans" in the county, except the few men he could enlist, the waffling Captain Blair's troop, and Texas Rangers armed only with their grim story. Of special interest were the tales of heroic "American" deaths, contrasting sharply with actions of perfidious "Mexicans."

Kerber forced Solomon Schutz to equip twenty-five men with pistols and carbines, though the latter were "nearly valueless."

News of the surrender reached General Pope that afternoon. Though Blair lacked confirmation of a rumor that the Rangers were en route to Franklin, he was "preparing to give them a *cordial* reception." He also telegraphed [Fort] Selden and Mesilla to hurry the troops en route "forward with all possible haste." (Nothing in the surviving record indicates that Blair dispatched any "hurry up" messages prior to the Rangers' surrender.) Pope instructed Blair, "Notify the Leaders of mob that sufficient force of U.S. Troops are on march to El Paso, and will be there almost immediately. The leaders of the mob will be severely dealt with, unless they cease their outrages at once. Use all your official and personal influence to stop the troubles, as well as what force you have." The cavalry was indeed closing in. The 9th Cavalry's Company L passed Mesilla that afternoon and was expected at El Paso shortly after daylight. Another company was close behind.[31]

The captain wired Pope at 6 P.M. that the Rangers had finally reached Franklin and that "The mob has dispersed." Shortly before 11 P.M., Pope dispatched another telegraph to Sheridan in Chicago, "*Blair is instructed as soon as Company L reaches him to move down to San Elizario*. There seems little doubt that it will require all the force that has been ordered there to arrest the ringleaders & participators in this outrage under the orders rec'd." Pope repeated Blair's claim that "the persons composing the mob were *entirely Mexican born* citizens of Texas. There does not appear to have been any help given them from the Mexican side of the river."[32] [Italics added.]

Pope's assertion that Barela's force was composed of "entirely Mexican born citizens of Texas" must be a reference to their ethnicity. The message Pope probably intended to convey, the only one that makes sense, was that the rioters were Texas citizens of Mexican descent. This meaning matches Blair's early reports on the matter. The source of Pope's phrase is uncertain. Did Blair use this language in a now-lost telegram, or did Pope badly paraphrase Blair's reports? It first appears in the transcription of Pope's telegram made by the receiving Western Union office in Chicago. It is repeated thereafter in messages to Adjutant General Townsend, the secretary of war, and ultimately the secretary of state. Earlier telegrams using the phrase, whether sent from El Paso or transcribed at Fort Leavenworth, Kansas (Pope's headquarters), are absent from the usual source collections. The phrase, obviously misleading, was altered when released to the press in Chicago. The altered language then became a focal point in the argument between those who wanted to cross the

Rio Grande to chastise Mexico and those who sought to avoid war with that neighbor.

News of the surrender also reached Washington via Austin. The governor exhorted the Texas delegation to "urge immediate investigation by United States authorities as to the part taken by citizens of Mexico in the late El Paso County troubles. Our information from State civil officers shows that Mexican citizens did largely engage in these outrages."[33]

Meanwhile, chaos reigned in San Elizario. Some insurgents or hangers-on degenerated into a looting mob. Even Teodora Ellis was not immune. The new widow's home was robbed of all valuable belongings, and she had to flee for her life. Others in San Elizario, whether insurgents or bystanders, were already beginning to scatter, "picking up their effects and moving them away in carts and wagons, some over the river and others in other directions."[34]

Could the Texas Rangers have avoided their fate? The answer is *certainly,* if on December 12, Tays had turned Howard around and escorted him back to the New Mexico line; but Tays, however unwisely, was fulfilling his initial, overarching instructions from Major Jones when he attempted to protect Howard while that citizen sought to uphold his legal rights. When he gave those orders, Jones had counted on Howard staying away until court, despite Howard's history of not staying away. In fact, Tays had advised Jones of Howard's visit to Franklin in late November. The record does not reveal that Tays ever asked for instructions should Howard want to go to San Elizario nor that Jones ever offered any. Could Tays have kept Howard away from trouble? It is hard to imagine that Jones would have failed to support his officer, even if that officer had exceeded the bounds of his authority in stopping Howard's progress. But could Tays have done it? Not without arresting the man, for Howard was bound by honor and shame—and fatalistic enough—to force the issue. Howard also had Sheriff Kerber on his side, and Tays was duty bound to support Kerber.

The answer is also *possibly,* if Blair had not turned around on the evening of December 12 or not hidden in Franklin on December 13. Whether Blair could have gotten through Barela's men depends on who one believes. The captain reported that he and his men were surrounded by no less than from 120 to 150 men, all well armed, with ropes stretched across the road to block their progress. Were the 150 Paseños united in a determination to resist the U.S. troops by force, come what may? *Possibly,* but Blair had no intention of finding out. Rocky Mathews, expressing the Ranger viewpoint, later testified that "Any person could have come in town that night, prior to half past eleven or a quarter to twelve. From my knowledge of the situation at the time, and from what I have heard since, I believe that a body of United States troops could have entered the town without meeting strong opposition. . . . I heard Chico Barela say that he wouldn't fire on the troops; they might have done it, but they said they wouldn't."[35]

Civil War veteran John Ball was also "perfectly satisfied that [Blair] could have come in, and perfectly satisfied that the State troops could have gone to him, because the mob was not then thoroughly organized. I was out through the town about midnight. Had Captain Blair then come in, I am perfectly satisfied that not a shot would have been fired." He based this opinion on Rucker's previous assurances made to the insurgents that the army would not fight. "The mob concluded from this that the government troops were their friends and sanctioned all they had done, and so expressed themselves."[36]

Neither Blair's view nor the Rangers' can ever be established as correct. It is possible that both they and Blair were right. Jesus Cobos had witnessed many men running up the Socorro Road late that evening. It is possible that Barela took all or the bulk of his still-organizing force up the highway in order to dispose of Blair first. The captain could have been faced by over a hundred armed and determined men just outside of town even as the Rangers noticed little insurgent activity.

But even if Blair had not been able to force his way through, he might have, and should have, taken aggressive military action after pulling back. Skirmishing action by Blair's detachment of highly mobile veterans could not have defeated the junt*a* army, but it would have given its farmers something to think about. Paseños were already split on the idea of revolt. The effect of intervention by Blair cannot be determined. Knowledge that the army, once reinforced, *could intervene powerfully* would work on the minds of men, possibly weakening the resolve of those *salineros* who did not want to be on the wrong side when that happened. (It might also have galvanized fence-sitters to support the insurgents.)

Could aggressive action by twenty to thirty Buffalo Soldiers convince a critical mass of Barela's men to give up the fight before December 18? Possibly not: They had displayed great tenacity and the ability to absorb substantial losses in fighting the Confederates in 1862. Yet, Sibley's army was just passing through, an already defeated force. The Paseños of 1862 were merely hastening their exit. The U.S. Army, should it ride to the rescue, would come to stay, and the potential for adverse consequences was far greater.

Might the Rangers have been emboldened to hold out longer if they detected a shift in insurgent morale or determination? Again, the answer is speculative. To triumph, the cooped up Rangers needed leadership, discipline, and fighting élan. Though Tays was no military genius, his sense of duty gave him the steel needed to fight to the bitter end. His mixed bag unit was another story. They were of uneven quality and not all were all imbued with Ranger traditions. Individuals within the unit, men such as Price Cooper and Miguel Garcia, not to mention Captain Garcia, were experienced and tough as nails. But as a unit, the Detachment of Company C lacked the shared experience and prior successes that might have built cohesiveness, élan, and trust in their leader. Their morale may have been flagging by Monday morning, but events would soon prove that the insurgent army's morale was even more brittle. When it broke, it broke quickly. Blair might have given it a push. He never tried, and so Howard, Atkinson, McBride, and the reputations of John Tays and his Texas Rangers were lost.

CHAPTER 18

"Half crazy from our troubles."

News of mob rioting and murder in Texas swept across the nation. Reporting largely reflected a newspaper's politics. In describing the "fusillade" of Howard, Atkinson, and McBride, *Galveston News* editor Alfred H. Belo published the "Texas view" of affairs: "No braver Texans than they have been sacrificed through the cruelty and treachery of the Mexican race in the forty years they have waged open or disguised war upon Texas." The paper predicted that Major Jones's "whole force will be sent to the Mexican border."[1]

The *New York Herald* felt no sympathy for Texas. "It is to be regretted that the riot has had this bloody end; but it was incited by the Texas authorities—if not willfully, yet stupidly—for the Governor was not ignorant of the nature of the dispute and of the force which with it could be supported. . . . But a riot even in Texas is only a riot. . . . That is the whole significance of the event, and all the howling that will ever be done over it will not give it any other character."[2]

The *Atlanta Constitution* was unimpressed by the Rangers' defense. "The Texans in the earthwork [*sic*] were not as gamey as their ancestors at the Alamo. . . . What if their ammunition did give out, could not they have given the enemy cold steel, or if their bayonets were not fixed, the butt end of their muskets, rifles and shot guns together with anything else available? The world never makes its heroes out of beleaguered men who surrender quickly, and San Elizario lost its first and probably best chance to become famous."[3]

The press noted congressional hearings and grandstanding on border tensions and diplomatic relations. The *Constitution* reported that Senator Conkling and several "Conklingites" on one committee constituted a "packed jury and [Secretary of State] Evarts and the administration in general are its victims." The House of Representatives adopted a resolution calling for the president to communicate information "referring to . . . any recent violation of the territory of the United States, by incursions from Mexico." Testifying before Congressman Gustave Schleicher's House Committee on Foreign Affairs, Brigadier General Ord called upon Congress to "agree by resolution that the order of the President [authorizing punitive pursuits into Mexico] is right and proper. . . . its moral effect in Mexico would be good."[4]

Texans were not the only Americans sharing a border with Mexico. In Tucson, Arizona Territory, talk of war met "with a great deal of popularity. Past atrocities committed at San Elizario seem to be considered by everyone a *casus belli,* and a multitude of impecunious immigrants who have come into the Territory would be glad of a chance to enlist for active warfare. Rumors are also current of orders having been received to keep the few troops in the Territory in readiness, and also to have supplies sufficient in case of any active operations." A dispatch to the San Francisco *Examiner* reported that *Lerdista* agents in Arizona were "jubilant at the prospect of war."[5]

Determined to keep the affair from escalating into war, the president and his generals were uniformly reluctant to apply a cross-border pursuit strategy against the insurgents. The *Chicago Journal* reported that the administration "entertain[s] no fear that the matter will become more complicated, but that when five hundred troops reach the scene of recent troubles everything will become quiet again." Indeed, in his first instructions on the subject of the uprising, Commanding General Sherman advised Sheridan, "it is deemed unwise to cross into Mexico and this must not be ordered as this case does not fall under existing orders in Texas for crossing into Mexico." Pope dropped Sherman's polite tone in ordering Hatch, "You are under no circumstances to cross the river into Mexico, unless you receive orders to that fact from here."[6]

In reporting the Texas Rangers' surrender, Sheridan's headquarters declared, "No help was given to the mob from the Mexican side of the river. The mob was composed entirely of native born citizens of Texas." The information reverses Pope's claim that only "Mexican born citizens of Texas" fought the Rangers. The change was substantial and as inaccurate as the original formulation. The prowar Galveston paper published the proclamation from Chicago, labeling it a "Highly Colored Dispatch from Phil Sheridan's Headquarters." The change begs the questions: who altered the words "Mexican born" to "native born," and what was their intention? The original version was obviously incorrect and did not correspond to Blair's previous reports. A correction was needed. Was Sheridan or his staff merely sloppy in what it released to the press, or was this new falsehood a purposeful attempt to eliminate the rioters' nationality as a causus belli? Did someone at Western Union or a Chicago newspaper make the change? Evidence to resolve such questions is lacking.[7]

One of those most startled by the news that the "mob" consisted entirely of "native born" Texans was Governor Hubbard. On December 20, he wired Kerber, "You officially advised me that citizens from both sides of the Rio Grande were in the mob. . . . Captain Blair, of Fort Bliss, now says all the 'mob were native-born citizens of Texas.' I shall rely on the truth of your dispatch." Within hours, the sheriff telegraphed back, "Captain Blair . . . has changed his opinion." No wonder, asserted the *Galveston News*. "Garcia and others with him also know the people, and it is beyond question this testimony is better and more credible than that of Lieut. [*sic*] Blair, who, twenty eight miles from the scene of the massacre, telegraphs that all the mob are native-born Texans. . . . It will be difficult for Lieut. [*sic*] Blair to explain how he could be in sufficient proximity to the Mexicans of the mob to ascertain that each one was a native of Texas, and having thirty American soldiers at his back, make no effort to prevent the butchery of Americans."[8]

The sheriff also advised Hubbard, "The mob is now at Ysleta plundering. Only one company arrived yet." This was Company L of the 9th Cavalry. Its seventeen troopers were led by Capt. Francis Moore, "one of the very best officers in the service." Moore shared a Scottish birth, blue uniform, and rank with Thomas Blair, but nothing more. He served throughout the Civil War, first as a noncommissioned officer with the 1st Colorado Cavalry, Kerber's regiment, and then as an officer with the 65th U.S. Colored Infantry, rising to lieutenant colonel. Moore's company did good service in Texas and New Mexico, but had been demoralized by the execution of one of its own in February 1877 for killing a white in a barroom scuffle. The unit's confidence was rebounding when it marched to El Paso.[9]

Outranking Blair by several years, Moore became the senior officer in El Paso, with over forty soldiers at his disposal. If he had specific orders to act, they are absent from the record. Hatch's own orders from Pope, dated December 16, instruct the colonel "to assemble at El Paso, at the earliest moment a force sufficient to insure the execution of all laws, both State and National, and to act in support of either U.S. or State officials as the emergency may demand." This command, issued while the Rangers were still holding out, carried less urgency than a second order, transmitted to Hatch (still somewhere on the road) no later than December 19. After receiving reports of the "mob" advance to the outskirts of Franklin, Pope instructed "the Commanding Officer in that section [i.e., Hatch] to occupy San Elizario and El Paso with his troops as fast as they arrive to keep the peace and to aid the sheriff in making the arrests of such persons as were directly connected with the mob." Pope advised Sheridan on December 19 that Hatch "will be at El Paso to-night or tomorrow by noon."[10]

Despite this second, urgent order to move quickly on San Elizario, the captain did not. There was no longer a military necessity nor a civilian authorization to do so. Armed combat between insurgents and the Rangers was over. The people who needed rescuing now were Paseños in the towns being looted. In the days after Howard's execution, "wagon-load after wagon-load of plunder was hauled away from town to the opposite side of the river. Dona Teodora, the widow of Ellis, was robbed of her jewelry, dresses, bed-clothing, furniture — everything; her house was stripped." She had to hire twelve men for four days to protect her home and Ellis's plundered store and mill against further destruction.[11]

Only the army could end this disorder, and only in support of the proper civilian authority, the sheriff. Nothing in the record suggests Kerber was overly concerned for his Paseño constituents. He had begged unrelentingly for federal and state support to rescue Howard and the Rangers. He had complained unceasingly about the delay in the arrival of help. Neither government records nor newspaper coverage contains one word from Kerber suggesting haste was needed to put down disorder in the lower towns after the surrender. Kerber was impatient to teach "Mexicans" a lesson, not rescue them. The army officers had no mandate to go down the valley to end civil disorder in advance of the sheriff's presence. General Ord had pointed out in congressional testimony on border troubles, coincidentally given on the first day of the San Elizario siege, that any army attempt to arrest depredators without the presence of a Texas Ranger or deputy sheriff would bring forth "pettifogging attorneys [to] cause the arrest of the officers and men." Should any soldier shoot a depredator, "their friends would have us arrested immediately and tried for murder." Although insurrection in the valley had been declared, and the army given authority to act, Captain Moore determined it was wiser to wait until Hatch or Kerber was ready to advance.[12]

By Thursday, December 20, the wait was approaching its end. Deputy Sheriff Harvey prematurely telegraphed Fountain, "The Rangers are rearming themselves and are going down this evening to get the bodies of the killed and attempt to recover their arms. . . . There will be hair flying towards night if the Rangers get started." On that day, twenty-two volunteers from Silver City passed through Mesilla. "John Kinney was said to be in command of the party." Colonel Hatch reached the town by coach that evening. Four companies of the 9th Cavalry, with two cannon and three howitzers, were a half day behind. A company of the 10th Cavalry marching from Fort Davis was expected to reach San Elizario by Friday night.[13]

Kerber, Blair, Pope, and Tays all used the lull to prepare reports justifying their actions. In a letter to George Zimpelman, Kerber advised the unfortunate father-in-law that he "ordered a nice coffin to be made, and will have friend Howard buried alongside his dear wife so soon as the coffin is done." Kerber recounted the recent events, from Blair's unkept promise to Tays and the false assurances of the "mob commanders" to the executions by "nine dirty greasers." Kerber related his efforts to rescue the trapped men. "All promised help, but the regulars travel slow. I could not get a posse of ten men here at El Paso. Mexicans I could not trust. Men one thousand miles from here can not imagine how hard it is to get help." In his mind, the situation was now more dangerous than ever. "At present the mob is at Ysleta, and I expect at every minute that my house will be sacked." The events had left Kerber "half crazy from our troubles." But noting the arrival of Moore's troop, and the approach of other soldiers and the Silver City men, the sheriff rendered a grim prediction: "I will drive the scoundrels ahead of me like sheep."[14]

"And whom can we thank for all this trouble?" he asked. "[E]ven on Sunday night certain white men from here were at San Elizario advising the mob, and these same men are officers of the State and United States." The allusion may have been to James P. Hague, laid up sick in San Elizario, but actually no longer an office holder, and to Justice Magoffin and revenue agent Slade, who had warned against pushing the Paseños too far. "My previous reports were not taken notice of at Austin, and considered exaggerated because I am only a poor man, and of course, reports of rich merchants [i.e., Solomon Schutz and others] are more reliable." Kerber charged that these unnamed Anglos "wanted . . . to have Howard killed or driven from the country." The sheriff complained, "I have no courts, no justices; nearly all were in the row, and I have to take it all in my hands. I know Texas will uphold me if no other State does. You will find I will not allow the greasers to pass unpunished."[15]

Indeed, two months earlier, Slade had described the arrest of Juarez and Gandara as a "high-handed and unjust proceeding" that exasperated the Paseño population. His report to Secretary of the Treasury John Sherman apportioned the greater share of blame to the Anglo side. Upon his return to San Elizario, Howard had "greatly exasperated and incensed the Mexicans by calling them 'Greasers' and other opprobrious epithets, and by inviting them to come and take him now."[16]

Blair wrote a self-serving and contradictory letter to Pope. He described in great detail how his relief column was prevented from reaching Tays. Blair did not mention Kerber's effort to get him to ride to San Elizario. In recounting his own meeting with the insurgent leaders on Saturday, he admitted that Barela lied in claiming the Rangers had suffered no losses but made no similar admission in any other respect. Beyond noting that "Last night their videttes and pickets were up as far as old Fort Bliss," Blair was unable to provide his superiors with any sort of useful current situation report. "The reports from [San Elizario] to-day are contradictory, some saying they have dispersed and gone to their work; others that they are fortifying the town." When all was said and done, Blair had great reason to pat himself on the back. "The rangers consider it very fortunate for them that troops did not appear while they were prisoners, as that would have been the signal for their destruction. . . . I believe that my talk with the leaders on Saturday was instrumental in saving the lives of the rangers."[17]

General Pope explained away the disaster. "[T]he wide dispersion of the small force in New Mexico, most of which was only returning from a campaign against the Apaches, made it impracticable that they should have been able to reach El Paso even an hour sooner." Hatch had been ordered "to occupy San Elizario and El Paso with his troops as fast as they arrive to keep the peace and to aid the sheriff in making the arrests of such persons as were directly connected with the mob." Pope saw the need for a continuing military presence in El Paso: "It does not appear to me that it will be practicable to enforce the laws of the State in El Paso County especially those laws which relate to the rioters, without the constant presence of a considerable military force. Whether such force is to be supplied by the State authorities or how far & in what manner the U.S. troops are to perform the service are questions concerning which I ask to be informed."[18]

Of particular interest to federal authorities was the role played by Mexican citizens and the Mexican government in the disturbance on American soil. "All the persons concerned in the outrage, according to the report of the officer at El Paso are Mexican citizens of Texas, and no Mexicans from the Mexican side of the river appear to have been present. The action of the Mexican authorities has been in the interest of peace, and orders were given by them forbidding any of the citizens of Mexico from crossing into Texas." Thus, no justification for war or even punitive military action existed.[19]

"There was against Howard peculiar animosity," Pope asserted, "first, because he seems to have been the principal person who set in motion and conducted the movement against the free use of the Salt Ponds which the entire population in that region had been accustomed from time immemorial to exercise, and the deprivation of which was considered by them a grievous wrong; and second, because Howard shot and killed a noted Mexican [a reference to Cardis], in this question, in the streets of El Paso only a few months ago. Toward the rangers there seems to have been no feeling of special hostility."[20]

John Tays pointed the finger at Atkinson, who, "to save his own life, had betrayed them." At several points, the lieutenant painted a picture of insurgent treachery and brutality, from the murder and mutilation of Ellis ("his scalp, eyebrows and beard taken, his throat cut from ear to ear") and the shots fired at Mrs. Campbell, to the promises made to let Howard and all the rest to go free. Twice, however, Tays made a point of stating, "The mob . . . but for Chico Barela, would have rushed in and killed us." Tays finished with a heaping dollop of sarcasm: "When I arrived here [in Franklin following the surrender] the greatest excitement prevailed and found Capt. Blair making preparations to start to my assistance with his command of 18 men some time next Spring."[21]

Tays wired a separate report of casualties to Jones: "Sergeants McBride and Mortimore and Miguel Garcia shot [and killed]; two missing. Got out by a scratch. They took our arms. One hundred and fifty from Mexico. Send arms and ammunition." The missing Rangers were those who escaped with Captain Garcia and were in hiding south of the border. Miguel Garcia, who had also been with this team, had not been killed, a fact not yet known to Tays.[22]

Interviewed during the siege by the *Houston Telegram*, Jones expressed confidence that the El Paso trouble would not erupt into "difficulties of a national character" because the federal government "does not desire and has no intention of entering such a conflict." He insisted that Mexicans had joined in the rioting and had been drilled by an ex-officer in the Mexican army. "The Mexicans on both sides of the river are virtually one and the same people; many of those on this side were born in Mexico. They are intermarried one among another, and I believe the Mexicans now surrounding my men at San Elizario are to a great extent Mexicans from Mexico." Jones described Tays as "cool and determined," and affirmed that he had gotten "twenty-two men to suit me," though he did have to accept "two [Paseño] men and a half-breed." Jones identified the leaders as Salcido, Granillo, Barela, and "another very violent Mexican by the name of [José María] Juárez, who has some influence with the mob."[23]

The insurgents' chief military leader, Francisco Barela, of course filed no "after-action" report. We do not know what he thought of the previous week, except that the uprising and the execution of Howard were morally and legally justified. As to how he conducted the war, we have only the reports of his opponents, but these tell us a great deal. The Paseños evidently knew that Howard could not stay away, and they planned to respond en masse and in an organized military fashion when he did return. When the moment came, the insurgents collected and organized quickly, creating concentric rings around the Ranger compound and the town of San Elizario, with streets blocked, rifle pits dug, and vedettes placed to watch the approaches. Barela could take pride in utilizing every tactic and stratagem at his disposal, in using his wiles to divide the enemy, weaken their resolve, and, most of all, keep federal troops out of the fray. He restrained violence as best he could, displaying coolness and control in his leadership of any angry people.

Barela was not completely able to restrain the insurrection's worst impulses; no one could. There is no evidence that Barela desired anyone's death other than Howard's. Tays testified to the insurgent leader's determined and successful effort to save the Rangers on both Monday afternoon and Tuesday morning, instances providing proof of his reluctance to kill unnecessarily. Blair also claimed that Barela saved his life. The executions of Atkinson and McBride seem to have been junta decisions. While outsiders such as Pope could not fathom why the "mob" exercised such a "particularly vindictive feeling" toward the two unfortunates, they were the obvious candidates to follow Howard. McBride, as Howard's salt agent, was virtually an extension of Howard, while Atkinson had made a career out of being loathed. It is unlikely that anyone could have saved these two. Still, their deaths and mutilation, inexplicable to outsiders, made the Paseños appear savage.

Neither the insurgent soldiers nor their leaders can be blamed for the looting of the Anglo stores. Even armies subject to the firmest discipline and highest *esprit de corps* have fallen to such temptation. The goods of Atkinson and Ellis could be considered contraband. An army several hundred strong needed food, drink, blankets, and ammunition, and the readiest providers were these rich merchants. But as the effective rulers of San Elizario during the week of siege, the junta leaders should have protected Dona Teadora's home, an obvious target for looters. To the extent that any homes were pillaged, the insurgent cause was embarrassed. The junta had enough citizens within their army to prevent this self-destructive activity, and it is not clear that they fully utilized their power to do so. Still, riding a whirlwind, Barela did an exceptional job of harnessing the unleashed storm and keeping its destructive power to a narrow path.

It was now Friday, December 21, one full week since Hatch was ordered to take every available man to El Paso. Most companies were still en route. Phil Sheridan complained to Pope, "I cannot understand the procrastination." Pope wired back, "I do not understand the delay myself and telegraphed this morning to Hatch to occupy San Elizario immediately."[24]

The first soldier to arrive behind Company L was the 9th Cavalry's colonel, Edward Hatch. He came by carriage, which enabled him to do his paperwork on the road. At age forty-five, Hatch was still a vigorous officer, disposed to act, but on the basis of sound intelligence and always with the wider military and political interests of the nation in mind. After years in Reconstruction Texas and New Mexico Territory, he understood the complexity of the issues and the competing motivations of interest groups. His regiment of "Buffalo Soldiers" was one of the finest in the army, capable of handling whatever opposition they were likely to face. Their discipline would serve them well in the task of restoring civil order. Hatch's Ninth committed no excesses during the Salt War.[25]

If Hatch had a weakness, it was reading the motivations of individuals. He had failed to see through Captain Blair and would continue to do so. Now it was the depth of Charles Kerber's racism Hatch failed to perceive. Although even readers of the *Galveston Daily News* were aware that "Lively times [were] expected when the sheriff acts," the colonel was not. Misapprehending Kerber's character, he failed to prevent the violence bound to occur once the aggressive sheriff had the unfortunate means to alleviate the bile eating away at him.[26]

Hatch arrived in El Paso at 7 P.M. "[T]he evening was exhausted" in his efforts to learn the state of affairs. The dispositions of the insurgent force were something of a mystery. Blair's reports included no significant intelligence following his trip to San Elizario on December 15. According to the colonel, it was a man named Martin, "an old scout for the Army in Texas," who brought news that the insurgents were "still in strength, and determined to resist." Hatch's strategic and tactical decisions required an understanding of the insurgents' strength, morale, and objectives. National policy concerns would have prompted him to ask about the presence of Mexican citizens within the insurgent force. The answers to these questions were inconclusive. Tays, an eyewitness, thought that perhaps 150 of 400 Paseño *soldados* were from the other side of the river. Others reported that the insurgents were "largely composed of Mexican citizens from the . . . west bank of the Rio Grande," who claimed "that their numbers can and will be increased to over one thousand men by sympathizers from the Mexican side of the river, if the necessity arises." From Rucker's reports and Blair's, Hatch already knew a great deal about the significance of the Guadalupe Salt Lakes and the impact of the Cardis killing. Hatch probably got an earful regarding the "mob." Some would have sought to convince him that it was the *mexicanos'* intention to restore El Paso County to Mexico. Others saw the salt question as "nothing but a pretext, the real cause [of the civil disturbance] is abhorrence of the American government and a desire for plunder." Kerber probably described insurgent intentions to kill every "American." Hatch would also have noted the report that the *jefe politico* of El Paso had called out the local National Guard, "purpose not known."[27]

Before retiring, Hatch telegraphed a brief situation report to Pope. He expected the troops from Forts Bayard and Stanton in the morning. "When here I shall have two hundred men, will move at once. The mob are well organized by officers who have served in the Mexican army. They number five hundred men, and claim three times as many. No matter what odds are, I shall attack at once. If fight (which they declare is their intention), I have no doubt of my ability to clean them out. Will occupy those towns at once as directed in your telegram."[28]

Tays used the time to reorganize his command. Dr. John Ball and Sam Frazer, "an abusive and insulting man," replaced McBride and Mortimer as First and Second Sergeants. The lieutenant enlisted three new Texas Rangers on the twenty-first of the month. Sergeant Frazer took an instant dislike to one, John C. Ford, a twenty-eight-year-old, black-haired, blue-eyed cowboy from New Mexico. Frazer may have had more in common with another new man, Jerry Barton, who gave his birthplace as Louisiana but was actually from New York. Barton, born sometime between 1837 and 1853, rivaled Price Cooper for his movable age. He was almost five feet, eight inches tall, according to Ranger records, with a fair complexion, light hair, and hazel eyes. More importantly, he was large and quick, capable of killing a man with his fists. What Tays's detachment regained in numbers, it lost in discipline and control. For some recruits, this was apparently an opportunity to go to war with Mexicans. It was reported "the Rangers are determined to pay the Greasers in their own coin, to which no one can offer reasonable objection." Whether their commanding officer shared this view, or simply failed to rein in the men's worst impulses, is unknown. To the *mexicanos* and pueblo Indians of El Paso County, it would make no difference.[29]

CHAPTER 19

"I will drive the scoundrels ahead of me like sheep."

Hatch had planned to wait until eight more companies arrived from Forts Bayard and Stanton, but the "many lawless characters" plundering San Elizario worried him. At daybreak on December 22, he marched for San Elizario with fifty-four mounted troopers and one howitzer. Blair himself remained behind to push the reinforcements forward "as fast as they arrive." News of the advance beat Hatch down the valley. His column found no rebellion to put down, for "all the Malcontents had rapidly crossed the river into Mexico." The colonel remained all day in San Elizario, questioning local citizens. Contrary to Blair's reports, Hatch learned that the local insurgents had been joined by four hundred Mexican citizens, "mostly bad characters who joined the mob merely for purpose of plunder." They had robbed all the American stores in San Elizario, and would have cleaned out those in Socorro, Ysleta "and even El Paso" had not the troops arrived. The governor's go-ahead for Kerber to enlist men in New Mexico had "alarmed and inflamed the Mexican people on both sides of the River, who firmly believed they were all doomed to destruction by the Texas State Troops, and led to an appeal to the Mexican population to arm and defend themselves." Hatch's conclusion that "it was not their intention to resist United States troops" begged one question. Was it ever?[1]

The Fort Bayard troops, sixty-three men in four companies, with one Gatling gun, two cannons, and three howitzers, all under the orders of Capt. Charles D. Beyer, arrived at Fort Bliss later that day. The Fort Stanton troops were still somewhere in New Mexico. From the east, Capt. Louis H. Carpenter and Company H of the 10th Cavalry approached the end of their 210-mile forced march from Fort Davis.[2]

The Texas Rangers also rode down the valley as far as Ysleta. They were under the orders of a sheriff who had declared his intention to "drive the scoundrels ahead of me like sheep." At Kerber's instruction, the Rangers searched homes in Ysleta for arms and "mob" leaders to arrest. Corporal Mathews recalled, "we surrounded the house of Crecencio Yrigoyen one of the

Sheriff Charles Kerber. Robert G. McCubbin Collection

leaders of the mob, and found him within, we also found in the house a large quantity of arms, some of our own among the number. We arrested him and . . . also arrested an Indian named Andres Colmeneros who had made himself very prominent in the murders, we all knew him well." Other witnesses identified Yrigoyen's fellow prisoner as Santiago Durand. The unfortunates were kept locked overnight at Price Cooper's home.[3]

Before the Rangers bedded down, they were joined by twenty-seven "Silver City volunteers." With John Kinney said to be in command, they reached El Paso that morning, took care of their horses, took in the local saloons, and that evening took the road to Ysleta. C. L. Sonnichsen called the New Mexicans "hard faced and battle scarred." Robert M. Utley wrote, "Most were freebooters, some even avowed outlaws, seeking whatever the state might pay or whatever they might gain in loot."[4]

At this writing, no roster has come to light, though evidence indicates one existed. Contemporary newspaper accounts identify two leaders: Grant County Deputy Sheriff Dan Tucker and on-the-lam killer John Kinney. While the notorious Kinney was on his way to becoming the territory's most successful rustler chieftain, "Dangerous Dan" Tucker was making his mark as the archetypal shoot-first, ask-questions-later peace officer, ultimately killing as many as eight men. Tucker, at five feet, seven inches, was of average height, while Kinney, standing almost 5 feet, 6 inches, was short, even for the time. The Canadian-born David "Dan" Tucker was slim, with fair hair and blue eyes; Kinney was barrel chested, ruddy faced, with brown hair and hazel eyes. At twenty-eight, Tucker was four years Kinney's senior.[5]

Just who led the Silver City "volunteers" to El Paso is open to question. Historians relying on testimony taken in 1878 by a military board of inquiry identify Tucker as the man in charge. The evidence from December 1877, however, points directly at John Kinney, with Tucker gaining leadership after the events that gave the volunteers their everlasting bad name.[6]

A remark by Adjutant General Steele provides indirect evidence that more than one man served as an "officer" in the Silver City contingent. In a letter written eight months after the fighting in San Elizario, he complained, "It is difficult to understand under what authority Mr. Kerber appointed officers and fixed the rate of *their* compensation [italics added]." This does not, of course, necessarily mean that Kinney preceded Tucker as one of Kerber's officers, but it supports the proposition.[7]

Whoever was the Silver City "officer" during the advance to Ysleta and Socorro, Kerber directed what happened in those places. He sanctioned riot and took unapologetic responsibility for it afterward. Kinney, a criminal chieftain with a veneer of respectability, was only the sheriff's instrument, exercising a felonious influence over other men fully in accord with Kerber's intention to teach the people of *la Isla* some hard lessons. Only a few months later, Kinney did pretty much the same thing in Lincoln County on behalf of Murphy, Dolan, and Sheriff Pippin.

Testimony taken in early 1878 identifies only four other "Silver City men" by name, all accused of crimes: Andy Boyle, George O'Brien, and two so-called "rangers" with the ubiquitous names of Johnson and Williams. Indirect evidence points to the presence of some of the most notorious "avowed outlaws" of New Mexico. The December 22 edition of the *Mesilla Valley Independent* reported that twenty-two men from Silver City had passed through town two days earlier. In that same issue, the *Independent* "reliably reported that Jessie Evans and his companions who recently escaped from the Lincoln County jail arrived in town to-day and that they intend to honor El Paso county with their presence." Outlaws drifted in and out of Evans's gang, and so it is impossible to precisely identify who might have ridden with him to El Paso, if indeed Evans did travel that way. Boyle may have arrived with either Evans or Kinney. Jessie's pal Jim McDaniels was of course already there. Other "avowed outlaws" known to pal with Kinney or Evans included Bob "Dutch" Martin and Charles Ray. The first was in El Paso in early May, and might have arrived before that. The second was a frequent associate of Kinney.[8]

Not all the Silver City volunteers deserved the appellation "Kinney men." Some were ordinary miners, freighters, laborers, even schoolteachers. They had never stolen a cow, robbed a stagecoach, or killed anyone. They were simply freebooting. Sheriff Kerber's invitation to adventure with a paycheck seemed a better deal than going back down into a mine or hauling freight across Apache lands. That they were inured to rough life and untimely death does not mean they had no qualms about robbing, raping, or killing "Mexicans." We lack the identities of most of these ordinary pioneers who signed up for a war not their own. William Rutherford is one who left behind a record. He was a twenty-seven-year-old Tennessean who had been mining in Arizona prior to "enlisting" at Silver City in the "Texas Volunteers." After many of his companions returned home, Rutherford stayed in El Paso, enlisted in Lieutenant Tays's detachment, and served nearly two years as a Texas Ranger. Another "volunteer" was a handsome six-footer, Asberry C. Ryall, age twenty-four. Ryall had been a schoolteacher in Georgia before moving west. By 1877, he was operating a stagecoach between Globe, Arizona, and Silver City, New Mexico, dodging bandits and Apaches. On December 12, 1877, Ryall drove into Silver City, bringing "the startling intelligence" of an Indian attack on a wagon train. That may have been just the incentive he needed to join the "volunteers" bound for El Paso. Ryall also soon joined Tays's Rangers, rising quickly to First Sergeant.[9]

The official status of Kerber's posse was a matter of some confusion then and since. The governor of Texas had authorized Kerber to raise a posse in New Mexico "to resist invasion. Put down insurrection and reinforce State troops." So were the New Mexicans a sheriff's posse or state troops? Kerber testified, "By virtue of my office as sheriff, and by authority of an order from the governor, I regarded Lieutenant Tays' command as subject to my orders, also some men from Silver City who came as volunteers." Kinney's "discharge certificate," signed by Kerber, identifies him as a "private" in the "Company of Volunteers of State forces." Lieutenant Tays, on the other hand, claimed "they had not been mustered into the service of the State." As late as March 1878, General Steele retroactively sought an opinion from Texas Attorney General Hannibal Honestus Boone as to whether "certain men called out by the Sheriff of El Paso County" should be paid as Texas Rangers "or as directed for troops called out by the Governor?" The best that can be said is that Governor Hubbard created an anomalous force authorized to act in service to the state of Texas. When the men were finally discharged by Kerber, the action was taken at the direction of Steele. After that, the state of Texas retreated from any responsibility for these men or their actions. This avoidance of accountability would ultimately leave several El Paso merchants holding several large, unpaid bills. Whether they were "Tucker's company" or "Kinney men," the boys from Silver City reached Ysleta that evening, spoiling for a fight against the "Mexican mob," eager to reestablish the superiority of "American" arms and manhood.[10]

For some citizens of Ysleta, retribution began even before the sheriff arrived. One afternoon "two rangers" came to the house of Salomé Telles, killed her dog and then shot at her chickens. Afraid the violence would escalate, she took herself and three small children to the nearby home of brother-in-law Juan de Dios Alderete. Her two older daughters, ages eighteen and twenty, she quietly sent on to a neighbor's home. That night, she and an unidentified man were at home with Alderete when the same two men knocked on the door. Alderete had fastened the door on the inside with an ax and held his head against it to hold it. But after the men promised "to do us no harm," he opened it. The men "entered, drew their pistols and pointed them at our breasts. They ordered me [Alderete] to open a trunk, which I refused to do, whereupon they threatened to break it open with an ax." The frightened homeowner opened the box and quickly grabbed two dollars. "One of the rangers then forced my hand open and took the money from me, while the other put the pistol to my side. They took two dollars more which were lying at the bottom of the trunk." After turning over the beds in a vain search for arms, the intruders ordered their two prisoners outside. Alderete made a break, but failed to find any help. Upon his return, "one of the rangers met me at the corner, putting a pistol to my breast, and asked me what I wanted. I told him I came back to my house. He ordered me to leave, and I left, and stayed over four hours. The rangers were at my house about two hours, when they left."[11]

The "rangers" were not content with stealing four dollars. Once inside, they asked for the whereabouts of Salomé's older daughters. Mrs. Telles told them that only small children were present in the Alderete home. It was at this point that the intruders ran Alderete and the other male Paseño out of the house. Then, Telles bravely admitted, "one of the rangers went out and stood guard on the outside, while the other, who remained inside, pointed the pistol at my breast and forced me to give up my person to him. They then left." Telles could only describe her attacker as "a large man with a black beard."[12]

According to Pedro Cavillario, justice of the peace at Ysleta, the two men who invaded the Alderete home were J. Williams and F. Johnson. The names may have been aliases. If so, they almost certainly did not belong to any of Tays's Rangers, who were then stationed in Ysleta. If Rangers, they could not have been local men or in the unit for long. The likeliest Ranger suspects, Sam Frazer, Jerry Barton, Jim McDaniels, and Frank Allen, were all light haired. Black-haired John C. Ford remained with the unit in Ysleta through the end of March, and by February would have been known to Mrs. Telles and Mr. Alderete. No other Texas Rangers on duty in December seem likely candidates, unless Mrs. Telles understandably misidentified her abusers. It seems likelier that "Johnson" and "Williams" reached Ysleta ahead of other "volunteers" to get a jump on the action. Unfortunately, neither Tays nor Kerber was asked about the crime or its perpetrators.[13]

On Sunday morning the sheriff of El Paso County was ready to enforce the law: not what was printed in the statute books, but what was written in his heart, law more in keeping with his sense of eye-for-an-eye justice. Kerber never admitted to a murderous intent; he was not stupid, but the lawman did express the view that full justice was unlikely to be obtained in the courts. No grand jury would ever be able to handle all the arrests he had in mind. He later testified, "After indictments were found the only way to try these cases would be by change of venue to another county." Furthermore, the El Paso J.P.s were, in Kerber's mind, "not very intelligent men; they understand very little about the laws of Texas, and they do not speak the English language.

In any other country than this they would be regarded as ignorant men. In this community they are a little above the average in intelligence." He could not trust "ignorant" judges or local juries to properly administer the law. Nor would he risk sending captured criminals to another county. There might be too many to escort, too many places where the escort might be ambushed. On this Sunday, therefore, the sheriff became judge, jury and executioner.[14]

Revenge required organization, and Kerber rose to the occasion. He placed Tays's Texas Rangers in the lead of the column bound for Socorro and San Elizario. Silver City men came next. In the rear were two wagons, each containing two empty coffins for the bodies of Howard, Atkinson, McBride, and Ellis. Ranger Price Cooper held the reins of the first wagon. The second was driven by brothers John and Isaac Campbell. Yrigoyen and Durand, the two men taken prisoner on Saturday, were trundled out of Cooper's home, their hands "pinioned with ropes" behind them, and assisted into the Campbell's wagon. Each sat on a coffin, facing the other. Sergeant Sam Frazer and Private Jerry Barton of the Rangers rode alongside the wagons or to the rear, assigned to guard the prisoners, who were to dig the graves of the executed men.[15]

Kerber led the way, with Tays at his side. Roughly thirty minutes out, the leaders were about one quarter mile ahead of the wagons when Tays heard shots to the rear. Captain Beyer in Ysleta "heard 12 to 15 shots fired in rapid succession in the direction of Socorro, a short distance from town." From the wagon, John Campbell "saw the guard dodging about some mesquite bushes about seventy-five yards distant. . . . I could not tell whether they were in or out of the road, as some bushes intervened where the road made a turn." John said to his brother, "They are shooting the prisoners." Riding back, Kerber and Tays met Sam Frazer and Jerry Barton, who reported that the prisoners "had jumped out of the wagon, attempting to escape and were shot" by them. Tays ordered his the dead men left by the side of the road, to be picked up on the column's return trip from San Elizario.[16]

Kerber and his company rode on to Socorro. If the sheriff ever had any intention of making arrests by the book, he gave it up by the time his men reached town. On the outskirts, the column allegedly took fire from a "Mexican" house. Frazer was the only man hit, stung by a spent ball that dropped inside his boot. The riders returned fire and Tays ordered them to push the door open and search for arms. Inside, the Rangers saw a man lying dead inside the door. By his side was "an old-fashioned revolver," which had been discharged. Inside also was a wounded woman, "screaming and crying." Dr. Ball, who arrived on the scene moments after, saw Frazer, who "wasn't hurt much," pulling off his boot to pick out the spent ball. Inside the house, Ball saw "a woman crying who was suffering from a flesh wound made by a gunshot." The physician picked her up and carried her into another room for treatment.[17]

The woman, Maria Núñez, told a different story. She and husband Pablo had been in her kitchen when about a dozen Rangers burst in. Mr. Núñez had been in Las Cruces, not San Elizario, during the siege of the Ranger quarters, returning home three days after Howard's execution.[18]

> As the rangers entered the front room she and her husband closed the door of the kitchen. The rangers then commenced firing through the door at her and her husband, who were holding the door from the inside. The first ball struck her husband in the forehead, the second in the pit of his stomach; then more balls struck him, all of which were fired through the door. One ball struck her in the right side. Nobody had any arms in the house before the rangers attacked it, and no one but herself and husband were in the house. After her husband fell, the rangers pushed open the door and shot her husband twice more, and then thrust his body to one side. They then searched the whole house, trunks, beds, etc., but [she did] not know what they were looking for. They took away nothing from the house.[19]

A third version was offered to the *El Paso Times* some sixty years after the event by Toribio G. Rubio, who either witnessed the event as a thirteen-year-old boy or recounted a time-altered translation of it. According to Rubio, he was at Gregorio Holguin's home when "[s]uddenly, we heard the sound of galloping horses. I ran to the door and saw many mounted men riding up to the house." No sooner did Rubio bring the door bar down, locking it, than men began "pounding on the door and commanding us to open it. Then they fired through the closed door and Holguin fell with a bullet wound in his head. I was scared out of my wits. The men kicked the door in and I ran behind Mrs. Holguin. Someone fired again, and she staggered. She was hit in the side, and the blood soaked through her clothes onto my hands. The men knocked her out of the way and then fired several times into the body of Holguin and asked Mrs. Holguin many questions. Finally, they left, and everybody in Socorro went to Mexico for safety."[20]

Tays testified that a shot was fired from inside, through the front door, hitting Frazer with a spent ball that dropped inside his boot. Sergeant Ball confirmed that Frazer complained about getting hit and was digging for something in his boot. Was it a bullet or perhaps a small stone? Only Hatch testified to having seen the bullet, and his letter on the subject supports the testimony of Mrs. Núñez. Shown the spent ball by a Ranger named "Jack" later that day, Hatch asserted "The bullet was a pistol-bullet, .45 caliber, and I don't think was fired through anything." So what happened at the Núñez home? We will never know, but perhaps the Ranger column heard the report of a gun coming from somewhere and misidentified it as coming from the Núñez home. It would not be beyond Frazer to invent such a story. The evidence strongly supports an unwarranted invasion of the Nunez home and the murder of Pablo Nunez and shooting of his wife.[21]

The mayhem at the Núñez home was only the prelude to Kerber's occupation of Socorro. The Rangers had identified some of the town's residents as insurgents, and now the sheriff was out to find them and any weapons. The chief target was Jesus Telles, the man who had hacked his own foot with a machete while slashing at Howard's freshly executed corpse. Kerber assigned that job to Deputy Sheriff (and Ranger) John C. Ford and a squad of eight or ten Rangers under First Sergeant Ball. Several Rangers entered the home of Jesus Maria Padilla and, at gunpoint, ordered him to direct them to Telles's house. Hoisted onto Ball's horse and seated behind that Ranger, Padilla pointed the way. Before the party reached their destination, Telles was spotted riding toward them. "There was no trouble recognizing him," said Rocky Mathews. "When we found him he was on his horse preparing to skedaddle for the other side, as we approached he fired at us and then ran; we shot his horse first, and then him." Deputy Sheriff Harvey reported, "Jesus Telles . . . fought desperately: he was riddled with bullets." The Rangers broke down the door of the dead man's house, then "broke the furniture and carried the saddle away with them."[22]

As the sheriff's posse approached the house of saddler Antonio Cadena, he retreated inside, shut the door, and put his family into a rear room. Going back to his front door, Cadena was assured that nothing would happen to him. Within moments, one man stepped inside the house and grabbed him. As many as ten others followed. A pistol to his chest, Cadena was forced outside. Meanwhile, his wife and children watched in fear as the intruders ransacked the home, turning beds upside down, opening trunks and throwing clothing and other contents around. From the wall by the front door, they took the Paseño's revolver, gun belt, and sixteen cartridges. Others confiscated his saddle, bridal, spurs, and his horse, as well as one belonging to a Denicio Guerra. Cadena identified Lieutenant Tays as one of the men who entered his home.[23]

Cruz Chavez was walking toward the home of Juan Armendáriz when a man identified as a "ranger" stepped away from a larger group of sheriff's men and walked across the street to confront him. The "ranger" demanded his pistol. Chavez refused the order twice, prompting the man to shoot him twice. The first ball hit him in the right breast, exiting under his right arm; the second grazed his face. Despite the wounds, Chavez ran from the scene. Cosme Pérez, on his way to church with his family, witnessed the event. He did not know if the assailant was a Texas Ranger or a New Mexico man. Among the sheriff's posse, Perez identified Price and Santiago Cooper, as well as one of the Campbell brothers.[24]

Kerber ordered one Ranger to arrest a Paseño who was running, about three hundred yards from Armendáriz's store. "Some of the men hallooed out, 'There goes one of the men who was in San Elizario, Bernabel Candelario.'" Ranger Cpl. Martin Oberst approached the man, who turned and exchanged shots with Oberst. When the Paseño's revolver misfired, he ran, making his escape as Oberst's horse "began to jump and plunge." Tays thought Oberst managed to wound the fleeing Candelario. At least one more incident took place when eighteen men came upon Noverto Pais. After pistol-whipping him, they forced entry to his house, where they confiscated a rifle belonging to his father.[25]

It was still morning. The people of Socorro had taken on the Confederate army in 1862; now they again prepared to defend their town against these invaders. While women and children headed for the safety of Mexico, the men began to arm themselves and rally for the fight. Someone dispatched a boy to the home of JP Demetrio Urtiago, alerting him to the mayhem. Urtiago saddled his horse, preparing to ride hard to San Elizario and alert the U.S. Cavalry there. Just then, Colonel Hatch drew up in his carriage. Urtiago explained the situation, and Hatch instructed him to climb in. The two men "went uptown and met the rangers in front of Simon Cortez' house."[26]

Having left his column at San Elizario under Captain Moore, Hatch, accompanied only by an orderly, was on his way back to Ysleta to retrieve his late-arriving troops. When he heard "sharp, desultory firing," he hurried to Socorro. Here he saw the citizens preparing for battle. To this Civil War veteran, "a serious collision was imminent." The colonel asked the JP to convince the people to disperse while he dealt with the sheriff himself. In his own reports, Hatch downplayed his confrontation with Kerber. The sheriff, he reported, "informed me that the men killed about the town were killed resisting arrest, and learned that two men were killed, two wounded, and one woman shot through the lungs." Hatch let the sheriff know the lawless shooting of citizens had to stop, that in future, "nothing of the kind will be tolerated." The colonel also cautioned him not to advance farther "because there were at least 250 Mexicans on both sides of the route in ambush." Tays and Kerber marched their men back to Ysleta.[27]

Hatch and Urtiago drove on toward Ysleta. Near town they came upon the bodies of Yrigoyen and Durand. The latter was "lying directly in the middle of the road, as though he had fallen either from a horse or wagon." Yrigoyen was found at the road's edge. The justice identified the corpses as those of Ysleta residents arrested by Rangers the day before. Hatch began to "examine the men carefully. One had some five or six bullet-holes in his head, shot through the hat-band; the other was shot in the forehead and in the side. So near was the shot in the side it was evident the gun had been near his clothing." According to Urtiago, "Santiago Durand was shot through the side, the ball entering one side and coming out the other, and Crescentio Iragoen [*sic*] was shot three or four times in the face. They were all pistol or rifle wounds. The bodies had the appearance of having been killed right there; there was no appearance of their having been dragged to the road."[28]

Just short of town Hatch met people who knew the prisoners. It was then he learned that the prisoners had been bound. The locals told the colonel they had overheard Rangers threatening to kill the two men. "I was then assured in my mind," said Hatch, "that these prisoners were killed without necessity." The colonel was "doubly assured" when Kerber lied that the prisoners had been shot seventy-five yards from the road while attempting to escape. Hatch, who had seen enough corpses in his lifetime, wasn't buying it. He "was aware that the bodies had never been moved." He looked at the Campbell wagon and was "confident that the stains . . . in the rear were blood stains."[29]

Kerber now received the full force of the colonel's rage. Hatch "denounced this inhumanity in no measured terms, and informed the sheriff it was his duty to arrest the murderers immediately; that the troops were sent there to assist the civil authorities in enforcing order and making arrests; that atrocities of this nature would not be for a moment tolerated." Hatch "told the sheriff that they had committed a cowardly murder and that if such proceedings were to be continued he would apply to the President for authority to place this county under martial law." The man who had mutinied against his own commander in Colorado was not one to be intimidated by the Ninth's colonel. "The sheriff did not arrest the parties who committed the murders or take any further notice of the matter."[30]

Maj. James F. Wade, who had just arrived with the companies from Fort Bayard, was "at once ordered to the scene of trouble, with orders to restore quiet and prevent further acts of this character." The major placed troops in Socorro and reached San Elizario at 3 P.M. At perhaps the same moment, Capt. Louis Carpenter arrived at San Elizario with H Company, 10th Cavalry.[31]

At least one other violent act was committed that day by Kerber's men. Pedro Candelario was rolling into Ysleta, on his way home from Las Cruces leading three wagons. As he passed by whatever building the Rangers were using as quarters, he was brought to a stop. "They pointed their pistols at my head and at my boy's head, asking who we were, and if we had any arms. I responded that I was justice of the peace of that town, and demanded to know by whose order they were acting." The gunmen, who Candelario thought were "Rangers" but who Kerber identified as "Tucker's men," told him they were under orders not to allow any man to come into town with arms. With that, said the JP, "They then took from me by force a Sharps rifle, an old infantry gun, and two pistols, the latter jerked out of my boy's belt." Candelario complained to the sheriff, who ordered the arms returned.[32]

As the day ended, the prisoners Yrigoyen and Durand, Mr. Núñez, and Jesus Telles were dead. Maria Núñez, Cruz Chavez, and Bernabel Candelario had also been shot. Salomé Telles had been raped and Noverto Pais pistol whipped. Juan de Dios Alderete, Jesus Padilla, Antonio Cadena, Pedro Candelario, and his son had all been threatened at gunpoint. Some accounts included Andres Colmeneros among the Ranger prisoners. His fate is unknown. There may have been others whose sad destinies went unrecorded. Families were left without fathers, sons, and brothers, almost certainly all of them providers. Others dealt with the trauma of rape and beatings. Guilty or innocent, all paid the price for the executions six days earlier. On that Sunday morning the injustice of American law took on a new face for the people of *la Isla*. It was no longer simply a matter of misrule by crafty lawyers and politicians. It became a thing to be feared for its naked violence applied capriciously and liberally.

Charles Kerber bears primary responsibility for unleashing the violence upon his constituents. With the exception of the shooting of Jesus Telles, virtually every incident was inspired by Kerber's order to disarm the population. Many western cities and towns had gun ordinances that forbade citizens from carrying arms on city streets. It was not illegal to own guns or have them in one's home. Kerber's effort to disarm the populace was unconstitutional under any standard but one. President Hayes had authorized the intervention of the U.S. Army on the basis of the existence of an insurrection or invasion. Constitutional protections of the citizenry's right to keep and bear arms would not apply. There was no invasion from Mexico taking place, certainly not inside Socorro, but an insurrection was still underway. The insurgent army was still somewhere near and might cause trouble. The sheriff could not afford to have an armed and angry population at his back. His posse would not be able to determine who was a "good Mexican" and who was "bad." Better to disarm them all. As an ex-Union officer now living in the old Confederacy, Kerber was well aware that disarming a rebellious people was sanctioned by recent history; and whatever the American tradition regarding the right to bear arms, it was an "American" tradition. Kerber, forever worried about armed "Mexicans," even those who were American citizens, now denied them that right. For the resulting shootings, the sheriff never publicly expressed the slightest regret.

The Silver City men have taken the blame for the legalized anarchy, but, except in the case of the rape of Salome Telles by "Williams" and "Johnson," Silver City men were rarely identified in the episodes of violence. This is not remarkable, since they were strangers to Paseños. That some of them participated in the sheriff's crimes appears probable from the reports of their subsequent criminal activities. What is obvious is the active role of the Texas Rangers in the disorder. The back-shooting of the two prisoners by Sergeant Frazer and Private Barton may have been a rogue act. Both men were capable of such homicides without prompting from anyone. That Kerber gave them license is clear, for he made no attempt to arrest them in spite of Hatch's expert testimony regarding the wounds and crime scene. But Tays, Ball, Mathews, Oberst, Price, and Santiago Cooper, and unidentified Texas Rangers participated in or were present at all the other shootings and home invasions (with the exception of the Telles rape). Frontier justice was by its nature rough and inexactly applied. The events of December 13 through December 17 had pushed the Texas Rangers past their limits, leaving them in no mood to be fair, go light, or follow the law.

The performance of Lieutenant Tays is troubling. In his defense, there was, in 1877, no Texas Ranger tradition of law enforcement, no Ranger police academy, no Ranger procedural guide, and, in this unit, precious little experience. Second, while Texas Rangers had quieted other Texas county wars, no Frontier Battalion unit had ever been asked to quell a popular insurrection. Texas Rangers *had* been asked to protect Texas from Mexico and Mexicans. To these Rangers, or many of them, the people of Ysleta, Socorro and San Elizario were Mexicans at war with Texas, plain and simple. Law enforcement, however understood in 1877, seemed less appropriate than the "laws of war." Third, Tays understood his orders from Major Jones to require his full support of Sheriff Kerber.

But whatever his authority to quell insurrection, whatever his duty to follow Kerber, at some point Tays should have seen the need to prevent excesses against the citizenry. He should have read the mood of his men and foreseen trouble. Before they set out, he should have strictly cautioned them against excessive force against civilians. Once he saw the bodies of Yrigoyen and Durand, he had a second chance to set the proper tone. The fiasco at the Núñez home was his third opportunity to pull back from mayhem, and his third strike as a leader. Whatever the legal premise behind the home invasions, they went south quickly. Tays may have been ruled by a sense of outrage at the death of his men and those they protected. One also gets the sense that he simply washed his hands of responsibility. The leader who had shown backbone at every stage of the siege and its dénouement disappeared, replaced on December 23 with a cipher who instead kowtowed to the will of Charles Kerber. If Kerber was primarily responsible for what happened that day, Tays was the man who might have stopped the riot, or at least measurably slowed it down. He did not.

Hatch received two visitors that Sunday, Jesus Escobar y Armendáriz and Ynocente Ochoa, leading citizens of Paso del Norte. They carried a message from Jesus Padilla. The *Jefe Politico* assured the general that his countrymen "lament" the troubles in San Elizario and considered it a matter for the United States to resolve. Second, he regretted Kerber's vow to disarm every Mexican citizen who passed to the American side of the border, as well as a prohibition against Texan merchants selling any arms or ammunition to Mexican citizens. Padilla wanted the sheriff's action revoked, "as having a hostile tendency" and giving cause for reprisals by Mexican merchants. Third, the *jefe* declared that Mexican authorities gave no aid to the San Elizario rioters. In fact, "the very contrary was done," witnessed by Padilla's refusal to loan the insurgents a piece of artillery. Padilla had warned the Mexican people to take no part in the hostilities. He also had no intention of permitting the war to spill over onto Mexican soil. He had "sent a small force, *after* the last events at San Elizario, with the sole object to watch the line and to prevent any force from passing over to persecute the families which had taken refuge on this side of the river."[33]

Hatch received the Mexican emissaries "with notable distinction, [showing] deference to the extreme." The officer completely agreed with Padilla's first point, and he assured his visitors that his opinion was likewise shared by the federal government. As to Kerber's warning, Hatch thought the sheriff had "acted without authority" but added that it would not be necessary to formally revoke the order. Hatch understood Padilla's third point. He expressed his intention to restore order, prevent further outrages, and to avoid provoking any international incident. He condemned the killing of prisoners "in the severest terms," and evidently expressed a few choice words about Sheriff Kerber. Finally, Hatch convinced his visitors "that he will not permit the designs by the civil forces of Texas, whose hostile intentions towards Mexico he understands and proposes to stop, if they should unfold them."[34]

Hatch dispatched two reports to Pope on the twenty-third of the month. The first, written that morning in San Elizario, advised that his "rapid march" had led "the mob" to disperse without fighting. The evening's telegram reported the deaths in Socorro and his success in restoring order. While the Paseño towns were now "reasonably quiet," he cautioned, "there are none too many troops here to carry out the orders for making arrests." Looking to the long term, the colonel noted that old Fort Bliss was no longer fit for occupancy, but rented quarters could be had at "reasonable rates."[35]

Hatch issued two more messages on Monday, December 24. The first advised Pope that "the Gov. of Texas should send some one here at once with authority & discretion to settle the matter some official not biased by local prejudices." The second was directed at the citizens of El Paso County. General Order No. 1 announced the army's occupation of the county, as far as the Quadrillo, as "authorized by the Constitution of the United States in case of insurrection and invasion." Hatch made it clear to Anglos and Paseños alike that his troops would maintain order and enforce both federal and state laws. "Outrages in the name and under color of the law, and by those who ought to be its representatives and guardians, will not be tolerated." Civil authorities meeting any resistance would receive prompt military assistance. He encouraged "good citizens" to aid the authorities, "but murderers, thieves and marauders ought not to be shielded or protected." Finally, he instructed his officers and men that, "Under no circumstances will the troops cross the Rio Grande without orders." With General Order No. 1, a semblance of legally constituted order was restored. Peace in the valley had not yet been achieved.[36]

CHAPTER 20

"I wish they would clear out of here."

Christmas brought little joy to the residents of *la Isla*. Scores of family and neighbors had been killed or wounded in the previous eleven days. Women had been violated. The fields had not been worked for weeks. Hundreds had fled to Mexico, creating armed camps south of the border; those remaining faced destitution. With stores looted of everyday goods, there was nothing to buy in San Elizario. Upriver, merchants were largely spared, but the disorder on *la Isla* was bad for business. Sales depended upon the economic health of the community, and a large sector of the economy had ground to a halt.

Shopkeepers could look to the Silver City men as a source of revenue. Thirty volunteers spending forty dollars monthly pay apiece was good for business. The Texas Rangers also needed new equipment and supplies. Unfortunately, many of the New Mexicans had no cash. The merchants would have to wait until the state reimbursed the amounts advanced, but they could anticipate that Texas would pay its debt.

Perhaps the chief consolation for Paseño and Anglo alike was the posting of "Buffalo Soldiers" in San Elizario, Socorro, and Ysleta, a deployment that spelled an end to large-scale violence. Enforcement of law remained out of the question. Sheriff Kerber had no suspects to arrest. The insurgents had fled, and he saw no wrong in the actions of his posse men. If anyone celebrated the holiday season, it would have been the Texas Rangers and Silver City men. The former had escaped death and rendered a bit of payback. The latter were on a paid lark.

As 1877 ended, armed bands separated by a dry riverbed and united by hatred and misunderstanding kept a watchful eye on one another. With so much anger and so many guns about, it was possible that some action by some irresponsible gunmen would spark a new round of violence. While immediate concerns, such as the next meal or the next bloody clash, weighed upon people at the Pass, the larger question of whether the county's disorder would trigger a new war with Mexico was again being debated by newspapers around the nation.

The Rangers returned to San Elizario on December 27. They found the remains of Howard, Atkinson, and McBride in a "deep hole" or old well about three-quarters of a mile east of town, to where they had been dragged. According to Rocky Mathews, "The body of Howard was much mutilated, the flesh had been cut away in strips from his side and leg and taken away to be dried; his bowels were cut out, the body had no clothing on it but a shirt and sock. Mortimer's body had been exhumed from where we had buried it. We took the remains of Howard to El Paso and buried them beside his wife. The other three we buried in one grave." Although Atkinson was blamed for the surrender, Mathews remembered him in the same breath as Howard. They "stood at their posts and acted like brave men; during our siege, they both fought bravely; Atkinson did not lie down for three days and nights, they made themselves general favorites by their cheerful manner, and met their fates like brave gentlemen fully believing that we would avenge them; *and we will.*" Putting words to the revulsion and anger felt by the Rangers, Mathews promised, "the boys are wild, woolly, and will yet be heard from."[1]

Ellis' body was never found, indicating someone may have taken the trouble to bury it in an unmarked grave, as Parra family tradition holds. Ellis's store had been cleaned out; Charles and Teodora's home had been thoroughly plundered of furniture and personal belongings. The widow Ellis's loss, in addition to her husband, was estimated at fifteen thousand dollars. Some three hundred sacks of ground flour were taken from the Ellis mill, although nearly a hundred *fanegas* of unprocessed wheat were left untouched. Atkinson's home had been stripped of its furniture, bedding, and "fine library," much of it loaded onto his wagon and horses. Hidden in the house, looters found some eight thousand dollars in "greenbacks" and checks, the fruits of his sale of stock to Ellis just weeks before. Merchant Nicholas Kohlhaus's loss was estimated at eight hundred fifty dollars. The horse and carriage that brought Howard to San Elizario for the last time were also in Mexico, supposedly taken there by Chico Barela himself. Howard, McBride, Mortimer, and Andy Loomis were thought to have lost over a thousand dollars among them. Frank Campbell, Ellis's miller, returned home to find that his own home had been emptied of everything except a stove and bedstead, the price of standing alongside the Rangers. Tays estimated the Rangers lost twenty-five hundred dollars in arms, ammunition, forage, and subsistence to looters.[2]

Hundreds of "free salt men" had fled to Mexico, taking their families and moveable property with them. Reports indicated that the insurgents regrouped on the other side, notably at Zaragosa, opposite Ysleta, where the

Rangers were quartered. One observer put their total at four hundred or more. They had nowhere else to go, but hope arguably lived among some. The absence of farmers from the fields benefited no one. With Cardis and Howard both dead, perhaps a negotiated settlement could be made. If not, their homes were just a few miles away, the Silver City men could not stay forever, and the U.S. Army might march away again. And so the insurgents maintained a watchful presence along the Rio Grande. The cohesion and purpose of this amorphous little army are open to speculation. Most Mexican citizens who had traveled to the Pass of the North for the religious festival and then joined in plundering on the American side probably returned to their homes. Others, spurred by the chance for continuing payback against *gringos,* probably remained under arms. Some Mexican outlaws hung about for the possibility of more loot.[3]

It was soon obvious that the U.S. army would not cross the international boundary. Would Mexican authorities break up Barela's force, arrest the "rioters" and turn them over to the Americans? Would they at least ensure that no organized force would cross over to attack the Rangers or retake their towns? From the *Mesilla Valley Independent's* perspective, the insurgents "evidently expect to be protected by the Mexican government from receiving the punishment their crimes merit. . . . it then remains to be seen whether Mexico is acting in good faith, or whether she will follow her usual policy of procrastination and delay so as to defeat the ends of justice."[4]

The latest headlines continued to mix in rumor and sensationalism. On Christmas Day, the Santa Fe *Weekly New Mexican* complained, "it is almost impossible to get facts." Through a thoroughly garbled report received from the *Mesilla Times,* via Chicago, readers of Denver's *Rocky Mountain News* learned of "a thousand rumors in circulation. . . . It is reported that the forces encountered and a fight followed, in which Lieutenant Mortimer was killed, and two soldiers captured, and that the invaders took a cannon from the United States troops. A pursuit of the Mexicans who crossed the Rio Grande, whither they were followed by Texas rangers. Lively work may be expected at any hour."[5]

In a page one editorial proclaiming "The Mob Triumphant," the *Galveston Daily News* called for "prompt and vigorous measures . . . to quell the insurrection, bring to justice its guilty leaders, and firmly establish in the whole El Paso region of Texas the dominion of law and order." The *Austin Statesman* countered, "The El Paso trouble teaches one thing very forcibly, and this is that neither Texas nor the United States require any more of these people as citizens. An army of 100,000 men would have to be sustained to watch and govern a conquered Mexico, and the evils to society would be incalculable."[6]

Beyond Texas, the nation's newspapers were solidly, though not uniformly, against escalating the El Paso disturbance into an expansionist war. The *New York Times* scoffed at any "attempt to magnify the affair into undue proportions. . . . Official dispatches to Gen. Sheridan's headquarters at Chicago, however, explicitly say that no Mexicans have crossed the border to join in the riot. . . . [T]here is no invasion."[7]

The *St. Louis Times* chimed in, "There is work for a grand jury in El Paso county instead of an army of regulars." The *Philadelphia Telegraph* assured its readers, "No one wants war but a handful of conscienceless speculators." Such talk, said the *Chicago Times*, was "infinite idiotcy [*sic*]. . . . There is ample testimony that the hostiles who attacked American magistrates and butchered American merchants and militiamen at San Elizario are Mexicans. . . . Their common aim was resistance to American law and authority. . . . [T]he fact that a considerable portion of that war party moved from Mexico into Texas for the express purpose of fighting with their countrymen against the execution of the laws of that country, puts a very different face upon the matter."[8]

The *Army-Navy Journal* declared the El Paso affair "a local brawl, magnified and misrepresented, with much persistence, as a Mexican invasion." The *Journal* explained, "Very likely, some Mexican citizens may have been on the ground or have crossed from El Paso to see it, as in the case of a riot in New York some persons might cross from Jersey City, attracted by the sensation. But to present the affair as an armed Mexican invasion, as the Texan authorities have done, is preposterous."[9]

One voice lacking any sympathy for Texas belonged to presidential hopeful James G. Blaine, who linked the event to an ultra-Southern conspiracy. "Those Texas fellows are responsible for the whole thing. . . . A fight with Mexico means nothing less than the acquisition of more territory in the South-west and this means more Southern States. . . . The Texans want a fight because they know they will get more territory to the west of them. . . . The people who would flock to the new territory to the west of the Rio Grande would go from the class of ultra Southerners."[10]

Despite his consistent claims of no Mexican involvement, Sheridan had received contradictory reports regarding whether the rioters in San Elizario included, as Colonel Hatch said, "four hundred Mexicans from the other side of the river," or, as Captain Blair reported earlier, "not one I believe from the other side." Directed to clarify matters, Pope asked Hatch for "answers that could not be impeached to the following questions: Were there any persons in the mob from the south side of the river? If so, how many and how did they come across? As a body or separately? Report the exact fact." Hatch dribbled his answers, reporting first that many crossed from Mexico, "but they did not come in large bodies or organized but simply in small squads, evidently for the purpose of plundering." Later, he estimated that "at least two hundred Mexicans from Mexico" took part in the siege and, after the execution, "almost two hundred more from Mexico" crossed over to join in the plunder. Unconvinced of Mexican involvement, Pope dispatched his own man to the scene, Lt. Col. William H. Lewis of the 19th Infantry Regiment, "a cool judicious officer whose reports can be relied on and who has served in New Mexico."[11]

War with Mexico, whether over troubles at El Paso or lower down the Rio Grande, was never entertained by the Hayes administration. The thornier problem forcing its way into the Oval Office was the state-sanctioned riot against the citizens of El Paso County. The army's presence had made the sheriff's vengeance possible. Coming just five months after the president had deployed troops to quash rail yard labor riots, the military's use as a shield for murder strengthened those in Congress who wanted to get the army out of the business of enforcing civil order. If his executive powers were not to be hamstrung in the future, Hayes would have to put a lid on El Paso's civil disorder without attacking civil liberties.[12]

As early as December 24, Pope ordered Hatch to "prevent the Rangers from murdering prisoners arrested by the sheriff & who could not have been arrested except for the fear of the forces under [the colonel's] command." Hatch was to use force if necessary. (Hatch issued his General Order No. 1 that day.) Pope warned Sheridan, "unless some responsible party is sent by the Governor with authority to act, it seems likely that outrages like these will be committed without even the color of law." This message was probably in the hands of Secretary of War McCrary before he met with the president and Secretary of State Evarts on the afternoon of December 26 to discuss the situation. Hayes directed McCrary to inform the Texas governor that federal troops could not be used to arrest citizens unless state authorities could ensure them due process of law. On the larger issue of relations with Mexico, the president postponed recognition of the Diaz regime.[13]

Kerber began his own campaign of "damage control." He reported to Hubbard, "The last fighting was on Sunday when entering Socorro, we were attacked by the mob, and of the mob, one was killed and one woman accidentally wounded. Two desperate prisoners who were under guard while en route to bury the remains of Howard, McBride and Atkinson, in attempting to make their escape, were pursued into the brush and killed. No United States troops engaged." Kerber's version was reported under the headline, "Rangers Exonerated."[14]

In a letter to the *Independent,* Deputy Sheriff H. H. Harvey deflected criticism away from his boss, pointing the finger instead at John Kinney's men, who killed four "Mexicans," including the two prisoners shot while trying to escape. "Others say that he shot them to get rid of them. General Hatch says it was a murder, and that Kinney would have been whipped if he had not interfered; for the mob were gathering very fast."[15]

There were other Anglo scapegoats. The *Independent* reported it "astounding to find among men who make pretensions to honesty and decency those who attempt to excuse and palliate the devilish crimes perpetrated by the El Paso county mob. And it is even more astonishing to find such men among the English speaking race. . . . these few are a disgrace to the race." J. P. Hague, Kerber advised Hubbard, had been "Lying sick" at Mauro Lujan's home before being "escorted by numbers of the mob to El Paso." Kerber failed to mention that Hague had been prostrated with a broken limb following a buggy accident before the siege of the Ranger quarters began. The injury was serious and transportation to Franklin "at risk of his life" required such an escort. Kerber also noted that "a train of S. Schutz & Bro. [was also] escorted by the mob, after having disarmed two Americans, who were traveling with the train."[16]

Corporal Mathews confirmed that "a gentleman" (i.e., Hague) had been escorted out of San Elizario by "rioters," including "a rear guard who would not be very popular even among the murderers." He went on to claim that, once Howard had been executed, "Chico Barela *sent an express to Franklin for further orders,* as he then and there stated; our lives hung in the balance until midnight when the *further orders* arrived; fortunately for us the verbal decrees (written ones are unsafe at such times) was favorable, and Chico Barela was ordered by his superior to let us go, he having accomplished all he desired."[17]

An "Eye-Witness," who turned out to be Juan N. Garcia, claimed that "leaders of the mob . . . told their friends in my hearing that 'their friends in Franklin had assured them that the U.S. troops would not interfere.' And these same leaders, especially Chico Barela, stated that they were simply obeying orders from their superiors at El Paso [probably Mexico]."[18]

Garcia later acknowledged under oath that he had not really seen or heard everything he had claimed, but he stuck to this part of his story. His account is not particularly revealing. Everyone in the valley had been given reason to believe that federal troops would take no part in the conflict under any circumstances short of an attack on the Custom House. Mathew's charge that Barela and his lieutenants acted under the orders from someone in Franklin, presumably an "American," was more troubling. His failure to name those who supposedly secretly pulled the insurgents' strings cast a net of suspicion over anyone who had expressed sympathy for the Paseños' cause, including Magoffin, Slade, the Schutz family, and of course Hague.[19]

Meanwhile, it looked like at least two men would pay for their crimes, Sergeant Frazer and Private Barton. Their troubles began when an "official dispatch" from Washington reached Austin, declaring, "U.S. troops arrested two Mexicans and turned them over to the rangers, for safe keeping, who took them out and lynched them." The report, not what the governor of Texas had previously heard, prompted Hubbard to order the arrest of the two Texas Rangers, who were put in irons for more effective "safe keeping."[20]

While some pointed fingers, others continued to point guns. They included men from New Mexico who had little if any interest in righting wrongs. Revenge was just an excuse. While called into service by Texas, they answered to no one but themselves. They were here for the sport, and if they had a leader, it was John Kinney. He may have only been a "high private" in Dan Tucker's "company," but he appeared to be the chief of their misdeeds. Silver City men were not the only creators of anarchy. A couple of "bad eggs" in Tays's Texas Rangers inspired or emulated the New Mexicans. Even a few U.S. soldiers from the 10th Cavalry engaged in misconduct. Across the border, Mexican border ruffians and angry Paseños joined the game of mayhem and reprisal as the last days of 1877 slipped into the first of 1878.

The chaos was random and untidy, part guerrilla war, part crime wave, and damned hard for anyone in authority to handle, if they cared. On the American side, the sheriff had few deputies and less interest in bringing anyone to justice who was not beyond his reach. Of soldiers there were plenty, but they could only act in support of local law, which returns us to Kerber. As for the Rangers, the tragic events of December had damaged the unit's usefulness.

One Ranger, Kinney's friend Jim McDaniels, instructed George Zwitzers to deliver a message to John Leahy "that he (McDaniel) had been informed that Leahy had spoken very hard of him to General Hatch, and that a good many of the rangers had made up their minds to kill Leahy at first sight, but that he did not want to kill an American, and that he (Leahy) had better leave the country." In testimony given weeks later, Leahy admitted to receiving a message that he might be "shot down like a dog" if he stuck around, but swore that McDaniels afterwards treated him "in an apparently friendly manner, as he had done before." Leahy chalked the experience up to envy on the part of Zwitzers. But Leahy also admitted to receiving a letter from a friend in Tays's detachment who advised him "to leave the country because the rangers might be mad with me about something I had done or said." He refused to believe such threats and claimed to have received "no harm or indignity" since receiving the letter.[21]

The army officers who questioned Leahy tried hard to get at the identity of the Ranger who conveyed the threatening letter to Leahy. His answers indicate a strong survival instinct: "This friend's name I decline to give. He belonged to Lieutenant Tays' company at the time. I think he did; but am not positive. He does not now belong to the company. This friend is not stationed anywhere but varies his time at one place and another. He is in El Paso County now, I presume. I have seen him, four days ago, in the county." Leahy appears to recognize he has already said too much. Friendship, he alleges, impels him to keep the man's name confidential. His answers indicate fear is the driver.[22]

Indirect evidence that McDaniels waged a private war on Paseños is found in this report by an anonymous "Observer": "Padre Borrajo, the Italian [*sic*] Jesuit priest, who grossly insulted Bishop Salpointe on the street when he was last here, has offered one thousand dollars for the heads of Jimmy McDaniels and Sam Frasier [*sic*], and has of late been issuing bullet proof bulls." Regardless if the report was true, it indicates that Borrajo lumped McDaniels in the same category with the murderous Fraser.[23]

Andy Boyle, like McDaniels a member of Jesse Evans's gang, was caught stealing a horse from George E. Angerstein, a wealthy American merchant in Paso del Norte. Boyle was arrested on the American side by Deputy Harvey and taken before the JP, but no one was eager to prosecute the case. Dist. Atty. James Zabriskie soon announced the matter settled: Boyle was released and the horse returned to Angerstein. Such men were dangerous enough acting on their own, but they were also part of a larger cancer. Kinney and his pals were hardened criminals masquerading as peace officers and Texas troops. They were a gang with badges, perhaps the first southwest borderlands criminals to be sworn in as a group to fight a local "war." The activities of the "Kinney gang" in the Rio Grande Valley established a dangerous precedent repeated from Lincoln to Tombstone.[24]

Their leader did not take long to establish bragging rights. "Capt. John Kinney now has an independent command of about 70 men and has issued a call for 30 more well armed and mounted to join him at or near El Paso. Pay, $40 per month and forage, by state of Texas. With 100 well armed and mounted men Capt. Kinney is liable to soon have a world wide reputation; and lay Lt. Bullis in the shade." Though this *Mesilla News* item was written with tongue firmly planted in cheek, it indicates that Kinney saw himself or was seen by others as a power separate from Dan Tucker's command. On December 27, the New Mexican raised a little hell in Old Mexico. According to Juan D. Ochoa, the Mexican acting vice-consul for El Paso, "in the square and streets of El Paso, Mexico, Mr. Juan Kyny and several others were riding horseback, firing guns, and disturbing the peace: they were not arrested." Even the Mexican authorities were afraid to touch the gang.[25]

Kinney's raid spurred the insurgents into counter-offensive action. Desidario Apodaca was elected captain "to capture any parties going down from [the American] side to the other, taking their arms and horses, if possible." On New Year's night, five Paseños repaid the favor of Kinney's hurrah. Crossing the Rio Grande, they "went to the house of a man furnishing forage to the troops and Rangers. They broke his wagon to pieces, stole his harness and broke open the door of the house. The owner recognized three of them as leaders of the mob. They also plundered several other houses." Such raids were said to occur nightly. An attempt to capture Captain Garcia on January 3 was foiled. That same evening, around 9 P.M., Cpl. Rocky Mathews was "returning from the quarters of Lt. Tucker's command" when two men stopped and opened fire. Though a half-dozen shots all missed their target in the dark, Mathews "thought it discreet to retire, which he did at double quick." Rumor had it that the insurgents intended to wage "constant war to the citizens remaining in Texas, until they kill them all, according to their desires." On or about January 2, some "rangers" (possibly Kinney's men) reportedly crossed the river near Zaragoza and fired upon the people. While small bands engaged in these desultory raids, the insurgents and Rangers still patrolled the border between Zaragoza and Ysleta, occasionally exchanging "a little sharp shooting" across the Rio Grande.[26]

Continuing insurgent-related activity inspired some purely criminal disorder. In neither camp did greed draw distinction between "friend" and "foe." It was recounted "on reliable authority" that three Mexicans were killed when a New Year's Day fight over division of spoils occurred at the house of Padre Borrajo. However accurate that particular hearsay information from the Mexican side, such stories make sense. "Throughout history," today's U.S. Army and Marine Corps officers learn, many defeated insurgent movements "have degenerated into criminality." Another lesson applies equally to the American side: "Fighters who have joined for money will probably become bandits once the fighting ends." That certainly happened among Kinney's men, two of whom found plundering Ysleta were chased (if not caught) by other Silver City men, evidently worried about their "good name."[27]

Anglos and Paseños were not the only people to engage in disorder. Some Buffalo Soldiers of Company H, 10th Cavalry, joined in by looting homes in San Elizario. One plundered domicile belonged to Father Bourgade, who had been driven out by the insurgents. One day, Anacleto Escajeda, while watching over the priest's home, heard a window break and pounding on the priest's door. After a boy opened the door for him, "four of five of them walked into the dispensary or storeroom and . . . they took some grapes and a pumpkin, when they left, taking nothing more." Escajeda reported the incident, but nothing could be done.[28]

More serious was the soldiers' ransacking of Pablo Mejia's home. The scrappy lawman was in quarters at Ysleta when two of his neighbors witnessed "ten colored soldiers" break in through a window. Minutes later, nine of them came out "with bundles under their arms." Investigating, Meregildo Montez stumbled across the tenth soldier, standing with a cocked pistol in one hand and tobacco stolen from Mejia in the other. Bravely, Montez ordered the cavalryman to leave. He did so. The soldiers had broken open trunks, stole clothing belonging to Mejia's wife and, in fact, "took everything out of the house except the furniture, and bedding and cooking utensils."[29]

While the robbery of Mejia's house was still underway, a neighbor grabbed John Clark and the two men enlisted Captain Carpenter. The three found Mejia's possessions broken and in disarray, with food spilled on the floor. Carpenter "had the men all fall in line, cooks and everybody, and told the man to walk up and down and see if he could point out the men who did it. He was able to point out but one. The captain had the camp searched, but could find none of the articles. The man recognized was punished by being forced to carry a log." When Mejia returned, he surveyed the damage. The stolen items came to a grand total of "about $80," a testament to the general poverty of the Paseños. Said the frustrated Ranger, "All I know I was robbed."[30]

Juan Mauro Lujan also had his house broken into by Carpenter's soldiers. Inside they found enough wine to make them "tolerably drunk." Again the captain made some transgressors "carry a log on their shoulders." When Lujan eventually returned, he discovered that some trunks had also been taken from his home.[31]

Some H Company troopers wanted more than wine. They invaded Jose

Perez's home one night, threatening to kill him. What they really desired was "a girl to marry." Perez or his wife had allegedly agreed to give them such a bride (probably to fend off an earlier threat), "and this was the night to finish the bargain and get the woman." Perez advised the men that the young woman wasn't home. Frustrated, the soldiers left, after one pulled his revolver and shot into the fireplace.[32]

It is notable that all the accusations made against the Buffalo Soldiers for rapacious acts in *la Isla* were placed at the door of the 10th Calvary's H Company. Carpenter was an excellent combat leader and H Troop had distinguished itself against the Cheyenne. Carpenter himself won the congressional medal of honor for his leadership of the company against Indians in 1868. Still, it appears the troop was undisciplined in camp.[33]

No similar accusations of criminal activity were made against any troop of Colonel Hatch's 9th Cavalry quartered in Socorro, Ysleta, or old Fort Bliss. But one authority on the African American regiments serving along the Rio Grande has observed the irony inherent in the Ninth's professionalism in El Paso. "Once again, black soldiers assumed the dual responsibility of protecting a developing social order that subordinated fellow people of color while trying to protect those same people from local whites."[34]

Meanwhile, depredations by the Tucker-Kinney men continued unabated. Ysleta resident Francisco Garcia reported that two "rangers" belonging to "Tucker's crowd" knocked on his door at night, demanding admittance. "Not being allowed to enter, they fired four shots through the door, then killed and took away four chickens which were roosting on the ladder outside the house." Ysleta carpenter Jacob P. Miller experienced far worse. One night in December, around midnight, he responded to a knock on his door. Raising the door curtain, Miller spied "two men with pistols bearing on me." Miller at first refused their demand for entry, but their pistols quickly convinced him otherwise. They marched him some twenty steps outside his door and then instructed him to "keep quiet and tell them where these women were." Miller begged for the women to be left alone. The intruders ordered him "to hunt the women or they would shoot me." Blinded in the darkness, they told Miller to strike a light, but Miller said there was no candle in the house. Nor were any women there. They had used the time to escape out the back to a neighbor's house. In the confusion and darkness, Miller broke free and went to "the soldier's camp." No arrests were made.[35]

If life in the valley was ever to return to "normal," this terrorizing of citizens had to be stopped. The flight of victims to the safety of "soldier camps" did not escape Hatch's attention. He cleaned house. Carpenter's Company H was transferred from San Elizario to Ysleta on December 26, exchanging places with one or more companies from the 9th Cavalry.[36]

Exhibiting complete frustration with El Paso's top lawman, the colonel instructed Captain Blair to notify Kerber that "the State troops under your command have been committing outrages of greater or less magnitude on the peaceable people of the valley, including women and children, who, for fear of your men, have fled from their homes, and are now perishing for want of food and from exposure to cold in and around the village of Saragosa, Mexico." It did not matter whether the men in these families were part of the "mob," Hatch declared. The women and children "should be protected from outrage and encouraged to return to their homes by assurances of protection."[37]

Treating the message with the respect he felt the messenger deserved, Kerber "respectfully" returned it "with the remark that it does not appear to have been indicated by the order of the commanding officer, and consequently, cannot be accepted as official." Blair tried again on January 5th: "Sir: I am directed by the district commander to inform you that one of your men, George O'Brien by name, is now in charge of the post guard here for demanding goods in a store in this town yesterday at the muzzle of a pistol. Please inform me what disposition you want made of him." Blair enclosed a copy of Kerber's undated certification that O'Brien was, "a private of a company of State forces raised by authority of the governor of the State of Texas." The slippery sheriff again "Respectfully returned" the message, pointing out that O'Brien "was discharged from the Texas State forces under my command January 4, previous to the act herein mentioned." Since the man under army lockup was no longer the sheriff's man, the sheriff had no interest in the alleged crime committed.[38]

Kerber was not entirely inattentive to complaints. He claimed to have personally reimbursed citizens one hundred and twenty-five dollars for damages done. But the evidence fails to reveal any step he took to stop the terror. Tays was somewhat more active. He did send a squad of Texas Rangers to arrest some Silver City men who had committed a rape. For the most part he just kept his men separate from the Tucker-Kinney crowd, complaining, "I wish they would clear out of here." Hatch had little use for any state forces he had seen, reporting, "Think order will be restored better without the Rangers than with them."[39]

El Paso merchants were also weary of the Tucker-Kinney bunch. Whether they presented vouchers for supplies and pay or demanded goods at the point of a gun, this pestilential crowd was a growing financial burden. Solomon

Schutz, Ynocente Ochoa, and Moritz Loewenstein submitted claims to the adjutant general totaling $1,140.97 for payroll and $1,121.06 for supplies taken from their stores by the Silver City men.[40]

Not until January 9 did the Hubbard administration learn that Kerber had sworn the New Mexico volunteers into state service eighteen days earlier. In Steele's opinion, the sheriff was "Much at fault" for exceeding his authority. The adjutant general later claimed that Kerber "asked for authority to raise men when the rangers were besieged," but had failed to report their arrival and use as a sheriff's posse afterwards. The criticism was disingenuous. Hubbard had authorized Kerber to "Raise one hundred (100) men at once to resist invasion. Put down insurrection and reinforce State troops." The governor's warrant was not limited to assisting besieged Texas Rangers. Kerber could well deem himself justified in keeping the so-called Silver City men in service after the Rangers had surrendered and were released. With an insurgent force remaining under arms on the Mexican side of the river, the threat of continuing insurrection hung over the county. While the cavalry garrisoning *la Isla* could easily beat back any conventional "invasion" without the sheriff's help, the army's behavior had not inspired Kerber's confidence.[41]

But the state had no money to spare, and Kerber exercised no control over his posse. On January 10 Steele wired the sheriff, "The emergency calling for the posse having passed, you will discharge them. With the [military] force now in El Paso, peace should be assured." Kerber dutifully released the men the same day. As far as the governor was concerned, the situation was now a police matter for the sheriff, supported by the Rangers and courts. Two men who would not face trial were Fraser and Barton. At the behest of Major Jones, who "was satisfied the report [of lynching] was not true," Hubbard instructed Kerber to release them from the manacles they'd worn for shooting prisoners Yrigoyen and Durand "while trying to escape."[42]

Wars sometimes wind down rather than end. The Salt War, which left the insurgent force "in the field," was no different. Desultory clashes continued, carried on mostly by marauding gangs. Two days after their discharge, the Silver City men made one last raid across the border. Ten or twelve men led by John Kinney crossed the river and skirmished with Mexicans. According to one source, it was not much of a fight. Deputy Sheriff Harvey claimed that the ten New Mexicans "left their arms here, with the exception of two pistols, which they took with them." The first report from the Mexican side had Kinney "shot in the shoulder," but it turned out that he had merely been thrown from his horse and injured. Badly armed or not, the Americans quickly concluded that safety lay in surrender. Five of "Kinney's gang" turned themselves over for protection to a Captain Montes, the *commandante* of the Paso del Norte city guards. Montes then arrested four more and locked the nine up for the night, together with seventeen Mexicans. All were released in the morning. None were fined. With this last hurrah to their dubious credit, many of the Silver City men had a final "lively" night on the town of El Paso, Texas, before starting on the journey home. The *Grant County Herald* described their arrival: "A considerable number of the Silver City contingent to the El Paso army of occupation returned to town last Saturday [January 19]. They expressed themselves as well satisfied with their lark. No expense, plenty of fun and a measure of experience which may be turned to advantage in future campaigns."[43]

CHAPTER 21

"Little if any crop will be planted this season."

By early January, Colonel Hatch was ready to return to his wider responsibilities in Santa Fe. He departed with the thanks of El Paso's leading politicians and merchants. Political opponents and feuding store owners forgot their quarrels long enough to joined in a "testimony of good will," thanking Hatch for rescuing them when "our lives and property were in danger at the hands of murderers and thieves." The citizens were also grateful that a "strong arm of absolute security and protection" remained in Maj. James F. Wade, who assumed command in El Paso on January 22, 1878. At age thirty-four, Wade had sixteen years of service behind him. The son of Ohio Senator and Radical Republican powerhouse Benjamin F. Wade, he was a fine officer whose distinguished career eventually brought him a general's star. The first troops to depart were Captain Moore's Company L and, on January 18, Captain Carpenter's H Company. These reductions left Wade with ten officers and 130 enlisted men, organized into seven companies stationed on *la Isla*. Blair remained as adjutant.[1]

The soldiers marching out of El Paso may have crossed paths with outlaws riding in. The Pass had long been a hangout for rustlers and killers of both nations taking advantage of weak law enforcement and open borders. The Salt War introduced a new and dangerous generation of criminals who had coalesced in New Mexico, where they learned to cooperate in what one historian calls "the rustler chain." Some had earned their reputations in the loose-knit "banditti" of Jesse Evans, as dangerous an outlaw as roamed these parts in 1878. Some looked for leadership to John Kinney, a young man with a genuine talent for organized crime. Kinney, in the guise of a rough-edged frontier "businessman," perhaps did more than anyone to make El Paso, Texas, a sanctuary for rustlers and hold-up men despite the presence of the Texas Rangers. When the Silver City men went home, the rustler element stayed on with Kinney. They established a base in El Paso and, allied with the sheriff, eventually reached for control of the county's government.[2]

The army's troop reduction in El Paso may have prompted the governor's decision to reinforce the Texas Ranger presence there. One rumor held that "Major Jones and four companies of Texas State troops are now en route to El Paso." In fact, Jones had simply authorized Tays to raise his detachment's strength to thirty privates, plus NCOs. The Frontier Battalion commander reminded his man in El Paso, "you will be particularly careful not to receive any but good men, such as you can control & manage in maintaining the law & restoring peace & quiet."[3]

Tays had to reinforce his unit in the face of fresh resignations. The wounded Miguel Garcia, Billy Marsh, Gregorio Garcia II, Sergeant Ball, Pablo Mejia, and Santiago Cooper all returned to civilian life. Tays discharged man-killer Jerry Barton but for some reason kept Sam Frazer on the roster. No fewer than seventeen new men were added in early January. Several recruits would earn the title "Texas Ranger," including Asberry C. Ryall, Mark Ludwick, and George Loyd. These were good men, but several bad ones, including habitual rustlers, also slipped into the unit. No Paseños were recruited to replace those who left, reinforcing the alienation of this police force from the people of *la Isla*.[4]

Jones expected to return to El Paso during February, in time for the opening of court. He instructed Tays to "get the names of as many of the Mexicans from the other side of the river, who were engaged in the fight at San Elizario as possible." He also alerted Tays that he would probably enlist another ten Rangers "so as to have a stronger force during Court to guard & escort prisoners etc. etc." It was Jones's plan to escort them out of the county. "[We] we will try to get a Grand Jury which will indict as many of the murderers as possible & then change the venue & later take them to Pecos or Presidio for trial, where we may be able to teach them a lesson they will not soon forget."[5]

Many of Jones's "murderers" remained under arms in Mexico, near Zaragosa. There they defied not only the Americans, but also the Mexican government. Reportedly, when authorities attempted to confiscate plunder brought from the American side to Guadalupe (downriver from Paso del Norte), a force of Paseños "resisted and repulsed the officers making the effort." The government did manage to seize Howard's buggy and team, as well as Atkinson's horses, at the town of Carrizal, on the road to Chihuahua. One "Observer" reported that "Masked parties of Mexicans are nightly plundering and murdering indiscriminately, on the Mexican side." Though this anonymous account must be treated with caution, it stands to reason that some armed men would turn their weapons on the nearest available victims, as well as each other. It certainly happened on the victors' side of the Rio Grande.[6]

The insurgents could not remain under arms forever. Some naturally wanted to go home. According to Hatch, "parties accused of being with mob now on the Mexican side of the river say they will surrender themselves for trial if the State troops are removed but will not return while those troops are here." While some waited for the right time to return, others realized they could never go back. And so their towns remained nearly "deserted." Returning to Mesilla from a visit to El Paso County, Judge S. B. Newcomb worried "that little if any crop will be planted there this season, unless the people return which is improbable." The *Herald's* "Observer" advised that "there is no prospect of Mexicans returning, although invited to do so by Sheriff Kerber, as they have already petitioned the Mexican Government to donate them lands on the opposite side. There will not be one twentieth of the usual crops raised in El Paso County this year."[7]

The *Eco del Rio Grande* feared, "The worst phase of the El Paso 'Salt War' has yet to be felt. This [Mesilla] valley is annually indebted to that section for not less than 16,000 fanegas of wheat, which, from present appearances, is not likely to be forthcoming the present season, for the reason that nine-tenths of the farmers of that county have abandoned their ranches and crossed to the Mexican side of the river, and are now idle and evince no disposition to return." The *Eco* feared "a greatly increased price" for wheat.[8]

Getting insurgents to return was of the highest priority to all concerned, whether one sought a judicial reckoning, a salvaging of the economy and lives, or the salvation of souls. Father Bourgade, forced to flee San Elizario, returned in mid-January with Bishop Salpointe. The clerics undoubtedly worked to facilitate the return of Bourgade's parishioners.[9]

The long-lost Mr. Phelan, perhaps the only Anglo drafted into the Paseño army, also returned home. Mary Phelan had received no word of her husband for weeks after his kidnapping from their Las Cruces home, and she naturally feared he was dead. He reappeared in the middle of an Indian attack on the Jerrell home. The photographer and a handyman fought them off until help arrived in the form of neighbors "rushing up with guns, pistols, stones and all manner of implements." Not for the first time, "the horror of the whole affair" was about to send Mary over the edge when she heard a voice saying, "My God, Mattie, where are you and the baby?" It was Mr. Phelan. "Worn, thin, and almost a wreck," he had escaped and returned in the midst of the battle. The couple's trials ended on a happy note, but decades later Mary Phelan recalled, "to this day I shudder when any one mentions that 'Salt War.'"[10]

The genesis for the president's military board of investigation into the Salt War was a telegram dispatched by Governor Hubbard to Congressman R. Q. Mills on December 19, 1877. Hubbard had pushed the Texas delegation to "urge immediate investigation by United States authorities as to the part taken by citizens of Mexico in the late El Paso troubles." From the Texan's viewpoint, the troubles were the latest of many Mexican border transgressions, and he expected the inquiry to focus on that fact. But already aware of the Kerber counter-riot against *Tejanos,* President Hayes instructed Secretary of War McCrary to order a wider-ranging investigation.[11]

McCrary's confidential message to General Sherman directed him to send "two discreet and competent officers" to El Paso to act in conjunction with a third commissioner to be selected by Hubbard in making "a thorough investigation in relation to the recent riots and murders in that vicinity." The board of inquiry was "expected to report as follows:

> The causes which led to disturbances, and the facts and circumstances of any outrages, riots, or murders that have occurred.
>
> A statement of the action taken by the civil and military authorities of the United States and the State of Texas.
>
> The number and character of persons from the Mexican side of the Rio Grande, if any, who took part in these unlawful and criminal proceedings, and the action of the Mexican authorities touching the subject, if any was taken, so far as can be ascertained.
>
> Any other information deemed by the board of value, and any recommendations they may see fit to submit."[12]

Colonel John Haskell King, 9th Infantry was appointed board chairman. A veteran of forty years in the regular army, King was an officer Sheridan trusted. King brought with him 1st Lt. Leonard Hay as recorder. The board's second member was William Henry Lewis, lieutenant colonel of the 19th Infantry and previously ordered to El Paso by Pope. A West Pointer, Lewis had also remained with the Regulars throughout the Civil War. He was an old New Mexico hand, and may have met Charles Kerber in the Glorieta campaign. Once in El Paso, the officers found that Hatch had already begun securing evidence, cautioning his officers, "Only reliable and unimpeachable testimony on this subject is of any value."[13]

The nagging question was how many insurgents were Mexican citizens? At least nineteen of Hatch's twenty-five witnesses offered wildly varying information or estimates. Telesforo Montes had seen fourteen armed Mexican citizens; Porfirio Garcia spotted one hundred, "crossing in small squads, and in different parties and places." Captain Garcia "saw at least forty men from the Republic of Mexico armed and aiding the mob," while "from his own knowledge," he knew of one hundred or more who crossed over from the Mexican side. During his travels outside the besieged Ranger compound, Pablo Mejia, "saw about two hundred citizens from the Republic of Mexico whom he knew by sight . . . all of them armed and engaged in the insurrection." Theodora Ellis insisted that "Mexicans from other side numbered about 300; that they did come over in organized bodies, but to assist their thieving friends on this side to rob and plunder; that they refused to give her any provisions to exist." Pedro Larraique, the parish priest of Ysleta, walked among the insurgents prior to the executions. He testified that Francisco Barela "personally told me he had under his immediate command three hundred armed men who came from the other side and were citizens of Mexico."[14]

Montes speculated that the "chief object" of Mexicans who crossed into Texas "was to assist their friends against Howard and his party, whom they considered their worst enemy, and to rob and plunder and seize all the arms they could. . . . they had no intention of offering any resistance to United States troops, or to become involved in any manner with the United States Government; . . . as soon as they heard of the United States troops approaching the town they dispersed and went over to the other side."[15]

Putting it all together, Hatch reported, "It is very probable that all the able-bodied men on both sides of the river visited San Elizario during the troubles there, either as spectators or participants. Anywhere from 300 to 1500 men could have been present. . . . [Yet] it was after all but a local affair, which can scarcely be looked upon as an act of aggression by the Mexican Republic, since the authorities on the Mexican side were active, but powerless to prevent the crossing of their own people. The robbery of stores following the taking of the law in their own hands by the mob was a result which usually accompanies such disorders."[16]

The people, a sympathetic Hatch advised, "are one and the same on the two sides of the river. . . . hence when you touch one you touch all, and when one is hurt all feel it. All had an equal share of interest in the Salt Lakes remaining public common property—and all were equally incensed against Howard for attempting to deprive them of what they consider their rights." Hatch recommended the reestablishment of a military post. "Had there been a garrison of even 100 men at Fort Bliss, it is not likely the present trouble would ever have occurred. . . . As long as the frontier remains as it now is, and there is little probability of it changing, troubles of a like nature or even more serious are likely to occur."[17]

The colonel provided a novel explanation for Blair's assertion that the insurgents included no Mexican citizens. "The insurgents were divided into four bodies or companies, each with its own headquarters. The headquarters visited by Captain Blair was that of Chico Barela, the Captain of one of the companies and when he asserted there were none but Citizens of El Paso County, Texas with him, he was probably asserting the truth to a limited degree and in a limited sense, meaning only those in and around his Headquarters and in his own company." This was nonsense. The insurgent army was certainly loose knit, but the evidence indicates Barela was the junta's chief military commander. Blair may have been too new to the area to personally identify Mexican nationals, but Tays had identified such river-crossers and had gotten that message through to Blair. The captain's reliance on the countervailing word of the "enemy" commander was a dereliction of duty driven by cowardice. It would be another nine months before Blair's superiors realized his estrangement from the truth.[18]

Another controversy lingered: whether or not Anglos directed the disorder. This and other charges laid by Corporal Mathews and Juan N. Garcia that Chico Barela was given orders by someone in Franklin was hotly denied by "American," who challenged "Mr. Mathews, or the prominent Mexican gentleman" to identify the alleged guilty parties. While the newspaper had published no names, "you can hear the names of the supposed guilty men whispered on the street-corners and in the bar-rooms of this town, and everybody knows who are referred to in . . . your paper. Now if the gentlemen who have furnished you with the information that the mob (as they call it) was directed and advised by prominent men in this town, are able to prove their assertions, let them do so; if they cannot, let them hold their peace and cease their insinuations."

As to the claim that Howard was attempting to legally assert his rights, "American" countered, "although a man may have a legal right to walk into a blazing fire, it would not always be good policy for him to exercise that right. . . . Howard knew that the people were determined to resist his 'legal right' and that they would resist it by force . . . The people . . . had fully made up their minds what course to pursue, and they carried out their views without asking the advice of anybody on this side of the river."

Finally, "American" defended Father Borrajo as the people's "friend and advisor for a quarter of a century . . . if he advised them to stand up for their rights, he no doubt thought he was advising them for their own good." For personal safety's sake, "American" hesitated to say more. Unless a man "devotes his time to denouncing the people of this county as a bloodthirsty mob, he is held to be an enemy of law and order, and is liable at any moment to fall a victim to 'mob' violence."[19]

In yet another epistle to the *Independent,* Mathews repeated his claim that "Americans at El Paso," including the unnamed Hague, had "aided and abetted the cut-throats." He called upon the newspaper to publish the names he had provided. The editors declined and pointed out that the unnamed American "present at the juntas at Mauro Lujan's house" (Hague) was there only because of a serious injury.[20]

Others attempted to sway public opinion. Fountain advised his readers that the people's December 4 petition begging the governor not to grant the salt lakes to Howard, "as a specimen of unadulterated assurance has no equal." Fountain castigated the "red-handed memorialists [as] no better than the men whose atrocious acts they attempt to justify." In the course of his editorial, Fountain revealed the motivation behind his scathing remarks. The people had long ago denounced him when, as state senator, he had attempted to secure those same salt lakes for them. The recollection of this political defeat still rankled.[21]

Before the reorganized Ranger detachment could focus on the task of bringing "murderers" to justice, a growing tension between its two sergeants needed resolution. Trouble between the two noncoms surfaced in early January 1878, when 2nd Sgt. Sam Frazer told Rocky Mathews "there were two men in the company he couldn't live with," First Sergeant Ball and John C. Ford. He threatened to kill Ford and "fix Dr. Ball." Mathews calmed him down, and Tays afterward told his sergeant, "You must behave yourself." Behaving was not within Frazer's personal code of conduct. The junior NCO had once bragged to Ball that the some graves the doctor had spotted along the San Antonio road near the Pecos were his handiwork. Frazer claimed to have killed the men for their money.[22]

Frazer's mood soured further when Tays named Ford as First Sergeant Ball's replacement. He again threatened to kill Ford. Rocky Mathews repeated Tays's earlier admonition. "Behave or no behave," snarled Frazer, "I will kill him." One evening Frazer repeated the threat to Ball, declaring that the unit was "not big enough to hold both of us." Mathews alerted Ford, cautioning "I don't think he will kill you, but he is a dangerous man." The even more dangerous Ford had no intention of making the coming fight a fair one.[23]

On Thursday, January 31, Frazer was dismounting in the Ranger corral when Ford appeared behind him, carrying a double-barreled shotgun held high at a forty-five-degree angle. With an "ordinary tone of voice," the First Sergeant called "Sam." Fraz turned around, spied the armed man he had carelessly threatened, threw up his hands, and screamed. Ford lowered his weapon and unleashed the contents of one barrel into the unwitting target's right side. The second load of buckshot missed as Frazer staggered and fell. The wounded man sat upright, crying, "Don't shoot me any more." Deaf to another killer's pleas, Ford walked up to Frazer "and emptied the six chambers of his revolver at Frazer, five shots taking effect in his head."[24]

Both a coroner's inquest and jury trial were held without delay. In a clear case of kill or be killed, the jury applied the Texas doctrine of "no duty to retreat" and found the homicide justifiable. Ford returned to duty that same day, no doubt assured that insubordination in the unit was a thing of the past. Ford's severe brand of discipline may have prompted Jimmy McDaniels to desert the Rangers two days later. McDaniels more than met Frontier Battalion standards for bravery, horsemanship and skill with a gun, but he remained at heart a thief and killer.[25]

Judge Blacker's return to El Paso on February 3 confirmed the state's intention to bring the "murderers and thieves" to justice. The *Mesilla Valley Independent* proclaimed, "Texas demands that her outraged laws shall be vindicated; the blood of the slaughtered victims calls for vengeance upon their murderers; the future peace of this frontier depends upon the prompt arrest and punishment of the criminals. . . . The Mexican authorities announce their willingness to extradite them upon demand; it remains to be seen, however, whether this announcement is made in good faith, or whether shallow pretexts will be resorted to for the purpose of defeating the ends of justice."[26]

Much had changed since the jurist had agreed to "look the ground over" before Howard's trial for murdering Cardis. Now the danger was not in putting one man on trial while four hundred shouted for his blood, but in trying to punish four hundred for shedding that blood and more. Before any proceedings could begin, the wanted men had to be apprehended. For that task, Blacker also served as extradition agent for the state of Texas. On February 4, he paid "a friendly visit" to *jefe politico* Padilla, who doubled as the local Mexican extradition agent. Padilla informed Blacker that the governor of Chihuahua had ordered him to turn over some horses once belonging to Howard, Atkinson, and McBride as soon as the estate administrators presented the necessary papers. Blacker and Padilla reached agreement on two other points. First, "all cases coming under the plain provision of the treaty to be acted upon without delay." Second, Mexico would have reasonable time to decide what to do about any of its own citizens demanded for extradition. Padilla also agreed to look into the matter of stolen Texas Ranger arms and ammunition.[27]

Blacker advised District Attorney James Zabriskie to file charges against those American citizens known to be leaders of the San Elizario mob and hiding in Mexico. Zabriskie, not relishing the assignment, had already submitted a letter of resignation. Blacker reminded Zabriskie that "he was district attorney still," and, urged him to begin making the "extraordinary efforts" needed to secure testimony.[28]

The judge spent nearly a week in Ysleta, the county seat, reconstituting local government and collecting evidence against seventeen insurgent leaders, all American citizens, he wanted extradited from Mexico. The evidence "to establish the crimes charged under the treaty" included statements from Lieutenant Tays and five other Texas Rangers. With "incontestable proof of their guilt" in hand, Blacker returned to Padilla's office on February 28 to demand delivery of the seventeen at a time and place of Padilla's choosing. Blacker was instead greeted by Rafael Barron. The *jefe politico's* secretary did not know where Padilla might be at that moment. All Blacker could do was leave his papers with the secretary and ask the interpreter to urge Padilla to take action as soon as possible.[29]

As the administrator of Atkinson's estate, U.S. Commercial Agent Solomon Schutz had personally experienced what he perceived as Padilla's lack of cooperation. "I know where [Atkinson's stolen] property is kept and in whose possession it can be found, but without the assistance of the authorities here I dare not try to recover the same as my life would no doubt be forfeited should I make such an attempt." Schutz judged that, "the only way to make Extradition in this case a success is for the Federal Authorities of the United states to make a demand for these criminals with a sufficient force of troops near at hand to cause just demands to be respected."[30]

The army's continuing force reduction at El Paso provided a clear sign that the Hayes administration was unlikely to threaten Mexico into extraditing indicted insurgents. Orders were issued on February 28 for two more companies to return to New Mexico, leaving Major Wade with just five officers and fifty men in three companies. A great deal of excitement was caused when rumor spread that the two companies were marching to Mexico to catch and bring back Paseños wanted for murder. Within twelve hours, an estimated two hundred Paseños "were under arms and posted along the river to repel invasion." They guarded the border until word came that the column of cavalry had passed El Paso and turned north for New Mexico. Captain Blair was also ordered to Fort Stanton, New Mexico, on February 28. Before long, his issues with integrity would come to the personal attention of William T. Sherman.[31]

CHAPTER 22
"Fearing that their statements might be published"

Governor Hubbard had delayed naming his representative on President Hayes's Board of Inquiry while the commission's scope of investigation and its financing were ironed out. When Washington refused to pay for Hubbard's man, the governor entertained the idea of withdrawing from participation. He wrote to Steele and Jones, then in Washington to testify before Congressman Schleicher's committee: "As the U.S. Government will not and Texas cannot pay a Commissioner for El Paso . . . would it be politic to send an officer of Texas forces? Please consult our friends—I don't want to be placed in the attitude of declining to appoint." Congressional "friends" urged participation, and before January was out, Hubbard designated Jones to serve on the El Paso board. His appointment could be said to have marked the opening of a new phase in the Salt War, one waged between federal and state officials to define the conflict and how it would be remembered.[1]

Jones's attitude of what had been at stake for Texas in its border troubles from Brownsville to El Paso was revealed in his congressional testimony. "Mexicans" of both nations "say that the United States have no right to that country anyhow; that it was stolen from Mexico. . . . [M]y opinion is that these Mexicans really look forward to the time when they are going to get the country back."[2]

Jones did not reach El Paso until after February 16. A small firestorm erupted when the major learned that King and Lewis had already collected evidence he and the governor considered well beyond the scope of the board's authority. One objectionable line of inquiry concerned the actions of Sheriff Kerber's posse as it swept down the valley. The most damning witnesses included Maria Núñez and Salomé Telles. Núñez recounted the invasion of her home that left her husband dead and herself lying on the floor bleeding from a bullet in her chest. Telles, supported by her brother-in-law, bravely admitted to being forced to "give up my person" to two brutes. Some Anglos were far less forthcoming on the atrocities. Isaac Campbell, who sat beside his brother John in the wagon carrying the unfortunate ranger prisoners Durand and Yrigoyen, claimed that "he knows nothing of the killing of them, but heard some shots."[3]

The board learned much, but gaps in evidence clearly existed. As Colonel King put it, "The Board met with noticeable reluctance on the part of many people [to] testify. . . . Some, fearing that their statements might be published after leaving the hands of the Board, and finding their way, through the medium of the press, back to this neighborhood, so excite local animosities as to lead to their ruin, refused to testify at all. Hence, also, the guarded language to be found in many of the statements made by intelligent parties."[4]

The board's examination into what the major considered the internal business of Texas put Jones in a difficult position. He telegraphed Steele, "Commissioners are investigating the conduct of county and State officials and Rangers since the massacre. Shall I join them in this?" In reply, Steele noted "The Commission was suggested and intended only to settle the question of invasion from Mexico or not. You will take no part with it in matters not relating to that subject or subsequent to order creating Commission." But if Jones held aloof from the investigation into the outrages allegedly sanctioned by Texas lawmen, the final report would inevitably be one sided and embarrassing to Texas. Since King and Lewis were determined to "investigate everything," Jones would "file protest but act with them and will introduce rebutting and explanatory testimony if possible."[5]

On February 27 he formally protested the taking and admission of testimony regarding events occurring in El Paso after December 17. The major identified four objections:

(1) Governor Hubbard had only asked for an investigation into the circumstances leading to the killing of five *Anglo* citizens of Texas by a mob, the subsequent carrying of property into Mexico, and the possible involvement of Mexico in these actions;

(2) the crimes committed by New Mexicans after the executions had no relevancy to the riots;

(3) civil courts were "fully competent to deal with all parties who have violated the laws" following the San Elizario riot, and Jones knew of "no tribunal or authority" the Board could officially report to that could do anything about such violations; and

(4) "the order of the President of the United States, dated December 31, 1877, which was made at the instance of the Governor of Texas, for the

purpose indicated above, does not contemplate or comprehend the investigation of anything which might occur subsequent to [December 17]."[6]

The colonels, granted broad powers of investigation and recommendation, declined to take action on any of Jones's points. King and Lewis were understandably interested in how the army's role in maintaining civil order at the Pass was complicated by the outrages committed by Texas lawmen and "volunteers" after the surrender. If Jones was to deflect the board's attention away from "the robbing of hen-roosts," as he sarcastically termed the colonels' focus, and back to rioting "Mexicans" and the failures of Mexican authorities, he had to do more than play impartial investigator. He had to argue the defense for Texas. He went to work immediately, contacting witnesses whose testimony would make his case.[7]

At the same time, Corporal Mathews penned another letter to the editors of the *Mesilla Valley Independent*. Two months earlier he had threatened revenge for the executions: "The Boys are wild, wooly and will be heard from." Now he publicly retracted these words expressed when he "naturally felt very sore. . . . We had been tricked into a surrender by a most shameful breach of trust: our comrades taken out and butchered and ourselves subjected to the greatest indignities." Now, "after calm deliberation," Mathews admitted that his threat was "wrong and out of place," since it tended "to convey the impression that we intended to resort to violent and unlawful acts." The Texas Rangers, he now asserted, never intended such acts. The letter's timing is strong evidence that Jones was behind Mathews' change of heart. The major needed the corporal's retraction to bolster his case that rioters, not Rangers, were the troublemakers who needed investigating.[8]

With Jones's protests out of the way, the board moved forward. On February 28, one of the most dramatic of the board's sessions was held on the outskirts of Socorro, at the home of Maria Núñez. Recalling Lieutenant Tays's testimony that the first shot in the "exchange" that badly wounded Mrs. Núñez and left her husband dead "was fired through the door," from inside the house, the officers examined the doors. "In none of the outside doors of the house could there be found any shot-holes, entering or coming out; but upon an inner door, across the *sala*, or opening into the kitchen, there are found about fifteen shot-holes, all piercing the door from the outside." Since no Ranger claimed that Núñez had fired a shot and then closed his door, there could be no doubt in the colonels' minds that the Rangers had entered the home and fired through the inner door without provocation, riddling Mr. Núñez's body with bullets, and wounding his wife as well.[9]

The board next took testimony in San Elizario and Socorro on a variety of points before reconvening in Ysleta on March 4. Prominent Anglos testified to the good character of Tays's Rangers. All witnesses agreed "that those offenses, such as ravishing women, taking people's property, and breaking doors and windows were committed by Lieutenant Tucker's men, and not by the Rangers under command of Lieutenant Tays." Merchant Moritz Lowenstein observed that Tays "keeps them under very good control."[10]

John B. Tays took the witness stand on March 5. His testimony smacks of strong coaching by Jones. The lieutenant presented his original orders as evidence in justification of his actions. When he took command, the major instructed him to "assist the sheriff in making arrests and escorting prisoners when called upon to do so." Tays testified that he and his men were acting under Kerber's orders in Socorro on December 23. "By his direction, certain houses were designated to be searched for arms and certain arrests to be made. . . . Neither myself nor my men, except when doing duty under the sheriff's instructions, had anything to do with the men from Silver City. . . . I ordered my men not to mix with them or have anything to do with them, because I knew a great many of them were bad men, that they were acting badly, and they didn't appear to be under any restraint." Tays's most difficult moment on the stand came as he was questioned about the invasion of the Núñez home. The lieutenant testified that much of what little he knew was only "what the men told me." Tays told the "Tucker men" to push the door open and search for arms, but he did not know "whether any men had gone in or not. I understood that all the shooting had been done through that door." Tays testimony makes little sense, unless he had given orders and then ridden off before they were carried out. If so, it was a dereliction of duty violating Jones's standing orders to "do all in his power to give protection to the lives and property of citizens against unlawful attacks or depredations from any source whatsoever."[11]

Kerber was next on the stand. Facing a panel in need of a scapegoat, the unrepentant sheriff hid behind no one. By the governor's grant of authority, he considered both the Rangers and the Silver City men "as subject to my order. . . . I had control of them and used them in making arrests and in searching houses for arms."As to the future, he urged that any rioters indicted be tried in another county, as not enough jurors could be found who could speak English. (Whether the accused would understand proceedings conducted in English was not his concern.)[12]

The board returned to El Paso (Franklin), Texas, where it focused on the questionable behavior of Capt. Thomas Blair. Blair appeared first, denying a claim by Caleb Miller that the captain had encouraged the insurgents on the evening of December 12 by telling them he had not come to fight them. Blair, said Miller, had shaken hands with a leader named Olguin and in the process lost his glove. Miller had not seen the glove, but he claimed his father-in-law had. Taking the stand, Blair admitted talking with some of the "mob," but in response to obviously pointed questioning, the captain repeatedly denied losing the glove. Finally, the interrogation got to the heart of the Rangers' charge against Blair. Had he shaken Olguin's hand, implicitly striking some kind of bargain that left Howard and his companions to their fate? The captain's response was ambiguous: "I cannot remember shaking hands with any of the mob at that time. I feel morally certain I did not."[13]

The next day, Tays was back under oath, this time regarding Blair's promise to come to his aid "in four hours time . . . without fail." The Ranger restated his unshakable belief Blair could have made it through to the barracks compound without the insurgents firing a shot. "Had he been there we could have held the place until assistance came, and more than likely would not have had to fire a shot." Rocky Mathews summed up the Ranger viewpoint: "I believe that a body of United States troops could have entered the town without meeting strong opposition. . . . I heard Chico Barela say that he wouldn't fire on the troops; they might have done it, but they said they wouldn't." The Rangers were certain: Blair was to blame for it all.[14]

A fresh item of evidence represented potential embarrassment for Major Jones. In September, before the insurgents' first arrest of Howard, a grand jury had brought to Judge Blacker's attention an allegation that Guadalupe Carabajal was not a U.S. citizen, "and therefore is disqualified from holding his present position as said justice of the peace." Carabajal was the JP whom Jones had prevailed upon to release Howard on bail without any examination. The board would revisit this issue.[15]

On March 9, the panel read into the record a lengthy letter from Judge Blacker, who took pains to defend Howard, Kerber, and the Texas Rangers. The first was a "gentleman of culture and good address" who exceeded the bounds of professional propriety only under great provocation and "many extenuating circumstances." As for Kerber, "a brave, skillful, and honest officer. . . . he was at first alone, and what mistakes he made were made in an honest endeavor to re-establish law and order." The Texas Ranger unit was necessarily created "in haste, under great excitement, and was necessarily just what it was. There was no other material." While the Silver City men "committed some fearful excesses and felonies," the Rangers were "orderly, disciplined, and law-abiding." The jurist also complimented Hubbard for dispatching "to the scene of trouble in its very incipiency one of his most prudent, sagacious, and cool of his military family," Major Jones.[16]

Blacker also addressed the question of *why* the Paseños rose up. They believe in their "inherent, inalienable right [to] 'to rise,' overthrow the government, and put in other officers." Blacker then fell back on the old standby argument: The people were sheep. "If they have a good man to lead them, there is not a more pacific, easily governed, and loyal people on the face of the earth than the Mexican people of El Paso County. If they have a bad man [he meant Cardis or Borrajo], they will be just as bad as he wants them to be."[17]

A brief session on March 11 was devoted to the issue of Howard's release on bail. Hague asserted that JP Carabajal had erred in allowing Howard to go free on bail without an examination held in the presence of an attorney for the state. Only days before, Blacker declared otherwise. In the judge's opinion, if the arrested party waives examination, "and the justice knows in any way that satisfies him that it was a sudden killing, the practice is to admit to bail. . . . This is law and the practice." After such conflicting opinion, the board let the matter die. After one last evidence gathering session on March 15, when Jones submitted several documents, the board adjourned, giving itself just twenty-four hours to issue its final report.[18]

The report was adopted on March 16 by a vote of two-to-one, Major Jones dissenting. "The causes of the disturbance," declared the colonels, "are local in their character, directly issuing from a disagreement or personal feud long subsisting between two prominent and influential men in the community . . . which grew into deadly hatred, culminating in the violent death of Cardis at the hands of Howard." In prose that should have satisfied Kerber, the board maligned the Paseños as "an ignorant and hot-blooded race," ready for vengeance. "These avengers of blood brought in their train a mongrel following of thieves and robbers, birds of prey scenting the quarry from afar, whose deeds of rapine entailed the most serious consequences."[19]

The exact strength of the insurgent force the board found "difficult to compute. . . . but there could not have been far from 400, of whom not less than one-third were Mexican citizens." The board found no evidence substantiating the report of organized drilling of insurgents by Mexican army officers, other than ex-Lt. Ferris Lermo. As for Mexican authorities, they had sympathized with their American counterparts, but were powerless to prevent their citizens from crossing over. The board noted that Colonel Hatch acted promptly to halt violence. Not knowing what to make of Captain Blair, they neither praised nor rebuked him, noting only that his actions were "circumstantially set forth in the report."[20]

The members were less kind to Judge G. N. Garcia. "The county judge, though an educated man, is addicted to drink, and is frequently unfitted for business by reason of intemperance." The board was spurred to make their critical remarks by Garcia's failure to answer their summons to testify on three separate days, "the assigned cause being drunkenness." Singling Judge Garcia out in this manner was inappropriate. The magistrate gave extensive testimony on March 8. Whether or not his courage came from inside a bottle of El Paso brandy, in taking the stand the judge showed more fortitude than the many people who feared to testify at all.[21]

Despite the conflicting legal interpretations offered by Hague and Blacker, and despite the fact that neither King nor Lewis was a trained lawyer, they averred that "the death of Cardis at the hands of Howard seems to have been a premeditated murder, a crime for which there is no bail provided by law. . . . Yet Howard . . . without being confined or examined in the presence of a prosecuting officer . . . was suffered to give bail and go free."[22]

After weeks of testimony and a small mountain of evidence, the Board made two recommendations. The first was to reestablish a permanent military post in the county, to be garrisoned by two hundred troops, including fifty cavalry, some of whom would remain in San Elizario so long as it took "to restore confidence to the well-disposed people now residing there." If Fort Bliss had not been abandoned, "the late deplorable events would not have transpired." The board also thought that a military force would be handy when growing competition between the two nations for scarce irrigation water led to "trouble beyond the reach of diplomacy to settle." Second, the Board recommended that the U.S. government press Mexico for extradition of criminals and reparation for damages caused by Mexican citizens. Colonels King and Lewis concluded, "the salt war is probably ended," and at 6:30 P.M., they adjourned *sine die*. Their business ended, King and Lewis left the county immediately.[23]

There was one more item of business. Jones was permitted to file a minority report and to add further testimony on the damages done by insurgents. Jones demurred on ten points, including the citizenship of several insurgent leaders, notably Sisto Salcido, J.M. Juárez, and Macedonio Gandara, whom he claimed were Mexicans. He denied that the October "mob" was only fifty or sixty strong, pointing to evidence that they numbered up to four hundred. To the board majority's finding that the December "mob" possessed an "imperfect organization and little discipline," Jones shot back, "The record shows that they had recognized leaders, and moved in companies and squads commanded by captains, sergeants and corporals; that they kept out guards and pickets, etc." And in the second and last time any Texas Ranger relied upon the word of Thomas Blair, the major quoted the captain's report of December 15: "I found their force to consist of what I estimated at 350 sober, well-organized, well-armed men." Jones vehemently disagreed that the Guadalupe lakes had been used for many years, citing testimony that they had only been discovered in 1862. He took exception to the board's finding that "civil law in this county is loosely administered." He also felt it his duty "to repel the extraordinary opinion of the majority that occurrences [of Kerber's advance] were not excelled by the atrocities of the mob itself, by simply saying that the parallel is not justified by the evidence, and is far-fetched and uncalled for." Finally, Jones disagreed with the board's opinion that Howard should not have been released on bail. In mentioning "the many secret plots and contrivances of Cardis against Howard's interests, and the efforts made by him to procure Howard's assassination," Jones alluded to Howard's innocence of murder under the so-called "Texas doctrine."[24]

Jones placed blame for the tragedy on Cardis, Borrajo, Blair, and the Mexican authorities. The major assessed the stolen property at $31,050. Finally, he concurred in the majority's recommendations, urging the U.S. government to take steps requiring the Mexican government to either bring to justice and punish the 150 Mexican nationals in the San Elizario mob or deliver them for trial in Texas. With that, Jones signed his report and forwarded it through military channels to Washington, DC.[25]

The nation lost interest in the fuss along the border. On April 8, 1878, John W. Foster, the state department's representative in Mexico, extended the U.S. government's formal recognition of the Diaz regime. Though border lawlessness and trade disputes would continue to engender friction, recognition ended any chance that Hayes would try to force Mexico to give Texan civil authorities the redress they demanded. As far as the United States was concerned, the El Paso Salt War was over.[26]

CHAPTER 23

"The general state of the county is not improving much."

The Hayes administration might close its books on the Salt War, but for Texans, Anglo and Paseño alike, uncertainty and complications remained. Angry men kept tensions high, their emotions barring the road to resolution. If estimates of the insurgent force were correct, over two hundred men were still on the Mexican side of the river. Some had families with them; others longed for the opportunity to rejoin loved ones on the American side. Meanwhile, officials looked for ways to bring these "criminals" to justice. In the March term, the grand jury presented 274 bills of indictment, including 119 for murder and the rest for accessory to murder. Should the insurgents be caught or extradited in large numbers, the sheriff lacked a jail to throw them in. Another peacekeeping problem arose from the wreckage of the December fighting. Some New Mexicans, "a band of thieves, . . . so strong as to defy all authority," remained in El Paso. They did as they pleased, intimidating all. Within the year, they threatened to become the law.[1]

Anglos feared that the Texas Ranger unit would be disbanded, but Hubbard decided to keep them in service. As General Steele advised Texas Supreme Court Chief Justice Oran M. Roberts, "The trouble is not yet ended and it may be necessary to shed more blood." Aware of the potential legal and political consequences if the Texas Rangers again went beyond the bounds of proper law enforcement, Steele expressed his concern to Roberts: "[If] called upon to use force to sustain the rights of our citizens, I would very much like to know that those rights had the full sanction of law."[2]

Only four survivors of the siege reenlisted when the detachment was reconstituted on April 1. The new First and Second Sergeants were ranch hand Tom Zickefoose and former schoolteacher-turned-stagecoach driver Asberry Ryall. The reformed unit included no Paseños, but it did include two new Rangers surnamed "Tays." El Paso left John Tays's nephew George unimpressed: "A very dreary looking country. Surrounding hills and high lands all bare and dead looking." Young Tays was "pretty lonesome at first," so joined the Rangers. He invested ninety-four dollars in a Winchester carbine, a Colt 45 revolver with belt and eighty cartridges, a .44-caliber pistol with belt and scabbard, a new hat, spurs, saddle, blanket, horseshoes, canteen, and rope. He saved sixty-three cents by splitting the cost of a shared comb and brush. George's "Uncle Jim" Tays took "Uncle John's" place in command. John Tays was about to marry the very eligible widow Amelia Antonetta Rohmann St. Vrain. She and her widowed mother, Amelia Rohmann, owned property that Tays would build into a small real estate empire. Major Jones accepted his resignation on March 25.[3]

As the first commander of a makeshift outfit, John B. Tays had initially done well. His defense of the Ranger position over five days revealed leadership, tactical ability, and courage. Then one colossal blunder cancelled everything. He allowed himself to become separated from his men, precipitating the entire company's surrender. Six days later, in blind obedience to orders, he allowed his Rangers to provide the muscle for Sheriff Kerber's outrageous tactics from Ysleta to Socorro. Afterwards, in testimony, he temporized with the truth about what happened. Major Jones expressed no fault with Tays. In his mind, any outrages committed by the state's peacekeepers, while regrettable, were understandable. Perhaps the best evidence that Jones was satisfied with the conduct of John Tays is his appointment of brother James Tays as the next commander of the detachment.

Major Jones departed El Paso once more, never to return. Before the summer was out, Jones scored a major law enforcement success by breaking up the Sam Bass gang. In January 1879 he succeeded William Steele as adjutant general of Texas. Afterwards, he continued to exercise his extraordinary administrative and strategic skills, effectively and efficiently commanding an overstretched force on a shoestring budget. Among his many successes, he could not count his handling of the Salt War. He had done his best to prevent local residents from slipping into violence, but it was not good enough. He had made errors, particularly in handling Howard with kid gloves and in not properly instructing John Tays. He was not the villain, but neither was he the hero.[4]

The existence of Lt. James A. Tays and his nine-month command of the Rangers in El Paso was for decades completely missed in Ranger histories. The new lieutenant was age forty-one, nearly six feet in height, with brown hair, brown eyes, and a fair complexion. He listed his prior occupation as "farmer," but he tended cattle rather than vegetables. His correspondence reveals a rudimentary ability to spell, an optimistic view of his worst men, and a tendency to make excuses. Whatever his flaws, Jones believed him ready to command.[5]

Tays's first assignment was to cross the Rio Grande into Mexico, opposite Ysleta, where an American had been killed. Considering it his Christian duty to see the man buried, Sheriff Kerber led ten Rangers, including an "advance guard in proper military style." They found the corpse stripped of clothing, with the head split open "by the stroke of an axe." The posse gave the dead man a hasty burial and then returned to the American side. The sheriff, Tays, and ten Rangers returned the following day with a coffin. They discovered the body dug up "and otherwise abused." The Americans collected the corpse and quickly headed back across the river, just ahead of sixty armed men headed, according to that atrocious speller Tays, "by the netorious Senorean who ledd the mob in this County." These dangerous forays into Mexico prompted Tays to ask Jones to "define the duty you want the detachment to do and if we are to be soly under the orders of the Sherif or follow the instructions in Special Order 110 of Nov 16th."[6]

C. L. Sonnichsen and some other historians have claimed that, once they finished rioting, the Paseño population meekly lined up to buy their salt. Not all did. The economy of *la Isla* was still a wreck. The El Paso county wheat crop, owing to the shortage of men to plant and tend the fields, was estimated in February to run eight to twelve thousand *fanegas* short, a condition that promised a continuing lack of income to pay the accumulating debts of Paseño farmers and families. Even if farmers and laborers were to return overnight, the half year of turmoil and neglect undoubtedly left their *acequias* in disrepair. It is likely that farm animals were missing, eaten, bartered for food or other needs, allowed to wander, or stolen by the outlaws now infesting the valley. By contrast, farmers in Dona Ana County, New Mexico, including some recent immigrants from El Paso, took advantage of the favorable weather and intact *acequias*, "devoting an unusually large portion of their lands to wheat plantings." It remained to be seen if the corn crop would go the same way. Because of the war, "the refugees are said to be in a starving condition on their side."[7]

Starving refugees could not remain in Mexico indefinitely. Unfortunately, any return at this time might lead to a prison cell or the end of a rope. On May 4, Padilla finally responded to Judge Blacker's request for the extradition of 157 Mexican citizens. He credited his "want of a competent translator" for the six-week delay. Now, he stalled again, referring the decision to state authorities, but he also signaled the eventual result. There were two hurdles, he advised. First, neither nation was obligated by the Extradition Treaty to extradite its own citizens. "[As] the individuals asked of me are of Mexican origin and it does not appear that these individuals have removed their nationality," Padilla believed extradition unwarranted. Second, the treaty did not contemplate extradition for "crimes or derelictions of a purely political character." This position echoed, from a far more sympathetic viewpoint, the Texan view that the Paseños had risen up against the American system of law and government. The *jefe* cautioned that the crimes "may be of this character."[8]

Padilla's response demonstrated the difficulty inherent in claiming that Mexico and her citizens were chiefly to blame for the "riots." The U.S. government had no intention of militarily or otherwise coercing extradition, and Texas could not expect her neighbor to voluntarily surrender her citizens when the treaty entertained no such obligation. Kerber and John Tays advised the *Mesilla Valley Independent* that "many of the rioters who fled . . . to escape punishment have taken the oath of allegiance to the Mexican government." The disgusted editor informed his readers, "Our extradition treaty with Mexico is a farce."[9]

On May 13, the Texas governor issued a proclamation offering a reward of $250 each for Barela, Salcido, Granillo, Apodaca, Luciano Fresquez, Jesus Garcia, and possibly Agatón Porras for murder. Kerber and Lieutenant Tays would have to catch "the scoundrels" before either justice or rewards could be meted out.[10]

For the sheriff, it was also a matter of self-preservation. Tays reported "quite an excitement" throughout the county as a result of "threatening communications being sent from the Mexican side that they were coming over to clean out the Americans." Large mobs were said to be assembling nightly, and "on the other side the scarcity of provisions among them is making them desperate." Tays identified Kerber as "the first man they will go for." Raiding parties twice fired shots through the windows of Kerber's Ysleta home. No one was hurt, but since Kerber didn't scare his luck would have to hold. He was not the only one threatened. The *Mesilla News* advised, "Word has been sent . . . for all Americans to leave, if they wanted to live."[11]

If Hubbard had issued a "most wanted" deck of cards, Desidario Apodaca would certainly have been an ace. On June 15 he was in hiding six miles below San Elizario, on the river-bottom *bosque,* among the tangled thickets and tall grasses. He had not counted on Pablo Mejia, now a deputy sheriff, who guided Sergeant Mark Ludwick and five other Texas Rangers to his hiding place. Apodaca was taken completely by surprise and without a fight.[12]

Word spread that friends would attempt his rescue. Kerber wired Judge Blacker that "1,000 Mexicans are en route to and drilling at El Paso, Mexico, and have petitioned the Chihuahua Government for permission to cross over and rescue Desidario Apodaca and disperse Texas State troops." Not content with arresting murderers, the sheriff intended to round up the parties charged with "False Imprisonment," including those who had detained him the previous October.

Disgusted at the "corrupt interest" he saw on every side, Father Bourgade was ready to stop playing the intermediary. But with Kerber's ham-handed approach keeping tensions high, the priest made one last plea to Jones. He begged the major to convince local lawmen not to make any arrests unless Mexican authorities first delivered wanted men from the other side. "Inasmuch as those who still live on this side for the most part are not the true culprits. . . . I want justice to be done, crime punished, but also the innocent saved and those who opposed or discountenanced [*sic*] the last local rebellion protected in their lives and interest, and I again call on you to exert all your influence and the means in your power to that effect."[13]

Heedless of the criticism, Kerber attempted to arrest suspects as they exited Sunday mass in San Elizario. Word spread and the effort failed. Tays advised Austin, "The Priest is highly displeased with the sheriff for the attempt to arrest his flock. Report says that another rising of the people is eminent [*sic*]."[14]

Meanwhile, Apodaca allegedly began to talk from behind bars, stating "positively that J. P. Hague encouraged them and promised to defend them if they got into trouble and that Shutz [*sic*] the Old Man [i.e., possibly Joseph Schutz] offered $300 and two kegs of powder if they would kill Howard the powder to blow up the Rangers Quarters." Kerber brought the accusations to Blacker's attention, but no charges were brought. One more "murderer" was handed over to the sheriff in June after a dramatic confrontation along the Fort Quitman salt road, reminding all that some *salineros* still refused to pay for "Zimpelman's" salt.[15]

That capitalist had hired John C. Ford to replace Howard as salt agent. In later testimony, Ford stated that "the Mexicans have applied to him for permission to get salt, and fully recognize the fact that the lakes are owned by said Zimpelman." C. L. Sonnichsen was the first to rely upon Ford to claim, "The salt question troubled no more . . . the Mexicans applied to [Ford] with great politeness for permission to haul salt." Others have drawn similar conclusions, ignoring the second part of Ford's statement, that "some Mexicans from the other side of the river had gone to the lakes and taken from them five wagons and one cart load of salt for which they paid nothing. This was about a week ago."[16]

The incident referred to by Ford was not isolated. On or about June 13,

Mexican citizens obtained a pass from the U.S. Customs House in Paso del Norte to cross the river and haul freight to Fort Davis. Instead of checking in with customs at San Elizario, they crossed farther down the river, fifteen miles above the abandoned Fort Quitman. From this remote point, they headed up the fort road to the salt lakes. Word reached Zimpelman, who obtained an arrest warrant. Corporal Mathews and four Rangers rode out to intercept the wagon train. Mathews caught the returning salt train on June 24, about eighteen miles past the old fort. Some *salineros* escaped, but the Rangers took four into custody, together with forty-one oxen and five salt-laden wagons. The oxen had been driven so hard that Matthews had to leave four wagons behind for later retrieval when he headed back to the abandoned fort to water men and beasts. Mexicans who escaped capture harassed the Ranger party as they slowly made their way to the fort. Here the corporal got word from a passing stagecoach that an armed party of Mexicans had stopped it, inquiring about the Rangers' movements. Mathews gave the driver a message for Tays to send help. He also freed one *salinero,* instructing him to deliver a warning that the other prisoners would be shot at the first sign of an attack.[17]

Tays dispatched five Rangers under Sergeant Ryall to relieve Mathews' group. Ryall, leading extra horses, made a forced march to Quitman. After linking up with Mathews, he pushed on to the abandoned wagons, hitched them up, and drove them back down to the old fort. Here the salt was delivered to Ford. The entire party, now under Sergeant Ryall, rolled up the sixty miles of river road to San Elizario. The trip took several days, during which time small parties of Paseños harassed the little column with gunfire. Exercising "the greatest vigilance," Ryall succeeded in keeping the *salineros* at a distance. He brought the train into San Elizario on June 29, losing a few oxen in the process. Once the salt was unloaded by Ford, the empty wagons were turned over to the local customs agent on charges of "breach of Revenue." One prisoner identified as one of the salt war "mob," Antonio Salazar, was turned over to the sheriff to be arraigned for murder.[18]

The seizures endangered the fragile peace. The Customs House officials were threatened by a crowd who attempted to take the wagons. The army refused to intervene. Tays reported to Jones, "From the excited state in which the Mexicans seem to be it is feared that we will have hot times up here but I am not afraid but we can give them a stand off for a while." The Rangers stayed close to San Elizario for the next few weeks, guarding the prisoners closely. "Since the Salt scrape there seems to be a Settled determination to have a row by the Mexicans."[19]

Seizures of salt trains did inspire some *salineros* to pay up. Under the terms of a handshake agreement, Ford was to collect fees for two years and turn half the proceeds over to Zimpelman. "Ford was taking his life in his hands every time he demanded and collected royalty on the salt. But Ford was all nerve and he stuck to his contract." At the end of one year, the two men met in El Paso to transact their business. Ford recalled, "I had a sock full of gold and silver and a roll of bills. And all of it was money I had collected from the salt haulers. There was, near as I can remember, about $2800 of it." When Ford presented Zimpelman with his bill for fourteen hundred dollars, the major choked, allegedly responding, "Well that agreement was not in writing and you cannot prove it." He then revealed that he had hired another man [Kerber] to collect the fees for just 25 percent. Not one to be maneuvered out of his earnings, Ford flatly told Zimpelman he would "keep all the money, half of it for each of the two years for which I had a contract to take care of the lakes, and I did keep it." Later on, so the story went, "Maj. Zimpelman laughed at what he called being held up by John Ford."[20]

The Salt War increased the popularity of the Pass as a hangout for murderers and bandits. One observer complained that local lawmen "are so afraid of their lives they dare not complain [about] drunken desperadoes of every State" flocking to the area. On May 11, Lieutenant Tays reported, "Horse stealing is of nightly occurrence in El Paso Texas there is a villainous mob gathered of the most desperate kind of men from all parts of the county all of them Americans and I fear trouble with them." Five days later, Tays complained, "The general state of the County is not improoving [*sic*] much. There is a mob of outlaws located around El Paso Texas & Mexico that threatens us with trouble." Tays labeled these bandits "the Kenny band."[21]

John Kinney had settled down in El Paso, taking on the cloak of a respectable businessman. His Exchange Saloon must have been a social and business center for the criminals exchanging "title" to stolen goods and information about new "markets." About this time, Kinney also became an El Paso deputy sheriff. The sheriff's continuing reliance upon Kinney's support was an unhappy development. Their alliance threatened to steer the county into a new and deadlier form of civic corruption, one that could menace the economic well-being and physical safety of Anglos and Paseños alike if left unchecked.[22]

While Kinney played lawman and saloonkeeper in El Paso, his associates kept to their old ways. Two "Kinney men" attempted a daring daylight armed robbery that was thwarted by the son of Civil War general Benjamin F. Butler. The leader of the two outlaws was Robert "Dutch" Martin. His criminal career included association with a veritable "Who's Who" of southwest bad men, including Kinney, Jessie Evans, Curly Bill Brocius, Jim McDaniels, Pony Deal, and, briefly, young Henry Antrim. He stole horses, robbed stagecoaches, and shot it out with posses. His companion in this holdup attempt was Curly Bill, then sporting the surname Bresnahan.[23]

The man who defeated the attempted robbery was 2nd Lt. Ben Israel Butler of the 9th Cavalry. His father had insisted on a West Point education and "a lieutenancy in a regiment of colored troops which was stationed on the Plains." "Every preceding generation in this country has had its war," the old general believed, "and there could be no war in which a son of mine especially would not take a part in his generation."[24]

On January 17, 1878, seven months after receiving his commission, Lieutenant Butler arrived at the Pass. After three months in the "dirty, dusty, Mexican-inhabited villages . . . monotonous and tiresome in the extreme," without comfort or society, he tendered his resignation, to take effect on June 1. Around 2:00 P.M., on Tuesday, May 21, Butler (looking "bright and happy"), Charles Johnson (a trooper in the Ninth), and driver George Shakespear (no *e*), boarded an army ambulance that would take them to Las Cruces. From there Butler would catch the stage to the railroad depot at Santa Fe.[25]

Lieutenant Butler's war occurred two hours later. About ten miles up the road, two men on horseback passed the ambulance. It was speculated that the two "probably imagined that [Butler] would have considerable money with him" for the journey home. As the ambulance turned a corner "on a sharp trot," two masked men jumped out, brandished their weapons, and demanded a halt. When the driver could not halt his mules, "the miscreants immediately blazed away," peppering Johnson's lungs, gut, and hip with buckshot, and putting a pistol ball through Shakespear's body. Other balls passed through the wagon's side, curtains, and door rail. "As the wounded men fell back in the ambulance, Lieut. Butler grasped a carbine from the hands of one of them and sprang to the ground, whereupon the cowardly assassins fled through the brush and made their escape."[26]

Sergeant Ryall and four Rangers, arriving a few minutes after the attack, gave chase into Mexico. Authorities there arrested Martin, Curly Bill, and a compatriot, Buckskin Joe—the last two described as "as ugly a pair of jail-birds as one would wish to see"—and on May 22 turned them over to the U.S. Army. For Tays, the crime was fresh evidence that "The Lincoln Co. outlaws are coming down pretty plentifull [*sic*]." For Lieutenant Butler, the adventure marked his last day on duty before he returned to Boston, carrying the harrowing tale of his "most narrow escape" from dangerous Wild West bandits.[27]

Within days, Kinney himself departed El Paso to enter the Lincoln County War on the side of the "Murphy-Dolan House." The disappearance of Kinney and his "Rio Grande Posse" was immediately noticed. Tays reported, "The mob of desperadoes that was round El Paso Texas have all left for Lincoln County," adding a month later, "Franklin is perfectly dead." The calm was temporary; Kinney would be back.[28]

Meanwhile, exiled Paseños still waited across the Rio Grande. Tensions were bound to increase as two potential flashpoints for violence drew near: the opening of district court and county elections. Sheriff Kerber worried about a thousand Mexicans crossing to rescue his prisoners, but he did little to ensure against the simplest of escapes. On August 27, his intrepid deputy, Pablo Mejia, arrested escaped convict Antonio Olguin. The outlaw was placed overnight in the old jail at San Elizario for transfer the next day to Ysleta. "The same night a band of Mexicans headed by Jesus Garcia crossed the river, broke the jail, tearing it completely asunder, rescued the prisoner, escaping safely to the other side." Garcia had risked much: he was one of the six insurgents with Hubbard's price on his head. According to Bryan Callaghan, the future mayor of San Antonio, the sheriff's negligence was to blame. Olguin had been "confined to jail and locked with a 50¢ padlock and left without guard."[29]

The daring delivery of Olguin left several important prisoners still in state custody, including accused murderers Desidario Apodaca and Antonio Salazar and highway robbers Bob Martin and Curly Bill. They were held in Ysleta under the watch of the Texas Rangers. The burden worried Tays. "The Mexicans have been more uneasy of late than I have ever seen them before they are buying all the arms and when all are assembled they will make a cleaning out of the Gringoes [*sic*]. This is rumor. I think it necessary to use double diligence for the Court term."[30]

The rumors also alarmed Paseños. Judge Blacker reported that "the remaining Mexican population" of *la Isla* had again fled "through fear." According to a newspaper tipster, "The county is almost depopulated as nearly all the Mexicans have crossed the river for fear of arrest during court term or else avoiding the necessity of giving evidence." Father Bourgade had earlier warned against the calamitous effect of threatening to bring everyone involved in the "last sad troubles" before the court. "The consequence will be a general stampede at the time of the year we need most all the hands to irrigate and Keep the water running in the acequias in order to sow corn. I do not know but our situation in San Elizario is very critical it seems as if this was a conspiracy by each and every one to ruin that yet so much distressed village."[31]

The exodus came just as the local economy was beginning to rebound. The *Independent* reported in June that "crops in El Paso county look well and more than an average yield to the acre may be expected. Owing to the late El Paso 'war' much less than the usual amount of land is planted, but notwithstanding this fact, a large surplus of grain will be raised." By the first week of September, fruit crops were reported to be "splendid." These might be the last bountiful crops harvested for some time.[32]

Despite the scarcity of potential jurors and witnesses, court opened as scheduled. The state's cases were handled by a new district attorney pro temp, twenty-four-year-old Tom A. Falvey. Highwaymen Bob Martin and Curly Bill were found guilty of assault with intent to kill. Blacker sentenced them to five years in the state penitentiary at Huntsville. The convicted men remained in Ranger custody pending their appeal. Mejia settled a two-hundred-dollar contract dispute with absent insurgents Salcido, Zambrano, Guillermo Gandara, Sotero Beltran, and Jesus Sierra by securing a writ of attachment against thirteen of Beltran's oxen.[33]

Of greater concern to the people El Paso County, watching from either side of the Rio Grande, were the fates of Apodaca, Salazar, Hipolito (or Pablo) Telles, and any other Paseños charged with offenses connected to the "riots." But one week into the session, Tays observed, "The murder cases cannot be tryed [*sic*] for want of Jurymen only 17 can be found." The same was true a week later. Finding twelve among this small pool who could survive the attorneys' challenges was proving impossible. And with court in session, a time when peace officers usually rounded up criminals to haul before the court, Tays reported, "The Sheriff makes no attempts at arrests."[34]

The year 1878 was an eventful year for Capt. Thomas Blair. In February he took command of the 15th Infantry's Company H at Fort Stanton, under Col. Nathan A. M. Dudley, "one of the stormy petrels of the U.S. Army." In June, Blair wrote to a young lady in Scotland named Rachel Nicholl, warning her that an expected reduction in the army's strength might force him to quit. He ended the letter, "I hope your mother is better now." Blair left Santa Fe in early September on yet another extended period of leave. From St. Louis he placed in the mail to Anne Nicholl of Glasgow his approved letter of resignation from the army. Blair appended a note to Anne, "What the future may bring I cannot tell any search for me will be useless. I will not give you my address but shall from time to time as I prosper send you something."[35]

Three months later, on September 17, Blair was in Lancaster, Kentucky, where he married Maria Letcher Granger, the widow of Blair's old colonel. Only thirty-three years old at her husband's death, Maria lived for a time in Washington, "and, being a lady of liberal education and fine personal appearance, soon became a favorite in social circles." She and her small son survived for a time on her small widow's pension and a job her "warm friend" General Sherman probably secured for her in the quartermaster general's office. Her future brightened considerably when Cornelius "Commodore" Vanderbilt died in 1877, leaving her railroad bonds valued between ten and twenty thousand dollars. The bequest "placed her in circumstances far above want." "Suitors for her hand [were] numerous," but it was the charming captain who "laid siege to the widow's heart [and] succeeded in winning her affections." The happy couple honeymooned throughout October, ignorant that their world was already collapsing. Mrs. Blair had no knowledge of Anne Baillie Nicholl. Their bigamist husband, who had never resigned his commission nor obtained a divorce, was unaware that Anne and the commanding general of the army were already corresponding.[36]

In the years since the Civil War, the ballots of El Paso County had been tossed, stuffed, ignored, multiplied, lost and, occasionally, counted. In 1878, the issue was not ballots, but voters. Would they show up at the polls? If they did, would they be intimidated, get arrested, or fight back? The offices up for election included Cardis's seat in the state legislature and key county positions, including sheriff. Following the deaths of Cardis and Howard, the struggle was wide open. With more than a hundred residents indicted for murder and the sheriff allied with the leader of the criminal element, the stakes were higher than they'd ever been for Paseño and Anglo alike.

Election campaigns were underway as early as May. Father Bourgade feared the political dispute "will be the cause of some uneasiness and unpleasant excitement as the contest will be between two extreme parties. Those who are most bitter against the Mexicans indiscriminately and the others who see their best hopes in that considerable element which was implicated one way or another in the last sad troubles." He expected campaign tactics to sink to their usual low level.[37]

The "bitter" party lacked an effective leader. Judge Blacker might have taken Howard's place, but he had recently relocated to Tom Green County. Zabriskie, a proven vote getter, already saw his future in the territories. Zimpelman was a natural leader but was now more interested in building his "portfolio" and in protecting his investments. That left the incumbent sheriff, but Kerber's excessive zeal made him a liability.

Cardis's shoes were even harder to fill. No Anglo could ever gain the complete trust that so many Paseños had placed in him. As for Paseño leaders, the men who had championed self-government in the autumn of 1877 were now exiles. Those who, like Captain Garcia, had adhered to a law-and-order stance still retained support within *la Isla,* but Anglos would never accept them as party bosses. James Price Hague came closest to filling the bill. He was sympathetic to the insurgents' cause and may indeed have offered them the advice that Mathews, Apodaca, and others accused him of doing. Now he was running for legislative representative. He had strong support on *la Isla,* but the district was large, and he would need votes from other counties. James B. Leahy, the man McDaniels advised to leave the state, was the frontrunner for county judge and could be placed in Hague's party. For county clerk, George Zwitzers opposed Kerber's friend G. W. Wahl.

The race for sheriff evoked the greatest interest. This heat presented a special opportunity for Paseños to redress grievances and provide themselves a measure of safety. Cardis had failed on several occasions to wrest this critical office from his enemies. Now Kerber was vulnerable. Bryan Callaghan, already showing the savvy that would make him an effective political boss in the Anglo-Tejano city of San Antonio, explained the sheriff's plight: "Mr. Kerber is a good man, but his acts of late have made him very unpopular—the Mexican people are unanimously against him and he is very much dislike[d] by great many of the Americans. Kerber committed grievous blunders in the discharge of his official duties."[38]

Campaigning was said to be "running high" by mid-October. Hague and Leahy looked like shoo-ins, while rumors of who would run against Kerber abounded. Tays heard that "a Mexican is going to oppose Kerber," but Blacker warned the governor that the lieutenant was "using undue influence with Mexican mob to become Sheriff of El Paso Co." Tays, it was said, planned to defeat the incumbent with a Paseño candidate, and then step in when that winner was disqualified. Affronted, Tays denied "mixing up in Politics [or] in favour of telling these people who ran away, return and vote."[39]

Meanwhile, Blacker advised Steele that some candidates were planning to flood the ballot boxes with the votes of Mexican nationals. Steele warned Tays, "you will be especially on the alert on that to aid in the arrest of such as may be indicated. San Elizario will probably be the place of voting." Tays thought the report "without foundation."[40]

Sheriff Benito Gonzales. Courtesy of El Paso County Sheriff's Office

The Paseños did find a candidate of their own to oust Kerber. He was Benito Gonzales of Ysleta, tainted neither as a "rioter" nor as an Anglo puppet. Gonzales was not a fence sitter however, but a man with a strong record of community service. He had been mayor of Ysleta and a private in Lt. Telesforo Montes's 1874 Texas Ranger company. Now, in his early forties, he sought to become the county's first Paseño sheriff since Charles Ellis replaced Juan Armendáriz in 1871. His supporters included James Tays.[41]

Though Kerber looked like he could be handily beaten, this was El Paso, where ballot counting was the least important part of any election. This year, it began to look more like a matter of counting guns, Kinney guns. The president's proclamation of martial law had made New Mexico too hot for desperadoes like Kinney, wanted on charges stemming from the Lincoln County War. Now he and his "Rio Grande Posse" were back at the Pass. They operated on both sides of the river, making life difficult for ranchers, farmers, and merchants alike. Tays reported "cattle thieves constantly round [and] make times lively for us," and, in another report, "the Jews [i.e., merchants such as the Schutzes and Moritz Lowenstein] are scared badly as the roughs along with Kinney are making threats on their stores." These were the same shop owners who had been taken for over twenty-five hundred dollars by some of these same men when they were "Silver City volunteers."[42]

On the evening of November 2, three days before the election, Kerber suffered an embarrassing setback. Murder suspects Apodaca, Salazar, and Telles, as well as convicted felons Martin and Curly Bill, escaped. Kerber had turned them over to the Rangers and then, said Tays, forgotten them. The men were kept in a one-room structure with one door and no windows. Families were allowed to visit their incarcerated relatives, and the lax security enabled someone to actually sneak the proverbial hacksaw inside. In the darkest corner of the dark room, the prisoners filed through their manacles and dug their way out under the adobe wall. They were missed within minutes but too late. The lieutenant blamed "the negligence of the Sheriff." Tays dismissed the notion that his own men were responsible, though "inside men" plagued his unit throughout 1878. The prisoners may have been helped by young Rumaldo Salazar. According to family tradition, the boy smuggled in a tunneling tool in the shape of a knife inside a lunch pail. Salazar made his way to Socorro, where a horse was waiting.[43]

As the election drew near, Steele warned Tays to be alert to an invasion of Mexican "murderers" seeking to cast ballots. The lieutenant properly saw Kinney and his thugs as a greater threat to a lawful and orderly election. He took seriously the rumor that "an attempt would be made by a party headed by Kenny [*sic*] to run the Election their own way." Kinney had in fact registered to vote in El Paso County, as did gunman George "Roxy" Rose and other Kinney allies. But the bullets these men might cast, not their ballots, promised trouble.[44]

The voting was held throughout the county on Tuesday, November 5. Tays personally watched San Elizario, stationing a man at each polling booth in that troubled town. Everything was quiet, "not any of the Murderers attempted to vote," although the lieutenant was certain that "some of them ware secreted around town but ware [*sic*] not game to show themselves." Tays also dispatched two men to watch the voting in Socorro and kept others in Ysleta.[45]

Tays may have positioned himself in San Elizario to assure his commander that he was personally watching for balloting "murderers," but it was in Franklin that he expected trouble. Here he sent the strongest Ranger contingent, eleven men. Their leader, Sgt. Marcus Humphrey Ludwick, had been a farm hand, an artist, a harness maker, and most recently a boiler operator. But his early life in Bates County, Missouri, had not been quite so prosaic. The county was in the heart of the vicious Kansas-Missouri border war. Ludwick had grown up in a world dominated by lawless Jayhawkers, Bushwhackers, Provost Marshals, outlaws, and revenge-seeking "Radicals." After living among those who were expert at doing their worst, he was not cowed by men of Kinney's kind.[46]

Details of what happened are lacking, but "the [Kinney] mob tried to drive [the Rangers] out of town. Officers of election conducted themselves in a lawless manner," in an effort to intimidate voters. "[W]hen Sgt. Ludwick arrived in town they made demonstration against him + his men but on finding him firm they cooled down." Ludwick's tactics on election day are unknown, but his firmness is evident.[47]

Circumstances indicate that Kinney had attempted to throw the sheriff's race to Kerber. His thugs were not there to support Hague or others whose stands were sympathetic to the insurgents and whose elections were foregone conclusions. Kinney was not there to coerce people into voting for John Gillett as constable for Precinct No. 2. There was only one race worth Kinney's attention, that of the trailing sheriff against his "Mexican" opponent, Benito Gonzales. Kinney, who had a *Nuevo Mexicano* wife, was not a complete racist. (Few men could match the sheriff in that quality.) He was a practical businessman, and a friendly sheriff was vitally important to a man of his business interests. Only five months before, he had answered the call of New Mexico DA Rynerson and Lincoln County Sheriff Pippin for mercenaries. It would have been good business for Kinney to steer voters in Kerber's direction. The sheriff's involvement in Kinney's voting day "demonstration" is unknown, but likely, given his own actions after the election. In any event, through the watchfulness of Tays and steadfastness of Ludwick, Kinney failed. When the ballots were finally counted, El Paso County could look back upon what was probably its first honest election since the end of the Civil War. It required Texas Ranger intervention to defeat the "contemptible means [Kinney] resorted to in this Election." But in the end there was neither ballot stuffing, ballot box theft, nor priestly interference to alter the results.[48]

Hague won a countywide majority but lost his race for Cardis's seat in the other counties. Paseños won all the precinct-level races in the three towns of *la Isla*. Captain Garcia was elected county commissioner for San Elizario, while his namesake became that town's Justice of the Peace. Benito Gonzales won the sheriff's heat. His 62 percent vote total was a clear repudiation of Charles Kerber. The margin of defeat convinced neither Kerber nor G. W. Wahl, who also lost his job as county clerk. Five weeks after the election, Callaghan advised Major Jones, "During the last contest . . . Wahl and Kerber were defeated by great majorities and yet, they were not willing to deliver the papers of their respective offices to the newly elected officers after they had duly qualified on the ground that the officers had not been commissioned by the Governor!! The people are outraged by such proceedings and the malice of sore heads is yet to be felt." The county commissioners court ordered Sheriff Gonzales to arrest Wahl for failing to turn over his books. The matter was resolved before Christmas. On December 21, 1878, the *Mesilla News* reported, "We are informed that some of the officers**** of El Paso county at first declined to turn over to those lately elected even when the laws, certificates of election, telegrams from the state capital, etc. were shown; but finally the 'constitution'—we heard it was 44 calibre—was 'presented' and the new officers had no further difficulty in assuming their respective positions."[49]

PART FOUR
Coda

CHAPTER 24
"With the Alamo and Goliad"

The El Paso Salt War was definitely over. With Charles Kerber's electoral defeat by Benito Gonzales, no longer did El Paso have a sheriff bent on exacting revenge. The insurgent prisoners' escape left no one for the courts to punish. Loose ends remained. The Paseño community had lost its bid for self-government, but continuing lawlessness produced a siege mentality among Anglos not yet able to benefit from the insurgent defeat. Old Salt War wounds festered in the civil and criminal courts during the years following the conflict, in some instances for decades. Open sores like these kept memories of the conflict alive for some, even as the county moved toward a heady future that took little notice of the past.

Postwar instability and Sheriff Kerber's malfeasance had left the county "infested by a band of thieves . . . so strong as to defy all authority [and] permitted to exercise their infernal profession without molestation." Lt. James Tays bears some blame for this. Despite defeating Kinney's attempt to control the election, he had taken rustler moles into his detachment, thus giving the Kinney band the upper edge. When Jones announced that budget cutbacks forced a reduction in the detachment's strength, Tays resigned, to accept "beter [*sic*] and easier work" as deputy collector of customs at Presidio Del Norte.[1]

Sgt. Mark Ludwick was the lieutenant's successor in command of the detachment. His achievements included arresting Desidario Apodaca and facing down Kinney's gang on election day. Major Jones was not yet sure that anyone would succeed Tays in permanent command of the El Paso unit, since the legislature had not yet funded the Frontier Battalion. Even without pay, Ludwick was expected to keep a lid on El Paso County. On January 11, a number of Franklin citizens met with Ludwick to discuss how best to fight the growing criminal menace. The sergeant obtained warrants against three outlaw leaders, including former El Paso Texas Rangers Fred Northcutt and "Red" Cartledge. The problem, Ludwick reported, was that, "The desperadoes never show themselves in daylight, but make their raids at night when unsuspecting parties are unarmed." But in daylight on January 16, Northcutt and Cartledge rode up to the Ranger quarters in Ysleta and opened fire. This was the act of men who had taken the measure of their former comrades and expressed their complete contempt.[2]

The unhappy record of the Detachment of Company C had not stopped with its surrender. During the ensuing year, one sergeant brutally killed another, two rangers deserted, and at least one went AWOL to fight alongside Kinney in the Lincoln County War. James Tays's questionable leadership got another Ranger killed during an Indian fight at Cornudas. Still another member of the detachment was shot by a revolver falling from the pocket of a fellow Ranger at a dance. Five prisoners in Texas Ranger custody escaped, a third Ranger deserted, and still another was discharged for drunkenness and insubordination. If this were not enough, Northcutt and Cartledge gave proof that El Paso's Texas Rangers were a joke to those who knew them best.[3]

Sergeant Ludwick gave orders to Rocky Mathews, George Loyd, and Calisto Benevides (the first Paseño recruited since the war) to saddle up. Spotting this activity, Northcutt and Cartledge turned their horses and made for Mexico. Ludwick was not deterred. He and his companions crossed the Rio Grande and chased the rustlers several miles. They drew near a large house as Cartledge was dismounting. "Red" was the first to open fire, but his shots missed. With the outlaws about to gain the safety of the house, Ludwick returned fire. The first round from his Winchester passed through the horn of the saddle and then struck Cartledge in the head. Ludwick turned his aim on Northcutt, who was trying to dismount. The sergeant's second shot passed through the outlaw's body, doubling the number of dead desperadoes.[4]

The excitement was not yet over. The Rangers were soon surrounded and detained by armed Mexicans. The local authorities allowed Ludwick's three men return to Texas but tossed the sergeant in the Paso del Norte jail until Solomon Schutz paid the bail. Once released, the sergeant renewed the pressure on the outlaws, stationing men in Franklin to make that town safe, and making a sweep of rustler hangouts, arresting any suspects he found. Ludwick's audacity and accuracy proved a turning point in the history of El Paso's Texas Rangers and the county's future. Crime along the Rio Grande certainly did not end with the deaths of Northcutt and Cartledge. But from this point forward, desperadoes could no longer defy El Paso's Texas Rangers with impunity. With two shots, Ludwick declared that law, not outlaws, would ultimately prevail.[5]

The deaths of Cardis, Howard, and others created a morass of legal problems for the administrators of their estates. John McBride left behind a few cows, a claim on the estate of Charles Howard, and three-year-old daughter Louise. Kerber won the guardianship, while administration of the estate was awarded to a Thomas McBride. The two representatives immediately clashed. McBride averred that John had paid Howard for real property that Howard had not conveyed before their deaths. Now McBride's estate wanted Howard's estate, also administered by Charles Kerber, to turn over the titles as agreed by the two executed men. The surviving file is silent on the outcome of that dispute, but two years later Kerber hauled McBride into court on behalf of Louise McBride. The child was still costing money, and, Kerber demanded that Thomas McBride sell John's remaining property (250 acres at Quadrilla) to pay for Louise's allowance.[6]

John Atkinson's estate netted a few hundred dollars for its chief creditor, Solomon Schutz, who also happened to be the administrator. Schutz managed to recover a few drafts taken south of the Rio Grande, but the thousands in cash stashed in Atkinson's trunk or hidden in his home were never recovered.[7]

Charles Ellis's estate administrators, Teodora Ellis and her brother Martin Alarcon, spent more than a decade fighting off the claims of merchants holding more than six thousand dollars in IOUs. Through the incompetence of her attorneys or the machinations of the court, Mrs. Ellis nearly lost everything except her home. Her store, steam mill, and other properties were ordered sold by Judge G. N. Garcia. She refused and the court battles continued. Ultimately, the Ellis store on Main Street and some acreage were sold. Teodora Ellis hung on to *El Molino* for a while longer, but in 1882, that building, Captain Garcia's bastion in the 1877 battle, was sold to Juan Armendáriz. The estate was finally closed in 1889, with Teodora Ellis still in possession of her home.[8]

Charles Howard's death triggered a race for the court house door. Kerber was appointed executor in early January 1878. At the same time, Zimpelman and the Austin firm of Foster & Ludlow began selling off the dead man's properties in San Elizario. Before long, Zimpelman was suing Howard's siblings in a struggle over ownership of properties in San Elizario, Quadrilla, and Pecos County. Also contested were 2,240 acres at the Guadalupe Salt Lakes and other lands in El Paso County purchased under Memphis, El Paso & Pacific Railroad script. The legal brawl continued for years. From time to time, Kerber obtained a court order to sell a lot or two to pay the estate's continuing debt.[9]

One heir was left out of the proceedings. Tiburcia Jurado, Howard's former lover, had died in July 1880, but she left a son, Jose Jesus Jurado, born July 4, 1876. The Jurado family kept him hidden, possibly to protect the boy. According to Jose, around 1895 he discovered, with some anger, that Howard had been searching for him prior to his death. Perhaps not realizing the estate was still unsettled in 1895, Jurado made no claim.[10]

In 1900, Kerber learned that the siblings had won their case and were entitled to whatever land remained. Armed with this knowledge, he went into court one last time to secure his long-awaited discharge as administrator. In the meantime, he discovered that, as to the salt lake properties, *"the Estate has no title to any of the 4 mentioned tracts, that no patents were ever granted to parties under whom these heirs and estate hold and that they all belong to the State"* (italics added). One can imagine Kerber's thoughts as he realized that many lives had been lost and countless others, including his own, disrupted by Zimpelman and Howard's insistence on payment for salt they never owned.[11]

Louis Cardis's estate passed through probate in 1878. Administrators Benigno Alderete and Ward Blanchard sold Cardis's stage line to pay estate bills. Through attorneys, Cardis's Italian relatives showed up in court on July 15. Joint heirs Guadenzio Cardis and Rosa Pavesi de Cardis sold Don Luis's property to his friend James Hague. For $2,750 dollars, and the payment of some debts, Hague obtained title to several properties in El Paso, Ysleta, and elsewhere, including his stagecoach stops at Van Horn Wells and the aptly named "Dead Man's Station" southwest of El Muerto Peak.[12]

On February 16, 1879, the Las Cruces newspaper *Thirty-Four* published an anonymous warning "that we may have a repetition of the San Elizario troubles here soon. The mob leaders return to this side whenever they choose and no effort is made to arrest them. It is said the sheriff is in sympathy with them. There are now four companies of troops here besides the 'Rangers,' and if another disturbance occurs it will be of short duration."

In the meantime, Sergeant Ludwick continued to do his job. The lawman was greatly concerned about the "murderers" in his midst. He had 117 warrants to serve, without any help from Sheriff Gonzales. On February 21 Ludwick and seven Rangers swooped into San Elizario, where they nabbed five "murderers" and Antonio Olguin, the penitentiary fugitive who had escaped from jail the previous August. Word of the crackdown spread, and Ranger dragnets came up empty at Quadrilla, Socorro, and again at San Elizario.[13]

The Rangers' arrests were often negated by the new sheriff's recurring inability to keep criminals behind bars. Gonzales allegedly allowed Antonio Olguin to walk about town, without any restraint and on his own recognizance, up until the moment he simply disappeared. "Roxy" Rose, a notorious Kinney man charged with shooting a soldier, escaped in March, while horse thief Leon Maximo followed in August. Most galling to many in the Anglo community was the escape on March 25 of the insurgent Beltran. This prisoner, charged with being an accessory to the Salt War executions, was allegedly released by Sheriff Gonzales with the promise he would not be molested. While this action may have been wise, it was beyond his authority. The pattern of escapes and releases prompted a criminal complaint against Gonzales on the charge of neglect of duty. Gonzales was charged with merely "pretending" to be sheriff, since his lack of English comprehension prevented him from understanding his job. The case never went to trial, and Gonzales continued his "early release" program.[14]

One alleged murderer who didn't escape before trial was Pablo Mejia. In what may have been the last violent killing related to the Salt War, the former Texas Ranger was charged with the murder of Higinio Loya at Quadrilla on August 19, 1879. Loya, one of Captain Garcia's Rangers in 1870 to 1871, was also identified as an insurgent leader. According to the Las Cruces *Thirty-Four,* Loya had been "the commander of the firing party who shot Howard and the other prisoners at San Elizario." Though other accounts gave that job to Desidario Apodaca, Loya's prominent role in the insurgency may have fed Mejia's readiness to shoot and kill. Mejia was soon released on bail.[15]

Major Jones had vowed to mete out justice to "murderers," but by 1879 Governor Hubbard wanted peace in the valley. Early that year, Jones began his search for someone of stature to take permanent command of the Ranger detachment. Though Ludwick "seems to have done very well," Jones judged it better to send in an outsider, "one of some established character and who could get in free from any local 'entangling alliances' or influences." Jones found his man in George Wythe Baylor. The former Indian fighter and Confederate colonel had experience of El Paso but carried no Salt War baggage. Baylor prided himself on his command of the Spanish language and his many intimate friendships with Mexicans. Governor Oran Roberts personally instructed Baylor "to use all the diplomacy possible to reconcile the two factions and settle the Salt Lake War peaceably. . . . what had been done could not be undone, and the restoration of order was now the prime requisite, rather than a punitive expedition against the members of the mob [*sic*]."[16]

Baylor arrived on September 12 with his mandate to "avoid further trouble." The major also gave Baylor instructions to reorganize the unit, with the particular advice to discharge Corporal Mathews, who "has been several times reported to me as having been guilty of bad conduct." With Mathews' departure, the detachment lost its final veteran of the San Elizario siege. Ludwick was given the option of staying on as First Sergeant, but the superseded detachment commander decided to exit. James B. Gillett stepped into Ludwick's position.[17]

It may be said that George Baylor, more than any other El Paso Ranger officer of his time, developed a rapport with his Mexican neighbors. It helped that Baylor was passionate about one task that put him in complete agreement with virtually the entire Paseño community: ending the Apache threat. In this important respect, life in the lower valley went on as before. Mescaleros continued to raid the settlements along the river, steal livestock, and kill isolated citizens and travelers along the corridors to El Paso. Ignoring the visiting insurgents and leaving outlaws to Sgt. James Gillett whenever possible, Baylor concentrated on Indians.

Just four weeks after assuming command, Baylor received a note from Captain Garcia, delivered by Mejia, stating that fifteen Apaches had attacked a small party of Mexican hay cutters. By midnight, Baylor was on the trail, accompanied by Garcia and Mejia, two of the Indians' most implacable foes. The Apaches crossed the Rio Grande into Mexico. Baylor "did not hesitate to cross the river as the trail was hot." Dozens of Mexicans joined the Ranger column. The joint pursuit force could not pin down their quarry, but the *ad hoc* alliance was first of many along the Rio Grande.[18]

In December 1879, Baylor's Rangers joined a hundred Mexican citizens in pursuit of Victorio. The Rangers discovered another veteran Apache fighter in their midst, "Old Chico Barela," as well as "many others" wanted for murder in Texas. Baylor "gave the old fellow to understand we were now fighting a common enemy and should act in harmony together," an arrangement he was glad to make in view of Barela's saving the Rangers' lives two years before.[19]

In late February 1880, Paseños dislocated by the Salt War attempted to disrupt "progress" threatening to dislodge them from *la Isla* for the second time in twenty-six months. The imbroglio began when El Paso County surveyor A. J. Wingo attempted to survey some land lying between San Elizario and the Rio Grande, land occupied by former insurgents who "hope[d] to keep any one from settling on the Island." An armed party of Paseños "presented their guns and threatened to kill him [Wingo] if he crossed a certain line." The surveyor asked for Texas Ranger protection. Baylor and his men rode down the valley, but turned back upon learning that Judge Blacker had ordered all the parties to bring the dispute into his court. Wingo, who did not get the news, returned with about seventy-five citizens of San Elizario to survey the site. These scattered when the squatters renewed their threats, but then the latter "decamped in haste as soon as Capt. Garcia and men rode up." As Garcia and his party from town were crossing the boundary, Antonio Garcia (no relation to the captain) shot at Mejia from ambush, prompting one of Mejia's friends to fire back. The shooting "would soon have become general," but the captain managed to stop his companions' fire. The other side followed suit. So ended the last armed conflict involving Salt War insurgents.[20]

Troubled by her husband's decision to resign and disappear, Anne Nicholl wrote to U.S. Army headquarters. General Sherman himself replied, "alas! if he be your husband he is now by the laws of God and man a criminal, for on the 17th of September 1878 he was married . . . to Mrs. Gordon Granger, a lady of undoubted respectability." In falsely marrying Maria Granger, Blair had sullied the widow of a distinguished general and a friend of William Henry Vanderbilt. In bringing public dishonor to the army, Blair made a personal enemy of William T. Sherman.[21]

The Blairs were in Santa Fe when they learned of the developing scandal. The "fearful charges" stunned them both. On December 22, the captain wired Sherman, tendering his resignation. One day later, the devoted Maria Blair begged Sherman, "don't accept it. He is almost frantic does not know what he is doing." Receiving permission to defend himself in Washington, the captain withdrew his request.[22]

The news hit Santa Fe like "a clap of thunder in a clear sky." In a city where Blair was regarded as "a gentleman from instinct" and his wife "beloved and respected," the *New Mexican* advised its readers to suspend judgment. Further south, the *Mesilla Valley Independent* sneered, "This is the young and gallant officer . . . who refused to advance for fear of death and said all would be annihilated, who retreated and allowed his men to be pulled off their horses by the mob, when within sight of the besieged men."[23]

While the Blairs were en route for Washington, incriminating documents, including a certified copy of the 1861 Scottish marriage certificate of Thomas Nicholl and Ann Baillie, reached Sherman. The marriage form revealed that Blair was the maiden name of Thomas's mother. Various letters sent to Ann were in the same handwriting as Blair's forged resignation. Sherman ordered Blair arrested "wherever found."[24]

The press everywhere began trumpeting "An Army Scandal" involving "A Brass-buttoned Scoundrel." Nationwide, newspaper readers were titillated by the lovers' triangle of Captain Blair, "Mrs. Granger Blair," and "Mrs. Thomas Blair Nichols." The Blairs changed their destination to New York, where the captain's wife and stepson would stay while he proceeded alone to the capital. Getting wind, Sherman ordered Gen. Winfield Scott Hancock to arrest Blair "and prevent him leaving the country." On January 9 the captain was arrested at the Sturtevant House on Broadway and immediately hustled off to Fort Columbus on Governor's Island. Blair requested leave to secure evidence in

Exhibit 5: Capt. Thomas Blair. National Archives

Scotland. Maria supported this request, but her brother, Ben Letcher, feared that Blair "might persuade his wife there to 'clear out,' and he come over here with some patched up evidence that might satisfy my sister." Blair's request to sail beyond the reach of military authority was denied.[25]

The trial began on February 10. Blair's "in-laws" procured a pair of prominent New York lawyers, Martin Thomas McMahon and Elliott F. Sheppard. Blair pleaded innocent to bigamy and forgery. General McMahon's objections to the submission of the damning evidence were overruled. The flood of documents had their effect on Maria Blair, who ceased her visits after the trial's second day. On February 18, Letcher made McMahon aware of a letter written by Blair to Anne Nicholl *after* his marriage to Maria Granger, giving Anne a choice. Either they would divorce, or he would "drop out of sight" and stop sending money. The captain declared himself legally and morally free to remarry (which he had already done). The letter would be submitted in evidence, Letcher warned, unless Blair pled guilty. He did. The judges deliberated for three minutes and found Blair guilty on both the specifications and the charge of conduct unbecoming an officer. He was dismissed from the army.[26]

The result, said the *Brooklyn Daily Eagle,* was "a ruined man and disgraced family." The *Waukesha Freeman* suggested it was time for Blair to see "how the inside of a penitentiary looks." After a grand jury in Lancaster, Kentucky, indicted Blair for bigamy, raising the horrible prospect of a salacious trial at home, Ben Letcher pleaded for his expedited release. President Hayes approved the court's actions on March 5. At 3:30 P.M. Blair was released. The disgraced officer had lost none of his charm. The *Eagle* reported "The officers of the garrison entertain a kindly feeling for him, and made his departure as pleasant as possible under the circumstances. He is said to be worn and thin, and to feel his punishment deeply." He was variously reported headed to either Scotland or Canada. It is likely that no one cared where he went.[27]

For independent reasons, neither the new sheriff nor the new Texas Ranger captain made any effort to bring in the accused murderers of Howard and the rest. Their nonfeasance cooled tempers and would have facilitated a return to normalcy, had normalcy been possible. But with the arrival of the railroad in May 1881, life in the valley would never be "normal" again. With the trains came explosive growth in population, wealth, and new problems. Change was felt everywhere, but most dramatically in the town selected by the Southern Pacific and other rail companies for their depots.

The decision was not clear cut. El Paso, Texas (city government was permanently reestablished in July 1880), was always the frontrunner, not only as the county's business center, but also because the former "Franklin, Texas" was the most "American" town in the valley. In September 1879 it boasted "five stores, three saloons, three hotels, a custom house, Masonic lodge, three physicians, half-a-dozen lawyers, a school, and no church." With the advancing tracks still hundreds of miles short, Ysleta, still the county seat and administrative center, made a case for its selection. Although labeled a "Mexican" town, it was reported "rapidly filling up with an enterprising American population." It was, said one publicist, "the rising town of the valley [showing] more improvement within the last three years than all the balance of the county. If the railroad goes there, it will be *the* town of northwestern Texas."[28]

Socorro was never in the running. San Elizario would have been an outside shot for the railroad depot in the best of circumstances, despite its relatively large population and long experience as an administrative and commercial hub. Its location far down the valley made less geographic sense than the crossroads at the Pass, it had lost the county seat, and it lagged far behind El Paso as a commercial center. Nevertheless, the community reincorporated to promote its candidacy. In 1879, Captain Garcia's name led the list of those petitioning for reincorporation. Son Tomas presided over the subsequent vote to incorporate, and twenty-four-year-old Gregorio N. Garcia II took office as mayor in February 1880. The council offered the Texas & Pacific Railway Company choice land through the center of town in exchange for building its depot there. Ultimately, the T&P was persuaded to build its depot in El Paso. Perhaps it was community opposition to a train running through the middle of San Elizario that persuaded the railroad to base its operations elsewhere. More likely the Salt War had made San Elizario an investment pariah.[29]

One year after the siege, executions, and looting, Judge Blacker declared that "San Elizario would be placed alongside with the Alamo and Goliad, with the belief that it would ultimately be avenged in God's own good way, as those sanguinary catastrophes were." The statement was more than an admission that the executioners of Howard and his ill-fated companions would never come to justice in his court. (He made no similar statement regarding the riotous Texas Rangers and Kinney men.) The murderous mob riots—which is how any railroad investor who read the newspapers viewed the uprising—placed San Elizario beyond the pale. Mayor Garcia's administration made another offer to the Galveston, Harrisburg, & San Antonio Railway in August 1882, but it was too late. The T&P already linked with the Southern Pacific in El Paso, establishing forever Anglo El Paso's dominancy. The proud little town of San Elizario was consigned to the backwater of El Paso's future.[30]

CHAPTER 25

"My enemies are all dead."

When the railhead came to El Paso, Texas, many of the county's men of power, those who stood to benefit most from this day, were already gone: Mills and Kerber, gone from their offices; Fountain, Zabriskie, Newcomb, and Kinney, gone from Texas; Cardis and Howard, gone from this earth. Salcido, Granillo, Barela, and others who resisted injustice were now exiled and in no position to exert *la Isla's* influence on the county's future.

A few Reconstruction era leaders remained to reap the benefits of El Paso's growth, joining the newcomers who swarmed into the valley. J. P. HAGUE remained active in El Paso's affairs as a politician, newspaper owner, and trial lawyer. His reputation for incorruptibility wiped away remembrance of accusations that he had prompted the citizens of *la Isla* to riot and kill "Americans." Hague died in 1895.[1]

ALLEN BLACKER was elected in 1880 to the Texas legislature. There he lobbied Governor Roberts on behalf of the El Paso merchants still seeking reimbursement for the twenty-five hundred dollars in goods and payroll they advanced to Kerber's Silver City posse years before. He died in 1905, having served as an El Paso alderman and city judge.[2]

JOSEPH MAGOFFIN, served as El Paso's first railroad era mayor, winning election elected four times. He died in 1923 in Washington, DC and was buried in El Paso.[3]

JIM MCDANIELS continued to roam the borderlands, keeping just ahead of the law. In March 1881, his death in Las Cruces was "suggested" by the district attorneys prosecuting him in absentia. A fabricated death is not beyond possibility.[4]

MAJOR JOHN B. JONES died in 1881, at age forty-seven, following surgery on a liver abscess. The Salt War had been a rare failure for the Frontier Battalion commander justly remembered for his otherwise effective leadership of the Texas Rangers.[5]

PABLO MEJIA died horribly in 1882. According to George Baylor, "The Apaches knew and hated Pablo Mejia, and they . . . placed his head on a stone and pounded it flat."[6]

TELESFORO MONTES, who resisted the 1872 school law and commanded El Paso's second company of *Tejano* Rangers, died in 1887.[7]

The youngest Salt War Rangers died young. Irishman ROCKY MATHEWS was killed by bandits on January 19, 1885, somewhere on the road between Durango and Jimulco, Coahuila. Syphilis or perhaps genes drove BILLY MARSH mad, and in January 1886 he was legally declared insane. Within days, he wrote out his will on the back of his father's Masonic Lodge certificate and then blew his brains out.[8]

JERRY BARTON drifted back to Arizona, where he ran several saloons, pummeling and shooting unruly customers at each one. Following a stint in Yuma Prison for breaking a man's neck with two punches, he and his wife ran a boarding house. Barton died in Tucson in 1904 of Bright's disease. An obituary told of how he had killed six Mexicans who "attempted to make a rough house" in a saloon he bartended. "They were all shot in the head." The newspaper did not note whether they got it in front or in back.[9]

Deadly gunman JOHN C. FORD became an "owner of extensive herds" and later a real estate agent. He died of pneumonia in 1914. "Captain John C. Ford" the *El Paso Times* remarked, "was never known to shirk, even when sent into the very jaws of death. . . . yet at the same time was as gentle as an infant." Sam Frazer might have taken exception to Ford's reputation for gentleness from his own well-deserved grave.[10]

In the early 1880s, JOHN KINNEY created the early Southwest's most organized criminal enterprise. His men were positioned along two hundred miles of the Rio Grande, and into Texas and Chihuahua. From these points they allegedly stole cattle by the thousands over a wide area from New Mexico to Old. Stolen beeves were relayed to the slaughterhouse and meat dressing operations at the Kinney ranch just south of Rincon. Beeves and packed meat made the short trip from the ranch to the train depot at Rincon, nicknamed "Kinneyville," on their way to buyers. With no middleman cutting into profits, Kinney reportedly deposited huge sums in El Paso's banks. He was, said one newspaper, "king of the rustlers."[11]

In February 1883, A. J. FOUNTAIN's New Mexico Volunteer Militia fanned out across southern New Mexico, scooping up Kinney's allies and lieutenants. In early March, the "boss rustler" himself was captured. Kinney's trial took just two days before prosecutor Fountain secured a guilty verdict from a jury that spent eight minutes to convict.[12]

Fountain worked for statehood as leader of the territory's Republican Party. He served as speaker during his only term in the territorial legislature, but in 1890 was defeated for reelection by Albert Fall. Fountain represented the interests of large ranchers in a mid-1890s range war. In 1896, while returning home by buckboard from the court session in Lincoln, he and his young son Henry disappeared somewhere between Tularosa and Las Cruces. Their bodies never turned up, and the men tried for their murders were acquitted. The Fountain murder remains one of the great mysteries of the West.[13]

Ever disputatious, but "possessing unfailing delusions of leadership," WILLIAM WALLACE MILLS served briefly as assistant U.S. marshal during the Garfield administration and later as American consul in Chihuahua. He told his brother Anson in 1900, "I do not know that my enemies are all reconciled to me, but they are all dead, and that is better, or at least safer." Mills expired in Austin in 1913.[14]

In September 1884, seven *salineros* were caught stealing 140 *fanegas* of salt from GEORGE B. ZIMPELMAN's "private lands." They were acquitted, prompting Zimpelman to bring a civil suit. The defendants were ordered to either return the salt to Zimpelman or pay him one dollar per *fanega*. Zimpelman served as postmaster in Austin under Presidents Cleveland and Roosevelt. He died in 1908. His obituary mentioned almost every noteworthy aspect of his life except Charles Howard and the Salt War.[15]

In 1879, COL. EDWARD HATCH wore out his men and animals in the fruitless chase of the Apache chief Victoria. The New Mexico press shouted for his removal, and in 1881 the 9th Cavalry was reassigned to Kansas. Hatch died at Fort Robinson on April 11, 1889, the army's senior colonel at age fifty-seven.[16]

In 1898, LOUIS HENRY RUCKER fought the Spanish, Filipino insurgents, and malaria. He was promoted to brigadier general in April 19, 1903, and placed on the retired list the next day. Bright's disease ended his life in Los Angeles on June 30, 1906.[17]

After a brief stint as the postmaster in Tombstone, THOMAS BLAIR lived out his days in Chihuahua. He ran a small store until one day in 1896, when he turned his back on the wrong customer, who picked up an axe and drove it through his head. It was a fitting end for one whose cowardice and self-interest cost Charles Howard and others their lives, deprived San Elizario of its future, and destroyed the happiness of the one woman he loved, though not more than himself. The unfortunate Maria Granger had already died in Lexington, Kentucky, at age forty-five in 1887.[18]

ANTONIO SEVERO BORRAJO died at age seventy-eight in 1896. A freezing rain accompanied his burial, as if to put his out his fiery temperament. Walter Prescott Webb labeled Borrajo "an evil spirit transplanted from the age of the Medicis." Borderlands historian Rick Hendricks sees instead, "First and foremost . . . a passionate defender of what he termed the 'Hispano-Mexican' expression of Catholicism. . . . a patriot, a fighter dedicated to the preservation of Hispano-Mexican culture and way of life."[19]

In 1879, the Frenchman PETER BOURGADE persuaded the Sisters of Loretto to build St. Joseph's Academy in San Elizario, which provided the Catholic education many parents desired for their daughters. In 1889, he was elevated to archbishop of Santa Fe. Peter Bourgade died in Chicago in 1908. He was described as a "special friend of the poor and lowly, and zealous for souls," a portrayal that also applied to Padre Borrajo.[20]

"PARSON" TAYS built El Paso's first Episcopalian Church (St. Clement's) and was co-founder of the *El Paso Times.* The reverend died of smallpox in 1884. Following a stroke, JAMES TAYS joined John in Ontario, California, where he died suddenly in 1892.[21]

JOHN B. TAYS was El Paso's postmaster in 1879 and its first city marshal in 1880 before turning to private enterprise. Building upon the inherited wealth of his wife Amelia, Tays engaged in a variety of businesses, most notably real estate. By 1884, he was one of El Paso's ten wealthiest individuals. Tays was said to have, "brought to bear in his various undertakings much business ability and indefatigable energy, and success attended his well directed endeavors."[22]

In 1883, the Tays family moved to the Ontario-Upland area of southern California. Here Tays became one of the region's foremost farmers, known for his use of "the best scientific methods." His last home, built in 1892, with its "grand entrance," "handsome parlors," and bathrooms with "Kentucky marble and silver fittings," was "in the luxury class." For Amelia Tays it was a long way from her first El Paso dirt floor adobe. For John Tays, a world apart from the embattled Ranger quarters in San Elizario. In 1892, Tays sold his acreage for forty thousand dollars, at that time reputedly the largest real estate transaction in the area.[23]

In 1894, Tays returned to mining, this time in the goldfields of British Guiana. On May 6, 1900, he was one of 140 killed in a spectacular boating accident when a river steamer and two boats went over the Tumatumari Falls. The El Paso *Daily Herald* reported his death, making no note of his command of the Salt War Texas Rangers. Amelia Tays died in Los Angeles in 1927.[24]

CHARLES KERBER served in a number of Ysleta city offices, including mayor, before his death in 1909 of heart failure. The *El Paso Herald* identified him as "a Soldier of Fortune, a Jurist and a Peace Officer" The *Herald* noted his enlistment in the prewar Army but not his desertion and name change before his "conspicuous" service in the Colorado volunteers. A thorough garbling of Salt War history prompted the paper to call him "the hero of the Howard salt riots." Kerber would have agreed.[25]

DAN TUCKER and John Kinney, the two man-killing leaders of the "Silver City Rangers" each died in bed. Deputy Sheriff Tucker returned to Silver City, where he also served as a deputy U.S. marshal. His health declined and he moved to California, where he died sometime after 1892, possibly in San Bernardino County Hospital.[26]

JOHN KINNEY went respectable. His conviction was reversed, and from then on he led an ordinary life. He operated a feed lot in Kingman, Arizona before moving to Prescott around the turn of the century. He died there in 1919. One obituary declared, "[As] a deputy sheriff and the chosen head of posses to run down criminals he attained renown that still endures." He died "one of the most generous and best loved men ever to grace the early life of the thrilling days of the border." No Paseños were interviewed on the subject of Kinney's generous nature.[27]

CAPT. GREGORIO GARCIA's participation in Baylor's 1879 campaign against the Apaches may have been the old soldier's last. He thereafter devoted himself to civic work until his death in 1898. Garcia had been San Elizario's greatest soldier of the post-Mexican era, never missing the opportunity to defend his community with his life. In the Salt War he took up arms against his own neighbors, because lawlessness was the enemy of the progressive future he saw for San Elizario.[28]

PRICE COOPER returned to freighting goods across Apacheria, more than once barely escaping death on the road. "Old Price" died in 1900. With jaw-dropping understatement, one former stagecoach driver recalled that Price "had his troubles with the Indians."[29]

Price's son SANTIAGO COOPER turned to cattle rustling before becoming an El Paso police officer. He passed away in 1919. The *El Paso Herald* accorded him a posthumous newspaper promotion to "captain of rangers" and noted, "He fought and was wounded in the battle of San Elizario."[30]

In 1920, nearly a dozen ex-*Tejano* Rangers or their widows filed federal military pension claims based on service against Apaches under Captain Garcia and Lieutenant Montes. (A 1917 federal law extended pension rights to certain veterans of Indian Wars who had fought in state militias or similar organizations under the control of the United States.) They included two veterans of the defense of *El Molino,* Captain Garcia's sons Tomas and Carlos. Benefits were also awarded to the widows of JESUS GARCIA and LINO GUERRE. In-

John Kinney and his daughter. Robert G. McCubbin Collection

triguingly, Garcia, who died in Ciudad Juarez in 1905, and Guerre, who lived in San Elizario at least from 1880 until his death in 1904, shared the same names as men once identified as Salt War insurgents.[31]

GREGORIO NANCIANCENO GARCIA II served as mayor of San Elizario several times between 1880 and 1911. He served in the U.S. customs service and the border patrol, was a deputy sheriff and later El Paso policeman. His federal pension claims were denied because he could only prove Salt War service against "marauding Mexicans," and not "against Indian depredations."[32]

Numerous *Tejano* names are absent from the list of Indian war pensioners. Foremost among these are insurgent leaders SISTO SALCIDO and FRANCISCO BARELA. By 1881, Barela had given up returning to Texas. In that year, he and his wife Miguela, then residents of the state of Chihuahua, sold the Ysleta home and land they had owned since 1873 for the sum of sixteen hundred dollars. In July 1887, these two, together with insurgent leader LINO GRANILLO, DESIDARIO APODACA, Salt War instigator JOSÉ MARÍA JUÁREZ, Garcia son-in-law RAMON ZAMBRANO, and a hundred other Salt War insurgents were ordered taken into custody by the sheriff on the still open charges of murder. Sheriff Benito Gonzales was gone by this time, but his replacement, Sheriff J. H. White, made no effort to round up anyone for the decade-old crimes. Slowly, quietly, some former insurgents finally came home as old grievances were forgotten. Who or how many is unknown. While a few like Lino Guerre, have surfaced in the examined records, the "chief offenders," men such as Barela, Salcido, and Granillo, apparently stayed in Mexico. Where they lived and how long they survived is at this point uncertain.[33]

Few Salt War participants survived to the 1930s. TOMAS GARCIA, one of Howard's four bondsmen and a defender of the Ellis store during the siege, passed away in 1934. Texas Ranger MARK LUDWICK moved to the Hill Country, married, and eventually became a San Antonio furniture dealer. The fearless and efficient lawman died in 1937.[34]

Eyewitness TORIBIO RUBIO, only twelve in 1877, was described by a local newspaper in 1937 as the "Only known survivor of the Salt War." Not so, said Gregorio N. Garcia II, who finally retired from the police force in 1935, at the age of eighty. Interviewed in February 1937, Garcia blamed the "Ambitions of Louis Cardis, youthful Italian, to control the political destinies of El Paso and Lower Valleys" for the bloody conflict. Looking back, he judged the truce with the insurgents "a fatal mistake, for no one can bargain with a mob."[35]

Thirteen months later, ex-Ranger JESUS MONTES was interviewed by the *El Paso Daily Herald*. Speaking from his wheel chair, "erect, white-haired, eagle faced," Montes recalled the Salt War, which had "ruined San Elizario." He admitted to missing the fighting, as he had gone to Chihuahua "to keep from joining the Texas Rangers to please his father." He did return in time to see the last moments of Howard, Atkinson, and McBride. "I stood on the bank of the acequia and watched the platoons shoot them down."[36]

Gregorio Garcia II died on December 28, 1938. The pallbearers included the mayor of El Paso, the sheriff, the district attorney, the police chief, two police captains, and four judges. Local newspapers recounted his many valuable services to San Elizario, the county, and the state. He was erroneously said to have "joined the Texas Rangers in 1879," after the Salt War, about which not a word was written. His passing left Jesus Montes as the oldest surviving resident of San Elizario. The latter died at the age of ninety on March 14, 1941.[37]

Ex-Texas Rangers THOMAS and ANDY ZICKEFOOSE lived in Colorado and Kansas before settling in Oklahoma. Andy died in 1928, leaving to his son the pair of steel knuckles he had carried during his months as a Texas Ranger. Tom's pension claim was rejected since his "entire service appears to have been against Mexican marauders." He died in 1942 at the age of ninety-three. He was probably the last "salt warrior" to have traded bullets in San Elizario over what Boundary Commissioner John Russell Bartlett had called "nothing but the saline incrustations of a dried up lake."[38]

EPILOGUE

On June 9, 1923, the *El Paso Herald* ran a feature article on "ancient San Elizario . . . a place where people lived in the quiet way of the Mexican aristocracy of half a century ago [and] now a mere spectre of what it once was." The article spoke of "High adobe walls . . . crumbling away; odd old gates that . . . have been nailed shut to keep them from falling from their rusty hinges." The valley's wine, once declared, "'the best to be had anywhere,' is no longer made. The El Paso onion that once grew near San Elizario grows there no more." The church had been "kept in good repair," but there were "more vacant buildings than . . . buildings with occupants." Los Portales, the former Garcia home, was still in use as a school, but across the *placita* Juan Mauro Lujan's large home, once used as an insurgent headquarters, was in ruins. The virtual disappearance of Rio Grande water had completed what the railroad began. Farming and clear-cutting of watershed timber in Colorado robbed the valley of both the water and soil nutrients previously making their way to *la Isla's* acequias. As the *Herald* noted, "The last support of the old town had been knocked from under it. Thus for two generations, San Elizario has been dead and still dying."[1]

The obituary for a town now in its third century was premature. As historian Rick Hendricks has noted, "Today, San Elizario and the ancient communities of Ysleta and Socorro are heroically struggling to preserve their histories." The town's listing in the National Register of Historic Places in 1997 provided the impetus for rehabilitation of surviving historic structures. Los Portales, the old home of Captain Garcia, nearly continuously in use since 1855, was renovated in 2000 and reopened as a museum under the care of the San Elizario Genealogy and Historical Society (SEGHS). Tourists can take a self-guided "Walking Tour" of historic landmarks, including the old Overland Stage station, Casa Ronquillo (the Ellis home), and the ancient *acequia*. Despite these improvements, the preservation challenges are immense. Key landmarks and community memories have eroded over time. In April 2002, I walked the historic district area with the society president, Sam Sanchez. We speculated on the location of key Salt War sites. We lamented the fact that so much history has become lost or muddied.[2]

Beyond preserving buildings, local historians have worked to halt and reverse this loss of the historic record. The society sponsors an annual confer-

Los Desperados take "Howard" prisoner, 2003. Courtesy of Ray Sanchez

ence devoted to the lower valley's history. In 2002, public historian Robert R. Archibald encouraged residents to use the special qualities of their community, including "the layers of Indian, Hispanic, and Anglo culture," upon which to build the future. In that spirit, recent conferences have included extended discussions of the Salt War by local historians such as Oscar Lozano and Nicholas Houser.[3]

As past society president Gus Lujan observed during the 2002 conference, those citizens who can trace their lineage to the first settlers are being joined by increasing numbers of fresh immigrants from Mexico, El Salvador, and other Latin American nations. Hendricks has noted the importance of imparting to these newcomers "a sense of the history of the place." In fact, the descendents of local Indians and presidial soldiers are intent upon assimilating the newest residents into the established community. One way the continuity of San Elizario's history finds expression in through its pride in its strong warrior tradition. Annually, Veterans' Day is marked by a parade led by old soldiers, sailors, and airmen, ending with patriotic speeches at the flag-adorned *placita*. The tribute offered on Veterans' Day 2003 to Ruben Estrella, Jr., a young Mexican immigrant who died an American soldier in Iraq, is a measure of the town's respect for the contributions of its newer families. Estrella joined a long line of Vietnam vets, World War II GIs, Texas Frontier Force Rangers, presidial soldiers, *salinero* insurgents, and other warriors who have defended their community and country at the cost of their lives.[4]

The annual SEGHS conference presentations precede an outdoor program. Dancers perform traditional dances, and actors in traditional costumes replay the first Thanksgiving held on U.S. soil, celebrated in 1598 by explorer Don Juan de Onate. Finally, "Los Desperados," San Elizario's own Wild West reenactors, appear. On a typically hot, late April afternoon, they stage several Western gunfights. The last is a much abbreviated reenactment of the Salt War. The complexities of the event are lost, but the issues and sides are clearly drawn. "Howard" hides in Los Portales, trading shots with angry Paseños until forced to surrender. Then, in front of the watching crowd, a firing squad executes him again for trying to steal the people's salt.

NOTES

Abbreviations Used in Notes

AGO	Adjutant General's Office, United States Army, NARA
CAH	Center for American History, University of Texas at Austin
EPC	El Paso County
EPT	El Paso Troubles in Texas (citation in Bibliography)
FBR	Frontier Battalion Records, Ranger Records, Texas Adjutant General's Department, TSLAC
HTO	Handbook of Texas Online at www.tsha.utexas.edu/handbook/online
LR/LS	Letters received/Letters sent
MBT	Mexican Border Troubles (see Bibliography)
MC	William Wallace Mills Manuscript Collection, CAH
MS	Microfilm Series
NAMS	National Archives Microfilm Series
NARA	National Archives and Records Administration
RBHLPC	Rutherford B. Hayes Library and Presidential Center
RG	Record Group
SSC	C. L. Sonnichsen Special Collections Department, Sonnichsen Library, University of Texas at El Paso
TAGC	Texas Adjutant General's Correspondence
TFT	Texas Frontier Troubles (see Bibliography)
TSLAC	Texas State Library and Archives Commission
WC	Walter Prescott Webb Manuscript Collection, CAH

Introduction

1. John Russell Bartlett, *Personal Narrative* I:121–22.

2. Pope, Captain John, *Reports of Explorations and Survey* II:30, 31; Report of Dr. John S. Ford, June 18, 1849, *Southwestern Historical Quarterly* 48 (October 1944), 266.

3. *MVI*, October 6, 1877, *EPT*, 15.

4. Thomas Jefferson to James Madison, 1787. http://etext.virginia.edu/jefferson/quotations/jeff0300.htm

5. McDougall, *Freedom Just Around the Corner*, 5.

6. For discussions of violence and its relationship to the concepts of honor/shame in the Old South, the so-called "Texas Doctrine," and the "Code of the West," see Bertram Wyatt-Brown, *Honor and Violence in the Old South*, Richard Maxwell Brown,

No Duty to Retreat, Gary L. Roberts, *Death Comes for the Chief Justice,* and Robert M. Utley, *High Noon in Lincoln.*

7. Border historian Oscar J. Martinez has noted, "Until the advent of modern transportation, the North [of Mexico] was practically cut off from the Mexican interior, and population centers in the region itself were isolated from each other." Martinez, *Troublesome Border,* 107.

8. Quotation taken from remarks by Dr. Walter Buenger during a panel on the state of Texas historiography, 110th Annual Texas State Historical Association meeting, March 2, 2006.

9. For discussions of the Apache impact on life in the Valley, see Griffen, *Utmost Good Faith,* and Sonnichsen, Mescalero Apaches. For overviews of Buffalo Soldiers in the Salt War, see Kenner, *Buffalo Soldiers,* and Leiker, *Racial Borders.* Salas, *Soldaderas,* discusses their role in Mexican military history. Eickhoff, *Exiled,* traces the history of the Tigua tribe.

Prologue

1. The number of men and carts taking part in the salt expedition of December 1877 was variously reported as nineteen or twenty men or eighty, with sixteen to twenty-four carts. I assume that a large number of men took part in the expedition to protect the train from Indians or Texas Rangers. Pierce to 2nd Asst. Sec. of State, March 11, 1873, March 17, 1873, Despatches from U.S. Consuls in Ciudad Juarez (Paso del Norte) 1850–1906, RG 59 Department of State, NAMS M184, Roll 2 (hereinafter Paso del Norte Despatches); *EPT,* 73, 126; Blair to Hatch, December 11, 1877, *Special Files of Headquarters, Division of the Missouri 1863–1885,* RG 393, NAMS M1495, Roll 14 (hereinafter *Special Files*).

2. *MVI,* February 2, 1878. Montes and Skaggs, "The Salt War." The use of "Pasenos" to describe natives of El Paso appeared in American newspapers as early as 1847. See Gettysburg Republican Compiler, June 7, 1847.

3. Bowden, "Magoffin Salt War," *Password* VII, Summer 1962, 95; Mills, *Forty Years,* 27.

4. The arithmetic is complicated but depends on the ratio of salt to unrefined silver ore, as well as the price of salt and refined silver. In 1877, Howard offered the going rate of $1.25 per *fanega* of salt. Bancroft, *Mexico* II:752, *EPT,* 112. A lengthy discussion of the "patio process" is found in Homer E. Milford, *History of the Los Cerrillos Mining Area* (New Mexico Abandoned Mine Land Bureau Reports 1994, vol. 2, 1996, vol. 1), www.cerrilloshills.org/mines/realdx1.htm. Inflation in price for salt based on unskilled wage rate inflation. See "How Much is That," http://eh.net/hmit/compare/.

5. *MVI,* August 18, 1877; *Grant County Herald,* September 29, 1877.

6. "Salt Industry," HTO; Bowden, *Magoffin Salt War,* 95–96.

7. In 1870, Granillo had a wife and six children at home, ages 5 to 18. See U.S. Census, 1870, San Elizario, EPC, Texas, 24. Whenever it was first used, "greaser"

was a pejorative term in common use among as early as the 1850s. See de León, *They Called Them Greasers,* 16.

8. Bartlett, *Personal Narrative,* 122. See also www.geoinfo.nmt.edu/staff/scholle/ep_csbd_roadlog.html for a geologic description of the lakes.

9. *MVI,* December 8, 1877; Gregg, *Commerce of the Prairies* I:211–212.

10. Davis, *El Gringo,* 212–213.

11. Pierce to State, November 15, 1872, Paso del Norte Despatches; Davis, *El Gringo,* 212.

12. "Animals in the West: The Ox and the Mule" Museum Gazette, Jefferson National Expansion Memorial, http://www.nps.gov/jeff/Gazettes/OxMule.html.

13. EPT, 70. Carpenter, "Search for the San Elizario Road," 141–162, presents a sketch map of West Texas (p. 143) indicating the Quitman road. An 1864 or later date of construction is indicated by the absence of this route on the topographical maps of west Texas produced by the U.S. Army between 1849 and 1864. See also "Map of a Route from Austin City to El Paso del Norte &c," (hereinafter 1849 Ford-Neighbors map), Texas General Land Office; "Map of Texas and Part of New Mexico compiled in the Bureau of Topographical Engineers," NA; "Map of the Military Department of New Mexico, Drawn under the direction of Brig. Gen. James A. Carleton," Southwest Parks and Monuments Association; "Map of the Military District of the Pecos, Department of Texas," National Park Service, Fort Davis National Historic Site.

14. Pope Report, 8, 30; "Hueco Tanks State Historical Park," HTO.

15. Report of Capt. R. B. Marcy, November 20, 1849, "Reports of the Secretary of War, with Reconnaissances of Routes from San Antonio to El Paso," Senate Ex. Doc., No. 64, 31st Cong., 1st Sess., 200; Julyan, *Place Names of New Mexico,* 100

16. This route is annotated "wagon road 20 miles" on the 1849 Ford-Neighbors map. Carpenter suggests that this route south of the Huecos "may have been more of a possibility than a common route. Early explorers were often concerned with identifying feasible routes, many of which proved to be impractical." Since the primary evidence indicates a "wagon road," and since in 1863 the Paseños succeeded in creating a successful upgraded direct road across the waterless tableland to the salt flats, this author accepts the presence of the route shown on the Ford-Neighbors map. Carpenter, "Search for the San Elizario Road," 157–158; Ward, *Salt War,* 23.

17. *NYT,* February 1, 1854, April 17, 1854.

18. Bowden, 108–110, quoting the *Texas State Gazette,* January 24, 1878; *NYT,* February 1, 1854.

19. According to Padre Ortiz, "When the people were dispossessed of the salt lakes of the San Andres by the State of Texas, although they were situated within the limits of New Mexico, they then opened a road to the Guadalupe Salt Lakes, and for two or three years following the people of both sides took salt from them without any oppositions whatever. At the end of that time the people of this side of the river were forbidden the use of the salt lakes, and soon thereafter those of the other side were

also forbidden. This was the origin of the lamentable occurrences which have taken place." *EPT,* 68.

20. *EPT,* 104

21. *O.R.,* vol. L:267; *EPT,* 127.

22. Bowden, "Magoffin Salt War," 110.

23. *NYT,* February 11, 1854.

24. Garcia was born in San Elizario in 1819, the son of Pedro Garcia and Josefa Ana Olguin. He married Maria de los Santos Abillar in 1841. Lockhart, "Gregorio Nacianceno Garcia," 84; U.S. Census, 1860, San Elizario, EPC, Texas.

25. Lockhart, "San Elizario Presidio History," *San Elizario Plaza,* ed. Peterson, Graves, and Hill, 58; Hendricks/Timmons, *San Elizario,* 84, 115; Griffen, *Utmost Good Faith,* 30–32, 56–57; "Ronquillo Land Grant," *HTO.*

26. Hendricks/Timmons, *San Elizario,* 57; *O.R. Series* I, 15:580, 50:267.

Chapter 1

1. HTO, s.v. "Paleo Indians"; Sonnichsen, *Pass of the North,* 13–14, 105; U.S. Census, 1870, EPC, Texas, NA.

2. Bartlett, *Personal Narrative* I, 188–89, 190; Emory, *Boundary Survey,* 92.

3. Bartlett, *Personal Narrative* I, 190.

4. Sonnichsen, *Mescalero Apaches,* 63.

5. SEGHS, "Walking Tour," 19–21; Lockhart, "San Elizario Presidio History," 55–56, 58.

6. Dr. Arnoldo De León applied this description of the Tejano "middling class" to certain residents in mid-nineteenth-century South Texas. It equally describes the *leading families* of *la Isla.* Hendricks and Timmons, *San Elizario,* 56–57; De León, "*Raza* of Middling Status on the Chaparral, 1880–1900," *Tejano Epic,* 59–70.

7. "Ortiz, Ramon," *HTO.*

8. Bartlett, *Personal Narrative* I, 152; *El Paso Times,* June 25, 1916.

9. Hendricks and Timmons, *San Elizario,* 50; *General Directory . . . El Paso,* 40, 42; *Letters of Ernst Kohlberg,* 64; Bartlett, *Personal Narrative* I, 186, 187.

10. *Las Cruces Thirty-Four,* January 5, 1881; Emory I, 91–92.

11. Bartlett, *Personal Narrative* I:184; *Letters of Ernst Kohlberg,* 64.

12. Edwards, *Campaign in New Mexico* IV, 90.

13. Pope Report, 32.

14. De Bow, "Progress of the Great West," 32; Pierson to State, November 26, 1872; Paso del Norte Despatches.

15. Pierson to State, November 15, 1872, Paso del Norte Despatches

16. James G. Bell, "A Log of the Texas-California Cattle Trail, 1854," ed. J. Evertts Haley, *Southwestern Historical Quarterly* XXXV (January 1932), 234.

17. *Letters of Ernst Kohlberg,* 38–39.

18. De Bow, "Progress of the Great West," 33; Pierson to State, November 15 and November 30, 1872, Paso del Norte Despatches.

19. Pope Report, 8, 30; *General Directory of the City of El Paso for 1886–1887* (C.A. Gould & Co.), 30–31.

20. Griffen, *Utmost Good Faith*, 71, 72, 91.

21. Edwards, 99–100, 178–79; AGO Records, LR 1878, No. 2348, referring to report by Captain W. S. Henry to AGO in July 1850; Dawson, *Doniphan's Epic March*, 120–135; *Brooklyn Eagle*, May 3, 1848; Sonnichsen, *Pass of the North* I, 132.

22. Hendricks and Timmons, *San Elizario*, 69; Sonnichsen, *Pass of the North*, 128–29; Bartlett, *Personal Narrative* I, 156–66.

23. Hendricks and Timmons, *San Elizario*, 72–73; "William Treyer," *Biographical and Historical Sketchbook*, 42.

24. Timmons, *James Wiley Magoffin*, 157–65; Bartlett, *Personal Narrative* I, 123–25, II, 436–37; Martinez, *Border Boom Town*, 10–11; Davis, *El Gringo*, 383.

25. Sonnichsen, *Pass of the North*, 141, 142; Hendricks and Timmons, 78–79.

26. Burdette, ed., *American Biography and Genealogy*, 295; U.S. Census, 1850, Galena, Jo Daviess County, IL; *El Paso Times*, June 25, 1916.

27. *El Paso Times*, June 25, 1916.

28. Strickland, *El Sabio Sembrador*, ESS-16.

29. SEGHS, *Walking Tour*.

30. Hendricks and Timmons, *San Elizario*, 75, 84; Paddock, "Hon. Maximo Aranda," *History and Biographical Record of North and West Texas*, 441. The chapel now standing on the east side of the town's plaza was not constructed until after the Salt War, with completion in 1882.

Chapter 2

1. Osorio, "Villismo," 97; Hendricks and Timmons, *San Elizario*, 51–61.

2. For a history of Mexican-Apache relations in Chihuahua, see Griffen, *Utmost Good Faith*. Sonnichsen, *Mescalero Apaches*, 62; Halla Jr., "Bi-ethnic Community," 47.

3. Mills, *Forty Years*, 22; *Brooklyn Daily Eagle*, March 3, 1859; Wadsworth, *Forgotten Fortress*, 261–62; Baylor, *Into the Far Wild Country*, 276.

4. Alberts, *Battle of Glorieta*, 9.

5. Wilson, *When the Texans Came*, 153; Mills, *Forty Years at El Paso*, ed. Strickland, xv-xvi, 37–52.

6. Alberts, *Battle of Glorieta*, 136; *OR*, series I, IX:714.

7. *OR*, IX:678.

8. Wilson, *When the Texans Came*, 311, 313–14. The index to Wilson incorrectly indicates that the fight occurred in Socorro, New Mexico.

9. *OR*, IX:722; Wilson, *When the Texans Came*, 311; *OR*, IX:585–93.

10. Wilson, *When the Texans Came*, 311–12; *OR*, IX:722; *OR*, IX:593.

11. *OR*, IX:566–67, 602.

12. Carleton to West, October 17, 1862, LS, Ninth Military Department, Department of New Mexico, and District of New Mexico, 1849–1890 (NA, RG393, NAMS

M1072, Roll 3) (hereinafter "LS New Mexico"); Baylor, *Into the Far Wild Country,* 272; *OR,* L, part 2, 109, 110–11.

13. Utley, *Frontiersmen in Blue,* 233; *OR,* XV:599–601.

14. *OR,* XV:576; Boatner, *Civil War Dictionary,* 122–23, 902; Heitman, *Historical Register* I, 282, 1020; *OR,* XV:580.

15. Carleton to West, October 19, 1862, Cutler to Garcia, October 20, 1862, LS New Mexico.

16. Indian Depredations Case File 4281, U.S. Court of Claims, RG 205.

17. Cutler to West, October 19, 1862, LS New Mexico; *OR,* series I, 50:213.

Chapter 3

1. Broaddus, *Legal Heritage,* 85; Mills, *Forty Years at El Paso,* xvii–xviii, 27–28; Gibson, *Life and Death,* 50.

2. Owen, *Two Alberts,* 9–12, 18; *OR* L,128–132.

3. *Rio Grande Republican,* January 30, 1891.

4. Gibson, *Life and Death,* 48, 50.

5. "Maverick, Samuel Augustus," *HTO,* http://www/tshs.utexas/edu/handbook/online/articles/view/MM/fma84.html; accessed October 4, 2002; "Texas Land Grants: The Land Grant Process in Texas after 1836," Texas General Land Office, www.glo.state.tx.us/archives (accessed October 4, 2002).

6. *EPT,* 127; Fountain Deposition; Fountain to Jones, March 4, 1878.

7. *EPT,* 128; Fountain Deposition.

8. Strickland, ed., Mills, *Forty Years,* xix.

9. Owen, *The Two Alberts,* 49–50; Gibson, *Life and Death,* 76. U.S. Census, 1860, Gilead Township, Branch County, MI, 642; U.S. Census, 1870, El Paso, EPC, Texas, 2.

10. *EPT,* 127–28; Fountain Deposition; Mills, *Forty Years at El Paso,* 146.

11. Hendricks, "Antonio Severo Borrajo," 1. His age is stated as fifty in the 1870 Federal Census for El Paso County, Texas.

12. Hendricks, "Antonio Severo Borrajo," 1–2.

13. "Ortiz, Ramon," *HTO;* Halla, "Bi-Ethnic Community," 216–17.

14. Sonnichsen, *El Paso Salt War,* 10.

15. Ibid. *EPT,* 127; Fountain Deposition.

16. *Chicago Daily News,* December 23, 1877; *Austin Statesman,* July 16, 1876; El Paso County Property Deed Books, C:205–209, 210, 297, 299; D:112; Internal Revenue Assessment Lists for Texas, 1865–1866 (NAMS M791, Roll 2), RG 058, NARA.

17. El Paso County Criminal Case No. 10, MF 524, Roll 1, UTEP. The episode was not Cardis's only run-in with the law that year. He was one of many arrested for "gaming" (Criminal Case 57), a charge typically brought by western communities in need of more revenue.

18. Cardis may have filed for naturalization in New York as early as 1864. Broaddus, *Legal Heritage,* 93; *Chicago Daily News,* October 23, 1877, and January 2, 1878.

19. *EPT,* 128.

20. Ibid.

21. Rick Hendricks correspondence with author, May 8, 2006.

22. Owen, *Two Alberts,* 55–67.

23. Owen, 68.

24. *EPT,* 128–129.

25. Ibid.

26. Ibid.

27. For Fountain's opposition to turning New Mexico's public schools over to the Society of Jesus, see *MVI,* January 26, February 9, and March 2, 1878.

28. Lopez, "La Expulsión del Padre Borrajo," based on the Jefe Politico's letter of June 3, 1866, found in the Archivo Municipal de Ciudad Juárez. Translation by Manuella Cárdenas-Oellermann.

29. Jose de Jesus Baca et alia, "Proposal for the creation of a bishopric of El Paso," September 21, 1871, cited in Hendricks, "Antonio Severo Borrajo," 3; Lopez, "La Expulsion del Padre Borrajo"; "Mexico," *Encyclopedia Britannica* 15 (1956), 391–92.

30. Lopez, "La Expulsion del Padre Borrajo."

Chapter 4

1. Owen, *The Two Alberts,* 70–71.

2. Gibson, 79; "Reminiscence of John D. Davis," El Paso Pioneer Society Files, El Paso Historical Society.

3. Gibson, 79.

4. For all crimes, Atkinson made the greatest number of recorded arrests (40), followed by Mejia (29). AGC, RG 401: General Corr., Record of Arrests by State Police 1870–1873, Boxes 1101–1102, TSLAC; Atkinson to A.G., April 13, 1872, Register of LR, AGC, Box 702, TSLAC;

5. Mills, *Forty Years,* 122–123.

6. Owen, *The Two Alberts,* 74; *Las Cruces Borderer,* January 1872; *Austin Tri-Weekly Statesman,* December 14, 1871, January 30, 1872.

7. *Flake's Daily Bulletin,* March 13, 1872, cited by Sonnichsen, *Pass of the North,* 191; "William Treyer," *Sketch Book of EPC,* 42; *Austin Daily Statesman,* April 30, 1872.

8. *Austin Daily Statesman,* April 30, 1872.

9. *Austin Daily Statesman,* April 30, 1872, June 15, 1872.

10. *Austin Daily Statesman,* June 15, 1872. Cardis himself had been charged with attempting to bribe Antonio Levan to kill Joseph Magoffin in April 1871. A date in district court was set for January 1872, but evidence of further judicial action is lacking. Obviously, Cardis suffered no significant consequences. El Paso Old Criminal Case File No. 10, MF 524, Roll 1, UTEP.

11. Rev. Tays settled in Indianola, Texas, until the yellow fever epidemic of 1867 carried off his wife, one brother, and a sister-in-law. Following this tragedy, he transferred to a new parsonage in Bryan, Texas, northeast of Austin. Here he instigated con-

troversy by marrying a white man and African American woman. "For this offence against decency," said the *Houston Times*, "many of his flock at Bryan refused to hear him preach." When Senator Fountain approached the Episcopalian bishop of Texas, Alexander Gregg, with a request for a pastor to establish a Protestant church in El Paso, Gregg suggested Tays. Fountain got Tays elected to the post of Senate Chaplain so that the politician could gauge the minister's effectiveness for himself. After the end of the 1870 session, Tays packed his belongings and with two sons made the long trek to El Paso. George H. Tays Military Pension File, No. XC2-291408, Indian War Service Records, Compiled Military Service Records, AGO Records, RG 94, NARA; Houston Times quoted in Bill Page, "Miscellaneous Brazos County Marriage and Divorce Notices, 1852–1889," *Brazos Genealogist* XVIII (2001): 3, http://user.txcyber.com/~bga/data/mar_div.txt; Senate Bills 206, 216 and 217, 12th Legislature, 1st (Regular) Session, Senate Journal, 335, 359; Lockhart, "Incorporations of San Elizario" 169–70.

12. Eickhoff, *Exiled*, 60–90; EPC Archives, Property Deed Book I:130–31; U.S. Census, 1870, Ysleta Township, EPC, Texas.

13. Owen, *Two Alberts*, 73; *EPT*, 129; Hendricks, "Antonio Severo Borrajo," 3.

14. *EPT*, 127; *OR*, series I, 15:229; California AG, Records of California Men.

15. The Ellis's hacienda boasted a sala, or ball room, with painted walls and ceiling, a private chapel, and twelve other rooms. Painted canvas cloth murals hung from the vigas (beams) and moldings. The hacienda's enclosed space included two patios plus a carriage house, stables, and a grape arbor. Mrs. Olga Trujillo de Hernandez, a descendant of Teodora's family, recalls the white grapevines that still grew in the patio into the mid-twentieth century. El Paso Times, August 4, 1939; SEGHS, *Walking Tour*, 20–21; Sonnichsen, *El Paso Salt War*, 22; Conversation with Olga Trujillo de Hernandez, March 1, 2004; U.S. Census, 1870, San Elizario, EPC, Texas, 5; *EPT*, 66.

16. Sonnichsen, *El Paso Salt War*, 22; *Chicago Daily Tribune*, January 2, 1878; *Sedalia Daily Democrat*, January 6, 1878.

17. Sonnichsen, *El Paso Salt War*, 22; John G. Atkinson v. Jacoba Alvorado, District Court, EPC, September 1868 Term (EPC Archives, UTEP); Records of Appointments of Postmasters, NAMS M841, Roll 123, RG 28.

18. *EPT*, 66; *Las Cruces Borderer*, June 22, 1871.

19. Sperfechter was discharged on May 10, 1860, and afterwards claimed this date as the end of his service. Army records indicate that he reenlisted on the same date, only to be reported as a deserter on August 22, 1860. His file reads, "No record of surrender or apprehension." Charles Kerber Military Pension File, Certificate No. 1051596, Civil War and Later Pension Files, Records of the Department of Veterans Affairs, RG 15, NARA (hereinafter "Kerber file"); Charles Kerber Military Records, Colorado State Archives and Public Records; Returns for Third Infantry Regiment, January 1860–December 1866, NAMS M665, Roll 33, Returns for Regular Army Infantry Regiments, June 1821–December 1916 (hereinafter "RA Returns"), NARA.

20. Kerber file; Gary Leland Roberts, "Sand Creek," 122–123, citing *Denver Republican,* April 20, 1890; *OR,* vol. IX, 536; Edrington and Taylor, *Battle of Glorieta Pass,* 73.

21. Kerber file; Monthly Returns, June 1864–October 1865, Fort Garland, Colorado, Returns from Military Posts 1800–1916; AGO Records, RG 94, NARA; *Saguache Crescent,* April 4, 1991.

22. *El Paso Herald,* July 16, 1909.

23. *EPT,* 66.

24. *Las Cruces Borderer,* May 15, 1872.

25. Telesforo Montes record, AG Records, Frontier Force, TSLAC. Montes's age was variously given as forty-three in 1860, forty-eight in 1870 (U.S. Census, 1860 and 1870, San Elizario, EPC, Texas), and forty-five in 1872 (Frontier Force Records).

26. Sonnichsen, *El Paso Salt War,* 23; *Las Cruces Borderer,* May 15, 1872.

27. Sonnichsen, *El Paso Salt War,* 23; Kerber to Coke, August 18, 1875, Gov. Coke Corr.; Atkinson to AG, June 8, 1872, AGC, Register of LR, RG 401, Box 702, TSLAC.

28. *EPT,* 66, 77; Owen, *Two Alberts,* 73.

29. *EPT,* 66

30. Jose de Jesus Baca, Jose Antonio Real y Vasquez, Antonio Severo Borrajo, and Juan de Jesus Trujillo, "Proposal for the creation of a bishopric of El Paso," September 21, 1871, quoted in Hendricks, "Antonio Severo Borrajo," 3

31. Baca, Jesus Corral, and Borrajo to Salpointe, San Elizario, November 24, 1872, quoted in Hendricks, "Antonio Severo Borrajo," 4–5.

32. *EPT,* 98–99; Catholic Church Records, San Elizario, Texas, MS 473, Roll 1, UTEP. Bishop Salpointe's misadventure with Borrajo are not mentioned in his memoir, *Soldier of the Cross,* ed. Odie B. Faulk (Diocese of Tucson, Arizona, 1966).

33. For biographical details regarding Bourgade, see Salpointe, *Soldier of the Cross,* 162–169. See also www.archdiocesesantafe.org, and "Clerical Leaders of Arizona," from "A heritage of Loving Service: The Sisters of St. Joseph of Carondelet in Tucson," at www.library.arizona.edu.

34. *EPT,* 98–99.

Chapter 5

1. *EPT,* 65; *Galveston News,* October 16, 1877, November 19, 1877; U.S. Census, 1860, Lavaca County, Texas, 24; HTO, s.v. "Charles Howard." http://www.tsha.utexas.edu/handbook/online/articles/view/HH/fhopy.html

2. Daniel Duval Family Patriot's Page, http://patriotspage.duvalfamilyassociation.com/civil_war.htm (accessed October 3, 2004); *Galveston Daily News,* December 27, 1877.

3. *Fort Worth Daily Democrat,* October 16, 1877.

4. U.S. Census, 1870, Belmont Precinct No. 3, 6, Gonzales County, Texas, and San Antonio, 2nd Ward, Bexar County, Texas, 166; *Galveston Daily News,* December 27, 1877; *EPT,* 65; Ward, "Salt War of San Elizario (1877)," 35, note 49. Thomas Jefferson

Devine (1820–1890), was the first president of the Texas state bar association. In 1870, Charles's sister Susan married William Lewis Davidson, who ultimately became assistant attorney general of Texas and later a justice on the civil court of appeals. HTO, s.v. "Thomas Jefferson Devine."

5. *EPT,* 122; *San Antonio Daily Herald,* March 24, 1872; *EPT,* 65; Fountain Deposition.

6. *San Antonio Daily Herald,* March 24, 1872, August 2, 1872, December 17, 1872.

7. Woods to Texas Secretary of State, August 8, 1872; Fountain to AG, September 11, 1872, AGC, Register of LR, vol. 401–703.

8. Atkinson to AG, October 2, October 9, 1872; Fountain to AG, October 11, 1872, AG Corr., Register of LR, Vol. 401–703.

9. Howard to AG, October 12, 1872, Cardis to AG, October 15, 1872, Fountain to AG, October 16, 1872, AGC, Register of LR, vol. 401–703; *Austin Tri-Weekly Statesman,* November 2, 1872.

10. Howard to W.W. Mills, December 30, 1872, MC.

11. Kerber, Aranda et al. to Registrar, EPC, November 8, 1872; Records of the Secretary of State, County Petitions and Protests, EPC Folder, RG 307, Series 84, Box 2-12-565, TSLAC (hereinafter 1872 Election Returns).

12. The original affidavits are not on file in the Archives. 1872 Election Returns.

13. Howard to Mills, December 2, 1872, MC; Mills to Newcomb, November 28, 1872; 1872 Election Returns.

14. Cardis to Mills, December 1, 1872, Howard to Mills, December 2, 1872, MC.

15. Ibid.

16. Mills to Newcomb, November 28, 1872; Newcomb to Mills, December 7, 1872, Petition of EPC Citizens re: Legality of elections, Box 100-394, TSLAC (hereinafter Petition).

17. Mills to Newcomb, December 9, 1872, Petition.

18. Newcomb to Mills, December 9, 1872, Petition.

19. Petition of Chas. Howard et al. to Secretary of State, undated, 1872 Election Returns.

20. Mills to Newcomb, December 14, 1872, 1872 Election Returns.

21. Affidavit of Charles H. Howard, December 25, 1872, 1872 Election Returns.

22. Affidavits included (1) Howard et al., December 25, 1872; (2) Socorro voters, December 25, 1872; (3) San Elizario voters, December 31, 1872; (4) Ysleta voters, January 1, 1873; (5) Telesforo Montes, December 25, 1872; (6) J. K. McCarty, December 25, 1872; (7) E. W. Bates, December 25, 1872; (8) A. H. French, December 28, 1872; (9) John Hale et al., December 29, 1872, 1872 Election Returns; Letter of Howard, Wahl, et al. to Governor Davis, December 23, 1872, 1872 Election Returns; French to AG, December 23, 1872; Fountain to AG, December 25, 1872, AGC, Box 401-703.

23. Howard to Mills, December 30, 1872, MC.

24. Mills to Newcomb, December 26, 1872, 1872 Election Returns.

25. Howard to Mills, December 2, 1872, MC; Broaddus, *Legal Heritage,* 102–103.

26. Petition of Residents of EPC to Legislature of the State of Texas via Gov. A. J. Hamilton and W. W. Mills, February 19, 1873, 1872 Election Returns.

27. HTO, s.v. "Reconstruction."

28. *El Paso Sentinel,* November 8, 1873, CAH.

29. *El Paso Sentinel,* November 8, 1873; *El Paso Times,* April 12, 1884; Middach, *Frontier Newspaper,* 7.

30. *El Paso Sentinel,* November 8, 1873. Text of poem courtesy the Center for American History, the University of Texas at Austin.

31. Howard to Mills, December 30, 1872 and May 7, 1873, MC; Howard et al., Letter to the Editor, *Austin Daily Statesman,* November 21, 1873.

32. Records of the Secretary of State, Elections, RG 307, Series 84, Box 2-12/576, EPC 1873; *Austin Daily Statesman,* December 12, 1873; *MVI,* November 3, 1877; *Chicago Daily Tribune,* January 2, 1878.

33. "William Treyer," *Sketch Book of EPC,* 42.

34. Fountain Deposition; *EPT,* 128.

Chapter 6

1. Broaddus, *Legal Heritage,* 118–19; El Paso *Sentinel,* November 8, 1873; Property Deed Books, vol. E:173–175, 223, I:98, EPC Archives, SSC; Campbell, *Gone to Texas,* 306, 308; "RECONSTRUCTION," HTO.

2. Broaddus, *Legal Heritage,* 119.

3. Sonnichsen, *Pass of the North,* 196.

4. Las Cruces *Borderer,* May 22 and June 5, 1872; Pierson to State, March 17, 1873, Paso del Norte Despatches; *Mesilla News,* June 6, 1874; El Paso *Sentinel,* November 8, 1873.

5. *Mesilla News,* May 9, May 16, May 30, 1874.

6. *Mesilla News,* June 6, 1874, citing *The Borderer;* Sonnichsen, *Pass of the North,* 196.

7. *Mesilla News,* August 29, 1874.

8. Newcomb followed Fountain to Dona Ana County, New Mexico, where he served successfully as a district attorney. Broaddus, *Legal Heritage,* 106, 117–18; *San Antonio Express,* April 17, 1874.

9. Broaddus, *Legal Heritage,* 119–20; Greene, *900 Miles on the Butterfield Trail,* 114; *MVI,* October 6, 1877.

10. *EPT,* 65, 122.

11. *Mesilla News,* April 18, 1874

12. El Paso County Deed Book F, 325; undated letters of Rafael Telles, Juan Armendáriz and Jose M. Gonzales to Governor Coke; Howard to Coke, March 29, 1875; Cardis to Coke, March 29, 1875; Hague to Coke, May 29, 1875; Governor Coke Corr.; *EPT,* 122–23.

13. Cardis to Coke, August 29, 1875; Howard to Coke, September 13, 1875; Governor Coke Corr.; Fountain Deposition.

14. Cardis to AG Steele, March 30, 1874, endorsing Letter of A.H. French, March 7, 1874, AGC.

15. Howard to Coke, July 2, 1875, August 16, 1875, August 19, 1875, RLR, Governor Coke Corr.

16. Kerber to Coke, August 18, 1875, Governor Coke Corr.

17. Coke to Kerber, September 4, 1875, Goveror Coke Corr.

18. *San Antonio Herald*, July 5, 1875, *Austin Daily Statesman*, July 17, 1875, August 19, 1875.

19. "Constitutional Convention of 1875," HTO; Hammon, "West Texas." excerpt located in "Louis Cardis" Vertical File, CAH; McKay, *Debates*, 176.

20. *Texas Constitutions Digitalization Project, Tarlton Law Library, University of Texas at Austin*, http://www.law.utexas.edu/constitutions/index2.html.

21. *Chicago Daily News*, December 26, 1877; Antoinette Duran Silva, "Notes regarding my Grandfather, Jose Jesus Jurado, his mother Tiburcia Jurado, and Charles Henry Howard." Shared with author June 29, 2006.

22. George B. Zimpelman, Compiled Service Record (NA Microfilm Pub. M323, Roll 53) (hereinafter Zimpelman Record); John M. Claiborne, "Brave George Barnhart Zimpelman," *Confederate Veteran* 19 3, 114, Online Archive of Terry's Texas Rangers, www.terrystexasrangers.org; *Biographical Encyclopedia of Texas* (New York: Southern Publishing Co., 1880), na.

23. Zimpelman Record; Hatley, *Bringing the Law to Texas*, 39–53; Registers of Felonies Committed 1866–1868, Department of Texas and 5th Military District 1865–1870, U.S. Continental Army Commands, RG 393, NARA.

24. *Biographies of Leading Texans* IV, 872–73; Brown, *Annals of Travis County*, 4–28, 30, 33–35, 38–39. Brown cites the *Austin Gazette* editions of January 16 and 17, 1874.

25. *Austin Democratic Statesman*, November 12, November 14, November 27, 1875; Ward, "El Paso Salt War," 35, note 49; *Austin Statesman*, January 1, 1876. Sonnichsen, *Pass of the North*, 444.

26. R. M. Keating v. C.H. Howard, El Paso County, Old Civil Case Files, No. 225.

27. A. J. Fountain Deposition; Williams, *Texas' Last Frontier*, 187; E. W. Bates Deposition.

Chapter 7

1. For Blacker's biography, see Paddock, *Biographical Record of North and West Texas* I, 501–03 at www.GenealogyMagazine.com. Other countywide winners included County Judge Joseph Magoffin, incumbent County Clerk G. W. Wahl, and County Assessor Charles Ellis. Pedro Candelano was elected county treasurer, while Tomas Maese took the surveyor post. Several lesser offices, including justice of the peace, county commissioner and constable, also went to Paseños. Contested Elections, EPC, 1876, Box 2-12/583, TSLAC (hereinafter "Contested Elections 1876"); *San Antonio Daily Herald*, April 21, 1876; Election Returns, 1876, EPC, Records of the Secretary of State, RG 307, Series 84, TSLAC (hereinafter "Elections 1876").

2. Contested Elections 1876.

3. House Journal, 15th Legislature, 1st Session, April 29, 1876, 97; Contested Elections 1876.

4. Treyer's testimony in the dispute over the district judgeship indicates that he had split with Cardis. Perhaps this occurred during the campaign, with the result that Kerber came out on top in the contest for sheriff.

5. Sonnichsen, *Pass of the North,* 310–44.

6. Case Files No. 207, 247, 248, 253, and 272, all State of Texas v. John G Atkinson, Criminal Docket District Court 1867–1877, El Paso County Archives, Record 1-5-3; *EPT,* 66.

7. See State of Texas v. John W. Campbell, El Paso County Old Criminal Case No. 275, Microfilm Series 524, Roll 1, EPC Archives, SSC; John G. Atkinson v. Juana Semienos [nee Campbell], Old Civil Case No. 215, EPC Archives, SSC. The 1870 federal census states that Juana's age was six, making her eleven or twelve at time of marriage. This is doubtful. See U.S. Census, 1870, San Elizario Township, EPC, Texas, 19, line 20.

8. I assume that the transcriber of Atkinson's statement erred in recording the warning as the question, "Que dado, John?" "Cuidado!" or "Look out!" makes far more sense here.

9. *EPT,* 122–123.

10. EPC Commissioner Court Minutes, Box 4E 139, 153–157, CAH; Mills, *Forty Years,* 180.

11. *Austin Daily Statesman,* June 27, 1877; Deposition of A. J. Fountain, Howard v. Zimpelman.

12. Hague to Newcomb, September 22 and October 24, 1877, James P. Newcomb Papers, Barker Library, University of Texas at Austin, quoted in Halla, "El Paso, Texas and Juarez, Mexico," 219.

13. *EPT,* 100; *MVI,* October 6, 1877; *EPT,* 69.

14. *EPT,* 120.

15. New Mexico's assassinations included that of Chief Justice John Slough in 1867 and the mysterious murder of A. J. Fountain and his young son in 1896. For the story of the Slough-Rynerson quarrel and a discussion of political violence in New Mexico Territory, see Roberts, *Death Comes to the Chief Justice.* Owen, *Two Alberts,* 69–72; *Austin Daily Statesman,* April 30, 1872; Fountain Deposition, Ninth and Tenth Interrogatories; *EPT,* 128.

16. *Chicago Daily Times,* January 2, 1878; U.S. Census, 1870, EPC, Texas, 410, line 38; Property Deed book D:271, EPC Archives; Early Marriages of El Paso County, at http://freepages.genealogy.rootsweb.com/~janarmstrong/elpmgs.html; Records of the Texas Secretary of State, Election Returns, EPC, 1873, TSALC; *EPT,* 69, 123.

17. *EPT,* 69, 120, 123.

18. *EPT,* 69.

19. The three sections in what is now Hudspeth County, were finally patented to

"George B. Zimpleman" [*sic*] on May 9, 1879. See Texas Land Title Abstracts, www.Ancestry.com; *MVI,* October 6, 1877; *EPT,* 78.

20. *EPT,* 123, 65.

21. Ward, Christopher, *War of the Revolution* I (New York: MacMillan Co., 1952), 8.

22. Halla, Jr., "El Paso, Texas, and Juarez, Mexico," 220; *EPT,* 123.

23. *EPT,* 123, 129.

24. *MVI,* October 6, 1877; *EPT,* 123; Kerber and Blacker to Governor R.B. Hubbard, September 14, 1877, AGC, Box 401-396.

Chapter 8

1. Deposition of Price Cooper, George H. Giddings v. U.S. Court of Claims, Indian Depredations Case File No. 3873 (NARA, RG 123); Texas General Land Office records. U.S. Federal Census, Ysleta, El Paso County, Texas, 1860, 1870, 1900; *El Paso Herald,* November 14, 1900.

2. Deposition of Price Cooper, George H. Giddings v. U.S. Court of Claims, Indian Depredations Case File No. 3873 (NARA, RG 123).

3. Deposition of Price Cooper, Giddings v. U.S., Case No. 3873; Austerman, *Sharps Rifles and Spanish Mules,* 84.

4. Price Cooper v. United States, Case No. 1651 (NARA, RG 123).

5. Utley, *Lone Star Justice,* 195. For similar broad-brush comments, see Sonnichsen, *Ten Texas Feuds,* 138, and Robinson, *The Men Who Wear the Star,* 225.

6. Sonnichsen, *Mescalero Apaches,* 132–34.

7. Winfrey and Day, eds., *Indian Papers* 4, 169–72.

8. The captain was to be paid one hundred dollars per month, the lieutenant eighty dollars, the three sergeants fifty-four dollars, the four corporals fifty-two dollars, and the fifty privates fifty dollars each. Wilkins, *The Law Comes to Texas,* 6.

9. Wilkins, 7; Garcia to A.G., December 3, 1870; Garcia to A.G., December 6, 1870; Return of Company N, Texas Frontier Forces, December 1870, AG. Records.

10. "Needle guns" were cartridge conversions of the muzzle loading Springfield musket. Garcia to A.G., January 7, 1871; Return of Company N, February, 1871; Garcia to A.G., March 31, 1871, A.G. Corr., Box 401-390.

11. Garcia to A.G. to April 17, 1871; A.G.'s Abstract of Articles transferred to Frontier Force, May 31, 1871, Box 401-390; Garcia to A.G., June 16, 1871; Wilkins, *Law Comes to Texas,* 16; Montes to A.G., September 13–14, 1871, *Indian Papers* 4, 384.

12. Garcia to A.G. March 31, 1871; Garcia to A.G., June 16, 1871; Garcia and Montes Payment Vouchers, A.G., Corr.

13. *Mesilla News,* March 7, 1874; Cardis to A.G., March 30, 1874, State Police Corr; Muster and Pay Roll of Company of Frontier Men, El Paso County, May 27, 1874.

14. Montes to A.G. William Steele, May 31, 1874; S. Schutz & Bro. voucher, June 15, 1874, Frontier Force records; Montes to Steele, August 5, 1874.

15. Montes to Steele, September 27, 1874, Box 401-704.

16. Montes to Steele, September 27, October 20, October 28, and November 27, 1874.

17. Montes to Steele, September 18, 1875.

18. Montes to Steele, October 21 and November 1, 1875; Montes to Steele, November 20, 1875; Montes to Steele, January 11, 1876.

19. Montes to Steele, April 21, 1876; Miguel Garcia to Steele, April 9, 1876.

20. Montes to Steele, April 21, 1876; Miguel Garcia to Steele, April 21, 1876.

21. Miguel Garcia to Steele, April 21, 1876; Montes to Steele, April 21, 1876.

22. Montes to Steele, April 21, 1876; Frederick Nolan, "San Augustine Ranch," *True West,* June 1999, 44.

23. Juan N. Garcia to Steele, May 22, 1876.

24. Juan N. Garcia to Steele, May 22, 1876.

25. *Texas Frontier Troubles,*105; *Kohlberg,* 40, 44; Rodenbough and Haskin, *Historical Sketches,* 697.

26. Besides Barela and Salcido, the Texas Rangers indicted for murder in the Salt War were Gabino Arias, Sisto Castillo, Lino Guerra, and Pedro Sierra (all Garcia's company), plus Dionisio Guerra, José Hernandes, Pomposo Paz, and Ysidro Sierra (all Montes' company). See Muster Rolls for Company N, 1870–1871, and El Paso Company, Frontier Men, 1874; EPC Criminal Case No. 282. The assumption that these Rangers are the same-named Salt War insurgents is based on their known willingness to fight for their community's safety and future.

27. The story of McDaniels and the Maese gang is taken from Howard to Coke, December 8, 1875, Governor Coke Corr.; *Mesilla News,* August 11, 1875.

28. "Muster Roll of Detachment in El Paso County, Texas State Troops, November 1877," A.G. Records, Ranger Records, TSALC; Howard to Coke, December 8, 1875.

29. Howard to Coke, December 8, 1875; *Mesilla News,* August 11, 1875.

30. *Mesilla News,* August 11, 1875.

31. Howard to Coke, December 8, 1875.

32. Cool, "Several Lives and Suggested Death of Jim McDaniels," *Revenge!,* 21–40.

33. Ibid.

34. *Mesilla News,* January 8, 1876; Devin to A. A. G., Dept. of New Mexico, January 5, 1876, Letters Received, Department of New Mexico (NARA, Microfilm Pub. M1088, Roll 27).

35. See Frederick Nolan, "Boss Rustler," *True West,* September 1996, 14–21; October 1996, 12–19; Rasch, *Desperadoes of Arizona Territory,* 33–39.

Chapter 9

1. The *Zone Libre,* or "Free Trade Zone," was placed into effect by the States of Tamaulipas and, intermittently, Chihuahua, in contravention of Mexican national policy, as a way of bolstering the local economies. The side effects included increased smuggling of goods into the United States. See Martinez, *Troublesome Border,* 112–13.

The contemporary official American attitude toward this policy is discussed in *Relations of the United States with Mexico,* xix–xxi.

2. Gregg, *Border Troubles,* 12, 69.

3. Bancroft, *History of Mexico* VI, 419–21; *NYT,* January 11, 1877.

4. *American Heritage Encyclopedia of American History,* s.v., "Stalwarts and Half-Breeds."

5. Sherman to Sheridan, November 29, 1877, Papers of William Tecumseh Sherman, Microfilm Roll 45, LC. See also Utley, *Frontier Regulars,* 66, 362–63 for the Texas delegation's key role in "saving the Army Bill" in 1877. For information on Gustave Schleicher, see http://bioguide.congres.gov and HTO, s.v. "SCHLEICHER,GUSTAV,"http://www.tsha.utexas.edu/handbook/online/articles/view/SS/fsc9.html.

6. Bryan to Hayes, August 24, 1878, R.B. Hayes Papers, RBHPCL.

7. Webster, "Texan Manifest Destiny," 181; Bryan to Hayes, October 26, 1877, R.B. Hayes Papers, RBHPCL.

8. *MBT,* 99–100.

9. Hagerstown (Maryland) *Herald and Torch,* July 11, 1877.

10. Gregg, *Border Troubles,* 43–44, 51–52.

11. *MBT,* 158; *EPT,* 135.

12. *MVI,* June 9, 1877; *EPT,* 135.

13. *EPT,* 136; *Letters of Ernst Kohlberg, June 5 1877,* 56.

14. *MBT,* 136, 167–68, 153–54.

15. *MBT,* 153, 154, 155, 219.

16. Louis Henry Rucker ACP file, OAG, RG 94, NARA; *OR,* series I, XLVI, part 1, 468.

17. Rucker File.

18. The others were the 10th Cavalry and the 24th and 25th Infantry Regiments.

19. Edward Hatch (1832–1889) needs a biographer. As a colonel, he took part in Grierson's famous raid south from La Grange, Tennessee, though he was detached at Columbus, Mississippi, to divert Confederate attentions from Grierson's main body. He commanded the 5th Div., Cavalry Corps against Hood and particularly distinguished himself at Nashville. Kenner, *Buffalo Soldiers and Officers,* 41; Hatch to Sec. of War, April 2, 1877, Rucker File.

20. *MBT,* 219–20.

21. *MBT,* 220–21.

22. *MBT,* 115, 116, 228.

23. *TFT,* 50; Webster, "Texan Manifest Destiny," 203.

24. *TFT,* 11–12.

25. *TFT,* 9, 50–51; Webster, "Texas Manifest Destiny," 204–209.

26. Webster, "Texas Manifest Destiny," 205, 209; Governor Hubbard Corr., September 1877.

Chapter 10

1. *Letters of Ernst Kohlberg*, 49.

2. Peterson, Graves, and Hill, *San Elizario Plaza Archeological Project*, 45–46.

3. *MVI*, September 22, 1877, October 20, 1877.

4. *EPT*, 54, 61, 72, 106, 118; *MVI*, October 6, 1877.

5. The record does not specify where Juarez was held by the sheriff. *EPT*, 106, 118.

6. *EPT*, 106–107.

7. For the role of law in the El Paso Salt War, see Harvard Law Review Association, "Law, Race, and the Border: The El Paso Salt War of 1877," *Harvard Law Review* 117 (January 2004), 941–63.

8. *EPT*, 54.

9. Kerber testified that they were joined on the roof by "Howard's sureties." No other evidence places Tomas Garcia, Juan Cobas, John Atkinson, and Charles Ellis in Ysleta on this night. *MVI*, October 6, 1877; *EPT*, 54, 59.

10. Lino Granillo, U.S. Census, 1870, San Elizario, EPC, Texas, 24; Francisco Barela, 1870 U.S. Census, Ysleta, EPC, Texas, 2.

11. *MVI*, October 6, 1877; *NYT*, January 5, 1878; *EPT*, 151–52, 142; Sonnichsen, *El Paso Salt War*, 31.

12. *MVI*, October 6, 1877; *EPT*, 151–52.

13. *EPT*, 151–52; MVI, October 6, 1877.

14. *EPT*, 140; *NYT*, January 5, 1878.

15. The author has found no information to identify J.R. Mariani. *EPT*, 60, 151–52.

16. *Atlanta Daily Constitution*, October 11, 1877; *Boston Daily Globe*, December 9, 1877; *EPT*, 71–72, 98–99.

17. *Galveston Weekly News*, October 8, 1877; *EPT*, 151. The quotation appears in EPT as, ". . . there was an order for my imprisonment to prevent me to risk (?) Judge Howard." [The "(?)" was evidently entered by the transcriber as an indication of his difficulty in making out the handwriting.]

18. Rucker to Hatch, October 2, 1877, Special Files, El Paso Troubles (NARA Microfilm Publication M1495, roll 14), RG 94, NARA (hereinafter Special Files).

19. *EPT*, 61, 99.

20. *EPT*, 152.

21. *MVI*, October 6, 1877; *EPT*, 151–52, 95.

22. *MBT*, 126–27.

23. *MVI*, October 6, 1877.

24. Ibid.

25. Bourgade did not testify as to whether the rites were in fact administered. *EPT*, 99; *MVI*, October 6, 1877.

26. *EPT*, 99, 151–52; *MVI*, October 6, 1877, December 22, 1877.

27. McBride is not mentioned in any account of this journey, but his separation from Howard at this point is questionable.

28. *MVI*, October 6, 1877; *EPT*, 151–52.

29. *MBT*, 126–27.

30. Wyatt-Brown, *Honor and Violence*, viii.

31. Ibid., 14; Freeman, *Affairs of Honor*, xix–xx; *Galveston Weekly News*, November 19, 1877.

32. Wyatt-Brown, "Honor's History Across the Academy," Historically Speaking, June 2002, http://www.clas.ufl.edu/users/bwyattb/honorhistory.html; Wyatt-Brown, *Honor and Violence*, viii.

Chapter 11

1. *EPT*, 140–41, 151; *Galveston News*, October 8, 1877.

2. *MBT*, 121; *Chicago Daily Tribune*, October 7, 1877.

3. *EPT*, 151–52. Note that the telegraph is dated October 5, which was a Friday, but relates events that happened on "Saturday."

4. *EPT*, 152–53.

5. Jos. Magoffin to Hubbard, October 5, 1877, WC.

6. *EPT*, 61–62.

7. *EPT*, 152; Kerber to Hubbard, October 10, 1877, WC.

8. *MVI*, October 6, 1877; Pope to Hatch, October 5, 1877, Special Files; *EPT*, 130.

9. *EPT*, 120.

10. *MVI*, October 6, 1877; Pope to Hatch, October 5, 1877, Special Files; *EPT*, 62.

11. *MBT*, 128–29.

12. *MBT*, 123.

13. *EPT*, 62–63.

14. *MBT*, 128–29.

15. Ibid.

16. Ibid.

17. Kerber to Hubbard, October 10, 1877. WC; *Letters of Ernst Kohlberg*, 69; *EPT*, 65–66.

18. *NYT*, January 5 1878.

19. *Grant County Herald*, October 13, 1877.

20. Fountain Deposition, Twelfth Interrogatory.

21. *MVI*, October 6, 1877.

22. Ibid.

23. *Galveston News*, October 29, 1877. How much land had been granted to the Texas and Pacific Railway was at issue because of the company's failure to complete tracks west of Fort Worth within the time stipulated by its charter. See HTO, s.v. "Texas and Pacific Railway."

24. *Galveston Weekly News*, October 8 and October 22, 1877.

25. *MBT*, 128–29, 120, 119–20.

26. *NYT,* October 8, 1877.

27. *Army-Navy Journal,* October 13, 1877, 152–53.

28. *Letters of Ernst Kohlberg,* 67.

29. Cardis to Hubbard, October 9, 1877, WC.

Chapter 12

1. *EPT,* 59, 62.

2. *MVI,* October 18, 1877.

3. *EPT,* 59–61.

4. *EPT,* 60–64.

5. White, *Out of the Desert,* 115; Howard to Mills, December 30, 1872, MC.

6. *EPT,* 59; Magoffin to Hubbard, October 10, 1877, WC; Fountain Deposition.

7. *Letters of Ernst Kohlberg,* 69.

8. Magoffin et al. to Hubbard, October 12, 1877, WC.

9. *MVI,* October 18, 1877.

10. *EPT,* 153–54.

11. *EPT,* 95.

12. *EPT,* 154, 156.

13. *MVI,* October 27, 1877.

14. *EPT,* 143; *MVI,* October 27, 1877; *Galveston News,* November 19, 1877.

15. *EPT,* 143.

16. Ibid. The other named insurgents were "Guillermo Gandara, Macedonia Gandara, Ambrosio Olguin, Nicolas Olguin, Alifonsa Zuñiga, Teofilo Astrada, Catamio Ballegos (from Mexico), Vidal Berrú (from Mexico), Nicolas Sierra, Jesus Peris, Sostenas Beltrán, Jesus Telles (an escaped convict from Mexico), Benigno Alderete, Juan Naranjo, Severriano Olguin, Antonio Garcia, El Sonoraman (a noted murderer and horse-thief), Juan Cordero, and Ricardo Cordero."

17. Ibid.

18. *EPT,* 143. In a letter to the *MVI* (October 27, 1877), a "wealthy American" claimed that Granillo, Juarez (misspelled "Ignores") and Salcido had never been citizens of the United States. While the 1870 federal census for El Paso County lists the three as natives of Mexico, the federal registration of male residents of El Paso County (taken 1867–1869) indicates that Salcido was born in "Texas," that is, in Mexican territory that became part of Texas in 1848. Francisco Barela was also born in Texas. The other insurgent leaders are not listed in this incomplete voters' register. Male residents of El Paso County, Headquarters, Fifth Military District, Office of the Secretary for Civil Affairs, Austin, Texas, April 16, 1870, republished in *Rio Grande Researcher* 19 (March–December, 2005), 1–4, http://www.rootsweb.com/txepgs/rgr_mar06.html.

19. Romero, "El Paso Salt War," 139.

20. Wasserman, *Capitalists, Caciques, and Revolution,* 10–11.

21. Martinez, *Troublesome Border,* 107–108.

22. *EPT,* 100.

23. *EPT*, 122.

24. Mary Romero concludes, "The opposition [of the Paseños to Howard's claims] was not an attempt to overthrow U.S. imperialism in the newly conquered territory; however, they were opposed to paying tribute as a conquered people. Protection of citizenship rights had been guaranteed by the Treaty of Guadalupe Hidalgo and therefore, Mexican-Americans resisted being relegated to an inferior position." Sonnichsen, *El Paso Salt War*, 20; Romero, "El Paso Salt War," 119–43.

25. Paust, "No Need to Revise," http://www.asil.org/taskforce/paust.pdf; Higgins, "Wars of National Liberation," www.jha.ac/articles/a132pdf.

26. *MVI*, November 3, 1877. Based on the similarity in content to other communications, the anonymous author was likely either Atkinson or Ellis.

27. Solomon Schutz to Secretary of State, October 21, 1877, U.S. Department of State Despatches from U.S. Ministers to Mexico, 1823–1906 (NAMS M97, roll 14), RG 59, NARA.

28. Pope to Hatch, October 13, 1877, Special Files.

29. *EPT*, 154.

30. *Galveston Weekly News*, November 19, 1877.

31. Ibid.; *MVI*, October 6, 1877.

32. *Army Navy Journal*, October 27, 1877; *Austin Democratic Daily Statesman*, October 18, 1877.

33. Pope to Sheridan, November 7, 1877, Special Files.

Chapter 13

1. Steele to Jones, October 24, 1877, WC; *EPT*, 154.

2. Webb, *The Texas Rangers*, 309; *Galveston Weekly News*, October 1, 1877.

3. *EPT*, 99, 101–102; *MVI*, November 10, 1877.

4. *EPT*, 26, 155; Kerber to Jones, November 8, 1877, FBR.

5. *EPT*, 26; *MVI*, November 24, 1877.

6. *EPT*, 26.

7. *EPT*, 26, 99–100.

8. *EPT*, 99–100, 155.

9. Frontier Battalion Special Order No. 109, November 12, 1877, WC.

10. *EPT*, 79–80. Texas Ranger historian Robert M. Utley censures Jones for his failure to appoint Garcia, "a capable leader." Utley, *Lone Star Justice*, 205.

11. Askins, "Texans, Guns & History," 166; Sonnichsen, *El Paso Salt War*, 37; Robinson, *Men Who Wear the Star*, 229; Webb, *Texas Rangers*, 367; Gillett, *Six Years with the Texas Rangers*, 139.

12. Joy, *John Tays and Jane Ellis of Nova Scotia*, Entry B2; Muster Roll, Detachment, Company C, November 1877, FBR; "John B. Tays," *American Biography & Genealogy*, ed. Burdette, 290, 293; Pierson to 2nd Assistant Secretary of State, October 12, 1872, Paso del Norte Despatches; Eickhoff, *Exiled*, 82, 84. EPC Deeds, Books E:440, F:549,

H:3, 58, 119, 129, 163, I:113, 117, 121, 136, EPC Records, Border History Manuscript Collection, MS 312, UTEP.

13. Howard to Jones, November 13, 1877, WC; *Galveston News,* December 20, 1877; *MVI,* November 24, 1877.

14. *MVI,* November 10, 1877.

15. Unless otherwise indicated, biographical details of Tays's Rangers are drawn from Muster Roll, Detachment, Company C, November 1877, FBR. Jones to Steele, November 19, 1877, *EPT,* 155 (This letter is dated the 18th in the transcribed *Webb Collection.* The original is no longer in its proper file in the Texas State Archives.); Sonnichsen, *El Paso Salt War,* 38; Robinson, *Men Who Wear the Star,* 225; Utley, *Lone Star Justice,* 195.

16. Howard to Jones, November 13, 1877, WC; Ward, "Salt War of San Elizario," 74; *Galveston Daily News,* January 11, 1878.

17. *MVI,* June 30, 1877; Nolan, *West of Billy the Kid* , 63, 67, 86; Enlistment Roll, Recruits for Company C, Frontier Battalion, El Paso, November 1877, FBR.

18. Tays to Jones, November 29, 1877, FBR.

19. *EPT,* 155; Jones to Steele, November 14, 1877, WC; Steele to Jones, undated, and November 19, 1877, WC.

20. *EPT,* 80, 130.

21. *EPT,* 80.

22. *EPT,* 63–64.

23. *EPT,* 26; Jones to Steele, November 19, 1877, WC;

24. *MVI,* November 10, 1877.

25. Ibid; *Mesilla News,* November 10, 1877; *Galveston Weekly News,* November 12, 1877.

26. *EPT,* 155, 124.

27. Special Order No. 110, November 15, 1877, WC.

Chapter 14

1. *EPT,* 121.

2. Thomas Blair ACP File, Records of the Appointment, Commission and Personnel Branch, OAG, RG 94, NARA (hereinafter Blair ACP); Blair Court-martial File, Record No. QQ978, Records of General Courts Martial, RG 153, NARA (hereinafter Blair Court-martial); *Washington Post,* February 14, 1879; *New York Herald,* February 19, 1879.

3. Shepherd to Col. Geo. A. Woodward, January 2, 1879, Blair Court-martial.

4. Ibid.; Portland *Morning Oregonian,* January 12, 1876.

5. Blair ACP.

6. *EPT,* 55–59; Sonnichsen, *El Paso Salt War,* 42; *New York Herald,* December 17, 1877.

7. *EPT,* 58–59.

8. *New York Herald,* December 17, 1877.

9. *EPT,* 121–126.

10. *EPT,* 72–73. Vidal was probably a close relative of Judge Garcia and his Ranger brother Miguel. In the 1860 federal census, Vidal was listed in the "Household" next to that of law student G. N. Garcia (the Salt War JP) and his "laborer" brother Miguel. See U.S. Census, 1860, San Elizario Township, EPC, Texas, households 809 and 810.

11. *MVI,* December 1, 1877, December 8, 1877.

12. *MVI,* December 1, 1877.

13. Fountain Deposition.

14. Tays to Jones, November 29, 1877, FBR; *Grant County Herald,* December 15, 1877.

15. *EPT,* 32–33.

16. *EPT,* 121–26.

17. *EPT,* 156; Frontier Battalion Special Order No. 110, November 15, 1877; *EPT,* 116.

18. Callahan, 220; *EPT,* 108–109.

19. Pope to Hatch, November 24, 1877, Special Files.

20. Pope to Hatch, November 26, 1877, Pope to Hatch, December 6, 1877, Special Files; *EPT,* 58.

21. *EPT,* 108. Blair's shortage of ammunition was an endemic problem. As Sherman explained to Sheridan only a month later, "the moment peace comes [civil officials] push the army to the wall. Pretending economy and a watchfulness over expenditures of money and property, they sometimes starve the army and make it impossible to even get cartridges for actual use, without such round about process that a modest officer would pay for them out of his pocket rather than be subject to the delay and forms of requisition." Sherman to Sheridan, January 10, 1878, *Papers of William Tecumseh Sherman,* Microfilm Roll 45, Manuscript Division, LC.

22. Pope to AAG, December 12, 1877, quoting Blair telegram, Special Files.

23. *EPT,* 144.

24. *EPT,* 78–79.

25. *EPT,* 80–82, 108–109.

26. Besides Salcido, Granillo and Juarez, the named defendants were Jose Peres, Jesus Arozes, Toribio Sanchez, Severano Olguin, Benito Caverario, Polino Chaves, Tomas Gonzales and Juan Lopes. El Paso Civil Case No.234; *EPT,* 80–82, 108–109, 121–26.

Chapter 15

1. Unless otherwise noted, information regarding the history and current use of the buildings described in this section comes from the SEGHS "Walking Tour."

2. Montes and Skaggs, "The Salt War."

3. *MVI,* January 5, 1878.

4. *EPT,* 73

5. *EPT,* 72; *MVI,* January 5, 1878; *EPT,* 80–82.

6. *EPT,* 73, 72.

7. *EPT,* 73.

8. There was no standardized time in 1878. Reported times must be viewed in that light. The exact sequence of events in the center and outskirts of San Elizario late on December 12 therefore cannot be determined with certitude. The narrative attempts to present the most probable order. *EPT,* 80–82, 119–20; *MVI,* January 5, 1878.

9. Anonymous letter (from Juan N. Garcia), *MVI,* January 12, 1878; *EPT,* 80–82.

10. *EPT,* 96; Trujillo de Hernandez, "Charles Ellis: Salt War Victim" (unpublished MS).

11. The cry, "Stop him!" was probably directed to Ellis. *MVI,* January 5, 1878; *EPT,* 80–82.

12. *EPT,* 119–20.

13. Pope to Townsend, December 13, 1877, Special Files. The name of 2nd Lt. Samuel Speece Pague is misspelled "Payne" in the printed version of *EPT,* a document that suffers from many transcription errors. Heitman (I, 765, 777) settles the issue. *EPT,* 55–57, 78–79, 105–106, 108–109.

14. *EPT,* 56.

15. Sonnichsen, *Pass of the North,* 206; *EPT,* 56.

16. Blair to Pope, December 19, 1877, Special Files.

17. *EPT,* 56.

18. *MVI,* December 22, 1877.

19. See Elizabeth Salas, *Soldaderas in the Mexican Military* (Austin: University of Texas Press, 1990).

20. Hatch to Pope, January 10, 1878, Special Files; Kearney and Medrano, *Medieval Culture and the Mexican American Borderlands,* 131–132; Encyclopedia Britannica Guide to Hispanic Heritage in the Americas, s.v. "Our Lady of Gudalupe," http://www.britannica.com/hispanic_heritage/article-9433088 (accessed March 24, 2007).

21. Deposition of Thomas Zickefoose, "Motin de Mexicanos contra las autoridades Americanas en Isleta y San Elizario, Texas," (hereinafter "Motin de Mexicanos") quoted in Callahan, 222–23; *EPT,* 80–82; *MVI,* January 5, 1878; *Grant County Herald,* December 22, 1878.

22. *EPT,* 80–82; *MVI,* January 5, 1878.

23. *EPT,* 80–82. Deposition of John B. Tays, "Motin de Mexicanos," cited by Callahan, 223. The printed account misspells Kohlhaus's name as Calhoun. The shopkeeper had already fled to the Mexican town of Guadalupe.

24. *EPT,* 80–82; *MVI,* February 2, 1878; *MVI,* January 5, 1878.

25. *MVI,* January 5, 1878; *El Paso Times,* February 14, 1937.

26. *EPT,* 80–82; *MVI,* January 5, 1878.

27. *EPT,* 56.

28. Blair to Hatch, December 13, 1877, Correspondence of the Secretary of State

(hereinafter CSS). The secretary of war kept the secretary of state apprised of the conversations within the army on the situation in El Paso. This and other telegrams were found in the records of the state department, but not in *El Paso Troubles in Texas* or the usual military records.

29. Blair to Hatch, December 13, 1877 (second telegram), CSS.

30. *EPT,* 156; *Grant County Herald,* December 15, 1877.

31. Deposition, Pablo Mejia, "Motin de Mexicanos," cited in Callahan, 220–21.

32. Pope to Sheridan, December 14, 1877, CSS.

33. *EPT,* 80–82; *MVI,* January 5, 1878.

34. Public Notice issued by Jesus Padilla, December 14, 1877, *EPT,* 76. The document appears in the original Spanish, 77.

35. *EPT,* 55–57; *Grant County Herald,* December 15, 1877; *EPT,* 78–79.

36. Kerber to Coke and Maxey, December 14, 1877, OAG/MBT, NAMS M666, Roll 203, RG 94, NARA (hereinafter OAG/MBT).

37. *EPT,* 156; *Galveston Weekly News,* December 17, 1877.

38. *EPT,* 144.

Chapter 16

1. *EPT,* 144–45.

2. *EPT,* 145–46.

3. *EPT,* 145; Sherman to Sheridan, November 29, 1877, Papers of William Tecumseh Sherman, MF Roll 45, LC.

4. *San Francisco Examiner,* December 17, 1877; *EPT,* 146; *New York Herald,* December 16, 1877.

5. Sherman to Sheridan, December 15, 1877, LR, AGO/MBT, M666, Roll 203, RG 094, NARA; Sherman to Sheridan, December 15, 1877 (second message), Special Files; Sheridan to Pope, December 15, 1877, AGO/MBT; Pope to Sheridan, December 15, 1877, Special Files.

6. Pope to Headquarters, District of New Mexico, December 15, 1877, Special Files; *Grant County Herald,* December 15, 1877; *Galveston News,* December 24, 1877.

7. *EPT,* 145.

8. *MVI,* December 15, 1877; *Grant County Herald,* December 15, 1877.

9. *Galveston Daily News,* December 15, 1877; *Brooklyn Daily Eagle,* December 15, 1877; *Eco del Rio Grande,* Las Cruces, New Mexico, December 27, 1877.

10. Phelan, "A 'Salt War' Experience," *New Mexico Highway Journal* VIII (December 1930): 23–24.

11. *Grant County Herald,* December 22, 1877; Tays to Jones, December 20, 1877, *EPT,* 80–82.

12. Tays says that Garcia and his band escaped, but both the captain and his son said they were captured. Miguel's name appears among the dead in a "corrected list of the killed and wounded," prepared on December 21 and published the following day in the *Mesilla Valley Independent.* The actions of Mrs. Campbell are difficult to

determine. Sheriff Kerber told George Zimpelman that she and her husband "while running to the rangers' quarters, with two children in their arms, had to stand a heavy fire." This statement contradicts other accounts, which have her bringing the children over before Frank Campbell ran to the Ranger quarters. *EPT,* 37, 81; *El Paso Times,* February 14, 1937; *Galveston Daily News,* January 6, 1878.

13. Trujillo de Hernandez, "Charles Ellis: Salt War Victim."

14. Ibid.

15. *EPT,* 56–57.

16. *EPT,* 56–57. The soldier had already been released.

17. Pope to AAG, Division of the Missouri, December 17, 1877, quoting telegraph from Blair, December 15, 1877, Special Files.

18. *EPT,* 146, 147.

19. *Galveston Daily News,* December 16, 1877.

20. *Denison News,* December 15, 1877, reprinted in *Las Cruces Eco del Rio Grande,* December 27, 1877.

21. *Brooklyn Daily Eagle,* December 16, 1877.

22. Pope to Headquarters, District of New Mexico, December 16, 1877, Special Files; *Army and Navy Journal,* December 22, 1877; Returns for U.S. Military Posts, 1800–1916, Fort Davis, Texas, M617, Roll 297; Pope to AAG, Division of the Missouri (hereinafter "Sheridan HQ"), December 17, 1877, Special Files; *New York Herald,* December 18, 1877.

23. *MVI,* January 5, 1878; "William Treyer," *Biographical and Historical Sketch Book of the Pioneers Association of El Paso County, Texas,* 43; *Galveston Daily News,* January 6, 1878; *Grant County Herald,* December 22, 1877.

24. *EPT,* 80–82, 115.

25. *MVI,* January 5, 1878; *EPT,* 57; Testimony of Juan S. Hart, William D. Marsh Probate File No. 74, EPC Archives, UTEP.

26. *MVI,* January 5, 1878; *EPT,* 59.

27. *EPT,* 57

28. Pope to Sheridan HQ, December 17, 1877, quoting Blair telegram, December 16, 1877, Special Files.

Chapter 17

1. Andrews to Carpenter, December 17, 1877, *Military Papers of Brig. General L. H. Carpenter,* vol. 2.

2. *New York Herald,* December 17, 1877.

3. *EPT,* 147, 156; *Galveston Weekly News,* December 24, 1877; *New York Herald,* December 18, 1877; *San Francisco Examiner,* December 18, 1877.

4. *EPT,* 102; Detachment of Company C, Monthly Return for December, 1877, FBR; *MVI,* December 22, 1877; *EPT,* 147.

5. Pope to Sheridan HQ, December 18, 1877, citing Blair's telegram of December 17, 1877, Special Files.

6. Mary Phelan, "A 'Salt War' Experience," 25. In February 1884, itinerant photographer Jerrell was shot to death by stagecoach robbers near Abilene, Texas. Mautz, *Biographies of Western Photographers,* 332.

7. *Mesilla Valley Independent,* November 10, 1877.

8. *Mesilla Valley Independent,* November 3, 1877, November 10, 1877. *Grant County Herald,* December 22, 1877. Kerber's telegram to Kinney hints at some kind of prior relationship, though no such evidence has come to light.

9. *EPT,* 80–82.

10. *EPT,* 73.

11. Ibid.

12. *EPT,* 157.

13. *EPT,* 158; *MVI,* January 5, 1878.

14. Gillett, *Six Years,* 140.

15. *MVI,* January 12, 1878.

16. *MVI,* January 5, 1878; *EPT,* 102.

17. *EPT,* 80–82. Corporal Mathews said that Barela, "the Commandante of the mob," entered the room before his men, "threw his arms around Howard and kissed and hugged him." When Tays refused to leave Howard, Barela's twelve men "pushed him into another room." *MVI,* January 5, 1878.

18. *EPT,* 114–115.

19. *EPT,* 102–103.

20. *EPT,* 73, *MVI,* January 5, 1878; Correspondence with Larry Zickefoose.

21. *MVI,* January 12, 1878.

22. *EPT,* 74.

23. *EPT,* 80–82; *MVI,* January 5, 1878; *EPT,* 74.

24. *MVI,* January 12, 1878.

25. *EPT,* 74, 80–82; *MVI,* January 5, 1878; *EPT,* 74.

26. Lt. Tays reported to Major Jones that the defenders of Ellis's made their dash to freedom after darkness on the day that Kent and Frank Campbell crossed over to the Ranger quarters. That was Saturday, December 15. This is contradicted by the testimony of the G.N. Garcias, father and son. Captain Garcia testified just two weeks after the event that his band was forced to surrender on the 15th. His son Gregorio told a reporter in 1937, "Right after this [the executions], my father and our friends in Ellis' store saw our chance to get through the lines of the mob." See *EPT,* 37, 82; *El Paso Times,* February 14, 1937. For the story of Captain Garcia's flight to Guadalupe, Mexico, see *MVI,* January 12, 1878.

27. *EPT,* 80–82; *MVI,* January 5, 1878.

28. Thomas Blair provided the lower estimate of insurgent deaths, as well as the number of wounded. "F" reported the higher number of deaths in a letter published in the *Herald.* The *Herald* later reported "about 25 men all told have been killed." Sergeant Mortimer, Ellis, the three executed men, and Paseño victims of December 23, 1877, accounted for nine deaths, leaving about sixteen deaths within the insurgent

ranks. According to a fourth unconfirmed report, also appearing in the *Herald,* "One of the San Elizario rioters admits that the mob lost from 30 to 40 men killed and wounded during the three days' attack on the ranger quarters." *EPT,* 57; *Grant County Herald,* December 22, 1877, January 5, 1878, January 19, 1878.

29. *EPT,* 158.

30. *EPT,* 80–82; Deposition of Price Cooper, "Motin de Mexicanos contra las autoridades Americanas en Isleta y San Elizario, Texas," cited in Callahan, 228; *Galveston Daily News,* January 6, 1878.

31. Pope to Blair, December 18, 1877, Pope to Sheridan HQ, December 18, 1877, 10:42 P.M., both Special Files.

32. Pope to Sheridan HQ, December 18, 1877, 11:56 P.M., Special Files. The phrase "the mob was entirely Mexican-born citizens of Texas" appears on the message transcription form used by Western Union Telegraph Company's receiving office in Chicago. This must be a mistake, as this completely contradicts the reports of the man on the scene, Captain Blair. The phrase occurs nowhere else in army correspondence. In fact, from this point forward, the press frequently regurgitated the phrase, "the mob was entirely composed of native-born citizens of Texas." Although no army telegram with *this* formulation has surfaced, this is undoubtedly what Pope meant to convey.

33. *EPT,* 147–48.

34. Ord to Sheridan HQ, December 20, 1877, AGO/MBT

35. *EPT,* 55–57; *MVI,* January 5, 1878.

36. *EPT,* 102.

Chapter 18

1. *Galveston Daily News,* December 19, 1877.

2. *New York Herald,* December 19, 1877.

3. *Atlanta Constitution,* December 20, 1877.

4. *San Francisco Examiner,* December 19, 1877; *Atlanta Constitution,* December 20, 1877.

5. *San Francisco Examiner,* December 22, 1877.

6. *San Francisco Examiner,* December 19, 1877, citing the *Chicago Journal;* Sherman to Sheridan, December 15, 1877, AGO/MBT; R. E. Platt, A.A.G. (hereinafter Pope HQ) to Commanding Officer, District of New Mexico (hereinafter Hatch), December 16, 1877, Special Files.

7. *Galveston Daily News,* December 20, 1877.

8. *EPT,* 148; *Galveston Daily News,* December 21, 1877.

9. Heitman, *Historical Register* I, 722; Kenner, *Buffalo Soldiers and Officers,* 199. See Kenner, 174-ff., for L Troop's story.

10. Pope HQ to Hatch, December 16, 1877, Pope to Sheridan HQ, December 19, 1877 (referring to the second set of orders), Special Files.

11. *MVI,* January 12, 1878; *EPT,* 100–101.

12. *Texas Border Troubles,* 11–12.

13. *MVI,* December 22, 1877; *Grant County Herald,* December 22, 1877.

14. *Galveston Daily News,* January 6, 1878.

15. Ibid.

16. *New York Times,* January 5, 1878 and January 29, 1878.

17. *EPT,* 58–59, 55–57.

18. Pope to Sheridan HQ, December 19, 1877, Special Files.

19. Ibid.

20. Ibid.

21. *EPT,* 80–82. This final comment is found in the Webb Collection transcription of the document but is missing from the published version of *EPT* (82, 158). Evidently, someone deemed the remark too hot for publication.

22. Magoffin had earlier wired Austin that three men were missing. The third was Captain Garcia, not a Ranger in 1877 and therefore not mentioned in Tays's report. See *Galveston Daily News,* December 20, 1877 and December 24, 1877.

23. *Galveston Daily News,* December 20, 1877.

24. Sheridan to Pope, December 21, 1877, AGO/MBT; Pope to Sheridan, December 21, 1877, Special Files.

25. For a biography of Hatch, see Kenner, *Buffalo Soldiers,* 29–50.

26. *Galveston Daily News,* December 21, 1877.

27. Hatch to Pope HQ, January 11, 1878, Special Files; *EPT,* 44; *MVI,* December 22, 1877; *Galveston Daily News,* December 21, 1877.

28. Pope to Sheridan, December 22, 1878, quoting Hatch to Pope, December 21, 1877, Special Files.

29. James B. Gillett, *Six Years with the Texas Rangers,* 140, is the source for John Ford's arrival from New Mexico. For Barton's life, see Rasch, *Desperadoes of Arizona Territory,* 134–41, and Nelson, "Trailing Jerry Barton," 36–42. Jones to Tays, December 20, 1877, Tays to Jones, December 21, 1877, Jones to Tays, December 22, 1877, FBR; Muster Roll for Detachment, Company C, December 1877, FBR; *EPT,* 102; Palo Pinto County *Western Star,* January 5, 1878.

Chapter 19

1. Hatch to Pope, January 11, 1878; Pope to Sheridan HQ, quoting Blair telegram, December 22, 1877, Special Files; Pope to Sheridan HQ, quoting Hatch telegram, December 23, 1877, Special Files.

2. *Grant County Herald,* December 22, 1877; Returns from U.S. Military Posts 1800–1916, Fort Davis Texas, December, 1877, NAMS M617, Roll 297.

3. *EPT,* 115–116; *Galveston News,* January 6, 1878; *MVI,* January 5, 1878.

4. *Grant County Herald,* December 22, 1877; Sonnichsen, *El Paso Salt War,* 58; Utley, *Lone Star Justice,* 201.

5. The median height for the fifty-one men who served as Texas Rangers in El Paso from November 1877 through January 1878 was five feet, eight inches, according to

the detachment's muster roll. Muster Rolls, Detachment of Company C, FBR; Alexander, *Dangerous Dan Tucker,* 9, 14; Mullin, "Here Lies John Kinney," 223.

6. Kerber and Tays each referred to "Tucker men." Kinney's "discharge" as a "Private of Lt. Tucker's Company of Volunteers of State forces . . ." provides further evidence of the deputy sheriff's status. The document was issued by Sheriff Kerber, who was in the best position to know who led the volunteers he had summoned to El Paso. On the other hand, the broadsheet giving the fullest account of the war, including the activities of the Silver City men, was the *Mesilla Valley Independent,* which attributed leadership to Kinney. By itself, this is not conclusive. The notorious Kinney sold papers in Dona Ana and El Paso; Dan Tucker did not. But stronger evidence comes from Tucker's hometown newspaper. The *Grant County Herald,* which displayed no prior interest in the activities of Kinney, consistently assigned recruitment and initial leadership responsibilities to him. Deputy Sheriff Tucker, the subject of favorable press in Silver City, was not mentioned by the *Herald* until January 5, 1878. The newspaper then announced that "D. Tucker has been chosen captain of the Silver City company *now* in El Paso county." (Italics added.) Taken together the evidence suggests that Tucker's late election to "captain" may have been an internal effort to clean up the Silver City act. *EPT,* 115, 117; Nolan, "Boss Rustler," *True West* (September 1996), 14–21.

7. Steele to Schutz & Co., August 9, 1878, TAGC Outgoing Corr., File 401-625.

8. *MVI,* December 22, 1877. Neither Evans biographers McCright and Powell nor the standard Lincoln County War authorities record any activities by Evans that would conflict with his participation in the Salt War.

9. William F. Rutherford, Indian War Veteran Pension Application No. 17049, Company C, Texas Frontier Battalion, Indian War Pension Files, 1865–1890; Records of the Veterans Administration, RG 15; NARA; "Life and Times of Asbury Crofford (AC) Ryall," unpublished MS; Detachment, Company C Muster Rolls, Ranger Records, AG Records, TSLAC; *Grant County Herald,* December 15, 1877.

10. *EPT,* 115–16, 117; Steele to Boone, March 7, 1878, TAGC Outgoing Corr., Steele to Kerber, January 10, 1878, TAGC General Corr., TSLAC.

11. *EPT,* 91.

12. *EPT,* 90.

13. *EPT,* 90; Detachment Muster Roll, Ranger Records, TAGC, TSLAC.

14. *EPT,* 117.

15. *EPT,* 84–86, 113, 115–16.

16. *EPT,* 84, 85–86.

17. *EPT,* 86, 88.

18. U.S. Census, Ysleta Township, EPC, Texas.

19. *EPT,* 83–84.

20. *El Paso Times,* January 31, 1937. The widow was named Maria Núñez, not Holguin. *EPT,* 83–84.

21. *EPT,* 94.

22. *EPT,* 89, 93–94, 117; *MVI,* January 5, 1878, December 29, 1877.

23. *EPT,* 93.

24. *EPT,* 85, 92–93.

25. *EPT,* 86, 94, 117.

26. *EPT,* 87, 89.

27. *EPT,* 79, 87; Pope to Sheridan HQ, quoting Hatch, December 24, 1877, Special Files.

28. *EPT,* 88, 89.

29. *EPT,* 87.

30. Ibid.

31. Carpenter was reported just twenty-five miles from San Elizario the night before. *MVI,* December 22, 1877.

32. *EPT,* 90, 117.

33. *EPT,* 75–76.

34. *EPT,* 76.

35. Pope to Sheridan HQ, December 23, 1877, quoting Hatch telegram of December 23, 1877, Special Files; Pope to Sheridan HQ, December 24, 1877, quoting Hatch telegram of December 23, 1877, Special Files.

36. Pope to Sheridan HQ, December 24, 1877, quoting Hatch telegram of December 24, 1877, Special Files; Headquarters, Military District of New Mexico, General Order No. 1, December 24, 1877, Special Files.

Chapter 20

1. *Chicago Daily Tribune,* January 25, 1878; *MVI,* January 5, 1878.

2. Trujillo de Hernandez, "Charles Ellis"; *EPT,* 30–33; *Sedalia Daily Democrat,* January 6, 1878, www.ancestry.com; *Grant County Herald,* January 5, 1878.

3. *Grant County Herald,* January 5, 1878; *MVI,* December 29, 1877.

4. *MVI,* December 29, 1877.

5. *Rocky Mountain News,* December 26, 1877.

6. *Galveston Daily News,* December 24, 1877.

7. Quoted in *Galveston Daily News,* December 31, 1877.

8. Ibid.

9. *Army-Navy Journal,* December 22, 1877.

10. *St. Louis Globe-Democrat,* December 21, 1877, Hayes Scrapbook for 1877, RB Hayes Collection, RBHPCL.

11. Pope to Sheridan HQ, quoting Hatch telegram, both December 23, 1877, Special Files; Pope to Sheridan HQ, December 17, 1877, quoting Blair telegram, December 15, 1877, Special Files; Pope to Sheridan HQ, December 24, 1877, AGO/MBT; Pope to Sheridan HQ, quoting Hatch telegram, both December 24, 1877, AGO/MBT; Pope to Sheridan HQ, December 27, 1877, quoting Hatch telegram, December 25, 1877, AGO/MBT; Pope to Sheridan, December 25, 1877, Special Files.

12. For a discussion of the federal response to the railroad strikes of 1877, see Bruce, *1877: Year of Violence* and Laurie and Cole, *Role of the Federal Military Forces in Domestic Disorders 1877–1945*, 29–55.

13. Pope to Sheridan, December 24, 1877, Special Files; *Galveston Weekly News*, December 31, 1877, *Army-Navy Journal*, December 29, 1877; *Galveston Weekly News*, December 31, 1877.

14. *San Antonio Herald*, January 1, 1878.

15. Harvey's indictment of Kinney may not have represented Kerber's views. The sheriff continued to have a close association with the New Mexico rustler. In fact Kinney replaced Harvey as deputy sheriff just a few months later. *MVI*, December 29, 1877

16. *EPT*, 78; *MVI*, December–February 2, 1878.

17. *MVI*, January 5, 1878.

18. *Mesilla Valley Independent*, January 12, 1878.

19. *EPT*, 96.

20. Jones to Tays, January 8, 1878, FB Records, Ranger Records, TAGC, TSLAC.

21. *EPT*, 84, 94–95.

22. *EPT*, 94–95.

23. *Grant County Herald*, January 19, 1878.

24. *EPT*, 64.

25. *Grant County Herald*, January 5, 1878; *EPT*, 91. The printed version of *EPT* identifies the border raider as "Yearn Kyng." An earlier handwritten version of Ochoa's testimony clearly reads "Juan Kyny." See *EPT*, 91; AGO/MBT *EPT*, NAMS M666, Roll 203.

26. *EPT*, 38; *Grant County Herald*, January 5, 1878.

27. U.S. Army, "Counterinsurgency," 1–9, 1–11; *Grant County Herald*, January 5, 1878.

28. *EPT*, 104.

29. *EPT*, 103.

30. *EPT*, 104, 103.

31. *EPT*, 104–105.

32. *EPT*, 104.

33. On September 27, 1868, Carpenter's company relieved George Forsyth's embattled scouts at Beecher's Island after a forced march. The following October 17, his troop held off several hundred Cheyenne for eight hours at Beaver Creek, Kansas. "Louis Henry Carpenter Heitman," *Historical Register* 1, 284; Utley, *Frontier Regulars*, 153–54.

34. Leiker, *Racial Borders*, 66.

35. *EPT*, 89, 95.

36. Hatch to Pope January 4, 1878, Special Files; Returns for U.S. Military Posts 1880–1916, Fort Davis, Texas, September 1854–December 1878, NAMS M617, Roll 297; *EPT*, 91–92.

37. *EPT*, 92.

38. Ibid.

39. *EPT*, 114, 117; Pope to Sheridan HQ, January 7, 1878, quoting Hatch telegram, January 5, 1878, Special Files.

40. As late as February 1882, they were still lobbying the governor and legislature to get their money. See TAGC, RG 401: Gen. Corr., TSLAC.

41. Steele to Messrs. Schutz & Co., August 9, 1878, AGC, Outgoing Corr., File 401-625, 197–199; *EPT*, 145.

42. *EPT*, 158; *Grant County Herald*, January 19, 1878; Jones to Tays, January 8, 1878, FBR.

43. *MVI*, January 12, 1878; *EPT*, 64; *Grant County Herald*, January 26, 1878.

Chapter 21

1. *Grant County Herald*, January 26, 1878; *Weekly New Mexican*, February 2, 1878; Heitman, *Historical Register* 1. 991; Returns for U.S. Military Posts 1800–1916 (hereinafter "Military Posts"): Fort Davis, Texas, September 1854–December 1878 and Troops in the field in El Paso County, January 1878, NARA RG 94, NAMS M617, Rolls, 297, 351.

2. Nolan, *West of Billy the Kid*, 53–54.

3. *MVI*, January 12, 1878; Jones to Tays, January 8, 1878, FBR.

4. Muster Rolls, Detachment of Company C, FBR.

5. Jones to Tays, January 8, 1878, FBR.

6. Pope to Sheridan HQ, January 7, 1878, quoting Hatch telegram, January 5, 1878, Special Files; *MVI*, January 12, 1878; *Grant County Herald*, January 19, 1878.

7. Pope to Sheridan HQ, January 9, 1878, quoting Hatch telegram, January 8, 1878, Special Files; *Grant County Herald*, January 5, 1878.

8. *Grant County Herald*, January 19, 1878, quoting *Eco del Rio Grande*.

9. *Grant County Herald*, January 19, 1878.

10. Phelan, "A 'Salt War' Experience," 25, 35.

11. *EPT*, 47–48, 148.

12. *EPT*, 47.

13. *EPT*, 5, 47–48; Heitman, *Historical Register* I, 599, 631; Utley, *Frontier Regulars*, 42, note 89; *Grant County Herald*, January 19, 1878; *EPT*, 6, 34–44; Hatch to Carpenter, January 8, 1878, Military Papers of Brig. Gen. L.H. Carpenter, II:28.

14. *EPT*, 34, 36, 37, 39, 42.

15. *EPT*, 39.

16. Hatch to Pope HQ, January 11, 1878, Special Files.

17. Ibid.

18. Ibid.

19. *MVI*, January 19, 1878.

20. *MVI*, February 2, 1878.

21. *MVI*, February 23, 1878.

22. *EPT,* 88, 102, 112.

23. *EPT,* 102, 112.

24. *EPT,* 86, 102.

25. Richard Maxwell Brown states, "it was in Texas—of all American states—where the no duty to retreat doctrine of standing one's ground became so deeply entrenched as to be referred to by legal scholars generically as 'the Texas rule.'" See Brown, *No Duty to Retreat,* 25–28. *EPT,* 102; Monthly Return, Detachment of Company C, February 1878, J.A. Tays to Jones, April 17, 1878, FBR.

26. *MVI,* February 9, 1878.

27. *EPT,* 124; *MVI,* February 9, 1878.

28. *EPT,* 124.

29. Besides Tays, the Rangers giving statements to Blacker were Second Sergeant Thomas Zickefoose, Corporal Mathews, Pablo Mejia, and Price and Santiago Cooper. *EPT,* 124; Blacker to Jones, February 24, 1878, FBR; Jones to Tays et al, Special Order No. 120, February 24, 1878, FBR.

30. S. Schutz to Second Assistant Secretary of State, March 6, 1878, Paso del Norte Despatches.

31. Ibid.; Military Posts, El Paso, Texas, February 1848–September 1916, NAMS M617, Roll 351.

Chapter 22

1. Hubbard to Steele and Jones, January 16, 1878, Governor Hubbard Corr., TSLAC.

2. Testimony of Maj. John B. Jones, January 19, 1878, House Committee on Foreign Affairs, 45th Cong. H.R. Ex. Doc. 701: *Texas Frontier Troubles,* 61.

3. *EPT,* 83–84, 85, 90.

4. Steele to Jones, February 8, 1878, TAGC; *EPT,* 13, 83–84, 85, 89.

5. Jones to Steele, February 19, 1878, Steel to Jones, February 19, 1878, Steele to Jones, February 23, 1878, Jones to Steele, February 27, 1878, TAGC.

6. *EPT,* 49.

7. Ibid; Jones to Fountain, February 27, 1878, FBR.

8. *MVI,* March 2, 1878. Jones apparently also put a stop to the corporal's war of words with "American." No more diatribes of any kind from the pen of H. O. Mathews appeared thereafter in the pages of the *Independent.*

9. *EPT,* 83–84, 86–87.

10. *EPT,* 113–14.

11. *EPT,* 115–16.

12. *EPT,* 117.

13. *EPT,* 105–106

14. *EPT,* 108–109, 119–20.

15. *EPT,* 119.

16. *EPT,* 121–26.

17. *EPT,* 122.

18. *EPT,* 11, 12, 125, 131–32, 138–39.

19. *EPT,* 13–18.

20. *EPT,* 13–18.

21. *EPT,* 13–18, 106–108.

22. *EPT,* 117.

23. *EPT,* 118. A law appropriating forty thousand dollars to purchase the necessary grounds and erect buildings for a military post at El Paso was signed by President Hayes on February 5, 1879. See Congressional Record, vol. 8, part I, 45th Congress, 3rd Sess., 1878–79, 774.

24. *EPT,* 18–19, 19–23. Macedonia Gandara was a Texas native, according to the 1870 Federal Census for El Paso County, Texas (*San Elizario,* 17). Salcido was born in Mexico, according to the same census (*San Elizario,* 23), but had demonstrated a commitment to his Texas home through service in the Texas Ranger companies of Captain Garcia and Lieutenant Montes.

25. *EPT,* 23–31.

26. Gregg, *Border Troubles,* 80.

Chapter 23

1. Petition of Blanchard et al to Jones, March 21, 1878, FBR; Criminal Case No. 296, Old Criminal Case Files, EPC Archives; Solomon Schutz et al to Jones, January 23, 1879; *MVI,* February 1, 1879.

2. Petition of Blanchard et al. to Jones, March 21, 1878, FBR; Steele to O.M. Roberts, March 25, 1878, AGC, Outgoing Corr., File 401-625.

3. John and Mary were married May 1, 1878, in the Ysleta home of Charles and Julia Kerber. Special Order No. 123, March 25, 1878, FBR; "Diary of George H. Tays" (hereinafter "Tays Diary"), Patterson, *John Tays and Jane Ellis and their Descendents. MVI,* May 11, 1878; Jones to Steele, March 25, 1878, FBR.

4. For Jones's career and accomplishments subsequent to the Salt War, see Utley, *Lone Star Justice,* 183–87, 207–22.

5. Cool, "Puzzling J. A. Tays," *The Dispatch* 7 (Summer 2001), at www.texasranger.org. Peter Brand first identified the Detachment's second commander in "Sherman W. McMaster(s)," *Western Outlaw-Lawman Association Journal* VIII (Winter 1999): 4–6.

6. Kerber to Jones, April 1, 1878, FBR; J. A. Tays to Jones, April 17, 1878, WC.

7. Sonnichsen, *El Paso Salt War of 1877,* 61; Robinson, *Men Who Wear the Star,* 231; Utley, *Lone Star Justice,* 205; *MVI,* February 16, 1878; J. A. Tays to Jones, April 17, 1878, WC.

8. Padilla to Blacker, May 4, 1878, Paso del Norte Despatches.

9. *MVI,* June 15, 1878.

10. In a May 7 letter to Kerber, Major Jones advised, "The Governor will offer rewards for Francisco Barela, Cisto Salcido, Luciano Frequez, Agaton Porros, Desidario Apodoca and Jesus Garcia and will send you the proclamation in a few days." The gov-

ernor's proclamation announced a reward of "Two Hundred and Fifty Dollars each" for just five men, "Cisto Salcido, Luciano Fresquez, Francisco Barila, Leon Granillo, and Jesus Garcia." Neither Porros nor Apodaca is mentioned in this document, yet a reward was offered for Apodaca. Jones to Kerber, May 7, 1878, FB Corr.; Governor's Proclamation, May 13, 1878, Records of Gov. R. J. Hubbard, TSLAC.

11. J. A. Tays to Jones, May 11, 1878, WC; *Arizona Silver Belt,* May 23, 1878.

12. The Zickefoose brothers fell ill that spring and, after slow recoveries, resigned. A.C. Ryall moved up to First Sergeant, while Mark Ludwick was promoted to Second Sergeant. J. A. Tays to Jones, April 17, 1878, WC; J. A. Tays to Jones, May 11, 1878, WC; J. A. Tays to Jones, June 18, 1878, WC.

13. J. A. Tays to Jones, June 18, 1878, WC; Kerber to Blacker, reprinted in the *Mesilla News,* July 20, 1878; Bourgade to Jones, May 18, 1878, TSLAC.

14. J. A. Tays to Jones, June 18, 1878, WC.

15. Ibid.; *Mesilla News,* July 20, 1878.

16. *EPT,* 71; Sonnichsen, *El Paso Salt War,* 61.

17. Tays to Jones, June 30, 1878, WC; Monthly Report for Detachment of Company C, Frontier Battalion, June 1878, FBR.

18. Ibid.

19. Tays to Jones, June 30, 1878, WC; Tays to Jones, July 15, 1878, FBR.

20. This version of Ford's experience collecting and keeping the salt fees is drawn from an El Paso newspaper article, "Collecting Salt Money Proves Dangerous Work," probably published in 1922. Another version, found in a typescript entitled "John Ford," was related by El Paso pioneer Frank Hunter, who misidentifies Howard as the man Ford "held up." Hunter also recalled that Ford set up a tent at the lakes to wait for any salt trains. This is doubtful, since Apaches still roamed the area, and collections could more easily be made in town, where witnesses helped ensure Ford's safety. In 1900, Charles Kerber submitted evidence in the Howard probate case that he met with Ford in San Elizario on fifteen occasions between May 1878 and May 1879. Both versions are found in the San Elizario Vertical File, Borderlands Collection, El Paso Public Library. See also Probate Case No. 5 (Charles H. Howard), EPC Archives.

21. *NYT,* June 10, 1878; J. A. Tays to Jones, May 11, 1878, WC; J. A. Tays to Jones, May 16, 1878, FBR.

22. *Arizona Silver Belt,* May 23, 1878

23. J. A. Tays to Jones, May 31, 1878.

24. Butler, *Butler's Book* I, 80.

25. *NYT,* June 10, 1878.

26. Ibid.; *MVI,* May 25, 1878.

27. The *Mesilla Valley Independent* reported that Corporal Mathews, and not Ryall, chased Martin and Curly Bill into Mexico. Tays was in a better position to know the truth. *MVI,* May 25, 1878; Tays to Jones, May 31, 1878, WC; *Grant County Herald,* June 1, 1878; *Army-Navy Journal,* April 27, May 25, 1878; *NYT,* June 10, 1878.

28. *MVI,* June 8, 1878; Tays to Jones, June 18, 1878, WC; Tays to Jones, July 15, 1878, FBR.

29. *MVI,* September 7, 1878; Callaghan to Jones, December 9, 1878, FBR.

30. Tays to Jones, September 1, 1878, FBR.

31. Blacker to Jones, September 5, 1878, FBR; Bourgade to Jones, May 28, 1878, FBR.

32. *MVI,* September 7, 1878; *Grant County Herald,* June 22, 1878.

33. Blacker to Jones, September 5, 1878; Broaddus, *Legal Heritage,* 131; State of Texas vs. Robert Martin et al, Case #300, EPC Archives; *Grant County Herald,* September 14, 1878; Tays to Jones, September 16, 1878; Mejia v. Gandara et al, JP Precinct No. 1, EPC, September Term, 1878 (Sonnichsen Collection, UTEP).

34. Tays to Jones, September 9, 1878, September 16, 1878, FBR.

35. Frederick Nolan, *West of Billy the Kid,* 125; Blair to Rachel Nicholl, June 7, 1878; Blair Court-martial file.

36. *Atlanta Constitution,* January 2, 1879; *Indiana Progress,* January 11, 1877; Blair to AG, October 1, 1878; Sherman to Mrs. Thomas B. Nicholl, October 18, 1878, Blair Court-martial File.

37. Bourgade to Jones, May 28, 1878.

38. Callaghan to Jones, December 9, 1878.

39. Tays to Jones, September 30, 1878, FBR; Wahl to Jones, October 16, 1878, FBR; Tays to Jones, October 17, 1878, FBR; Blacker to Hubbard, no date, Register of Letters, Gov. Hubbard Corr, TSLAC.

40. Steele to Tays, October 14, 1878, TAGC; Tays to Jones, November 1, 1878, FBR.

41. State of Texas v. Benito Gonzales, EPC Criminal Case File No. 256, SC.

42. Monthly Report, Detachment of Co. C, October, 1878, FBR; J. A, Tays to Jones, September 30, 1878, FBR.

43. J. A. Tays to Jones, November 6, 1878, FBR; J. A. Tays to Jones, November 16, 1878, FBR; Skaggs, Amelia Montes, and Dr. Samuel Ralph Skaggs, "Salt War Jailbreak," *The Bells of San Elizario,* nd, quoted in Harry Wilson Clark, "Traditions of the San Elizario Jail," 95.

44. Broaddus, *Legal Heritage,* 130.

45. J. A. Tays to Jones, November 6, 1878, FBR.

46. Cool, "Texas Ranger Mark Ludwick," *NOLA Quarterly* XXV:3, 42–55 and XXV:4, 33–45.

47. Monthly Report, Detachment of Co. C, November, 1878, FBR; J. A. Tays to Jones, November 6, 1878, FBR.

48. J. A. Tays to Jones, November 16, 1878, FBR.

49. EPC Elections, 1878, Records of the Secretary of State, RG 307, Series 84, TSLAC; *MVI,* November 9, 1878; Callaghan to Jones, December 9, 1878, FBR; EPC Commissioners Court Minutes, December 10, 1878, Box 4E130, 1:36, CAH; *Mesilla News,* December 21, 1878.

Chapter 24

1. Solomon Schutz et al. to Jones, January 23, 1879, appearing in *MVI*, February 1, 1879; J. A. Tays to Jones, November 16, 1878, November 26, 1878, FBR.

2. Jones to Ludwick, January 9, 1879, WC (the transcription erroneously gives the date as December 9, 1878); Ludwick to Jones, January 21, 1879, FBR; "Tays Diary," January 11, 1879.

3. Cool, "The Puzzling J. A. Tays," The Dispatch 7 (Summer 2001), www.texasrangers.org.

4. *MVI*, February 1, 1879; *Tays Diary*, January 16, 1879.

5. *MVI*, February 1, 1879.

6. Probate File No.10, EPC Archives, SC.

7. Probate File No.6, EPC Archives, SC.

8. Probate Case No.8, EPC Archives; Deed Books, I:187, and J:279, EPC Archives; Property Deed Records, Commonwealth Title Company, El Paso, Texas, Nicholas Houser Files.

9. Probate Case No. 5, EPC Archives.

10. Silva, Antoinette Duran. "Notes Regarding My Grandfather."

11. Probate Case No. 5, EPC Archives.

12. Probate Case No. 4, EPC Archives.

13. Las Cruces *Thirty-Four*, February 16, 1879; Ludwick to Jones, January 31, 1879, FBR; Monthly Return, Detachment Co. C, February, 1879, FBR; *Tays Diary*, February 20–21 and 24, 1879.

14. State of Texas v. Benito Gonzales, Criminal Case No. 256, EPC Archives.

15. State vs. Mejia, Criminal Case No. 319, EPC Archives; Las Cruces *Thirty Four*, September 17, 1879.

16. Jones to John B. Tays, February 1, 1879, Jones to Joseph Magoffin, et al., February 10, 1879, FBR; S. Schutz et al. to Jones, January 23, 1878, quoted in *MVI*, February 1, 1879; Jones to Blacker, May 17, 1879, FBR; Baylor, *Into the Far Wild Country*, 273; Gillett, *Six Years with the Texas Rangers*, 140–41.

17. Baylor, *Into the Far Wild Country*, 273; Jones to Baylor, July 26, 1879, FBR.

18. Baylor to Jones, October 10, 1879, FBR. Garcia's attitude regarding Apaches had not softened with the years. As late as November 1878, he lost forty or fifty head of cattle to raiders. Gregorio Garcia v. United States, Case No. 4231, Indian Depredation Case Records, U.S. Court of Claims, RG 123, NARA.

19. Baylor, *Into the Far Wild Country*, 287.

20. Gillett to Jones, February 29, 1880, FBR; Baylor to Jones, March 1, 1880, FBR; Baylor to Jones, March 13, 1880, FBR.

21. Sherman to Nicholl, October 18, 1878. All correspondence related to the scandal is from the Blair Court-martial File.

22. *Atlanta Constitution*, January 2, 1879; Blair to Sherman, December 22, 1878, M. L. Blair to Sherman, December 23, 1878.

23. *Santa Fe New Mexican*, December 26, 1878; *MVI*, January 11, 1879.

24. Townsend to Ben Letcher, December 31, 1878.

25. *Morning Oregonian,* December 28, 1878; *Brooklyn Daily Eagle* December 30, 1878; *Atlanta Constitution,* January 2, 1879; Sherman to Hancock, January 8, 1879; Janus B. Fry, A.A.G. to Sherman, January 10, 1879; *NYT,* February 12, 1879; Townsend to Blair, January 13, 1879; Letcher to Townsend, January 16 and January 18, 1879.

26. *Brooklyn Daily Eagle,* February 13, 1879; *New York Times,* February 12–13, 1879; *Brooklyn Daily Eagle,* February 19, 1879; Letcher to Townsend, February 23, 1879, quoting from letter of Blair to Nicholl, November 8, 1878.

27. *Brooklyn Daily Eagle,* February 19, 1879; *Waukesha Freeman,* February 27, 1879; Letcher to Townsend, March 2, 1879; Endorsement of R. B. Hayes, March 5, 1879, Blair Court-martial File; *Brooklyn Daily Eagle,* March 7, 1879; Rasch, *Gunsmoke in Lincoln County,* 152.

28. *Thirty-Four,* September 17, 1879; *MVI,* May 17, 1879.

29. Hendricks and Timmons, *San Elizario,* 89.

30. Blacker to Steele, December 4, 1878, TAGC; "Report of the Adjutant General of the State of Texas for the Fiscal Year Ending August 31, 1878," Galveston, 1878, TAGC.

Chapter 25

1. HTO, s.v., "HAGUE, JAMES PRICE."

2. Lowenstein, Schutz, and Ochoa to Blacker, February 19, 1882, endorsement Blacker to Roberts, February 22, 1882, TAGC, Box 401-401; HTO, s.v., "BLACKER, ALLEN."

3. HTO, s.v., "MAGOFFIN, JOSEPH."

4. For a discussion of McDaniels reported death, see Cool, "The Many Lives and Suggested Death of Jim McDaniels," *Revenge,* 21–40.

5. Utley, *Lone Star Justice,* 221.

6. Baylor, *Into the Far Wild Country,* 276–81.

7. Records of Old San Elizario Cemetery, http://www.rootsweb.com/%7Etxelpaso/oldsanelizariocem.html.

8. *El Paso Lone Star,* January 24, 1885; Probate File No. 85, EPC Archives; *Lone Star,* September 27, 1884; Texas v. William Marsh, El Paso Criminal Case No. 347, Microfilm Series 524, Roll 2, SSC.

9. See Nelson, "Trailing Jerry Barton," *NOLA Quarterly* XXVI:1, 36–42.

10. *El Paso Herald,* February 21, 1914; *El Paso Times,* February 20, 1914.

11. Owens, *Two Alberts,* 130; Nolan, "Boss Rustler," *True West,* September, 1996, 20–21, and October 1996, 12–13; *Santa Fe Daily New Mexican,* February 18, 1883, cited in Philip J. Rasch, "John Kinney: King of the Rustlers," *English Westerners Brand Book* 4 (October 1961): 10–13.

12. Nolan, "Boss Rustler," *True West* (September 1996): 21 and (October 1996): 13; Records of the United States Territorial and New Mexico District Courts for Doña Ana County, New Mexico State Records Center and Archives, Santa Fe, New Mexico,

Criminal Case Nos. 907, 911–915, 917–921, 953, 955, 991; Gibson, *Life and Death,* 125–27; Owen, *Two Alberts,* 93–95.

13. See Owen, *Two Alberts,* for the story of the Fountain-Fall feud. For the Fountain murder case, see Corey Recko, *Murder on the White Sands,* UNT Press, 2007.

14. Rex Strickland, "Introduction," to Mills, *Forty Years,* xx–xxi; Mills, *Forty Years,* 171.

15. *Lone Star,* October 8, 1884; *El Paso Morning Star,* December 13, 1884; *Austin Statesman,* January 2, 1908.

16. Utley, *Frontier Regulars,* 348–50, 369–74; Kenner, *Buffalo Soldiers and Officers,* 44–45, 46, 49–50; *Lone Star/Thirty-Four,* November 2, 1881.

17. Rucker ACP File.

18. Blair, occupation "Postmaster," is listed in U.S. Census, 1880, Tombstone, Cochise County, Arizona Territory. The facts of Blair's last years and death obtained in conversation with Dr. Ruben Osorio, November 11, 2006. For Maria's life after the general's death, see Portland *Morning Oregonian,* December 28, 1887.

19. "Temple and Pantheon of San Jose," www.geocities.com/santisimatrinidad dejuarez/historia.htm; Webb, *Texas Rangers,* 348; Hendricks, "Antonio Severo Borrajo: Hispano-Mexican Patriot Priest," 4.

20. "Sisters of Loretto," HTO; "Archbishop Peter Bourgade," www.catholic-hierarchy.org/bishop/bbourgade.html; "San Xavier: The White Dove," http://www.azstarnet.com/patronato_san_xavier/050-3664.htm; Jean Baptiste Salpointe, *Soldier of the Cross,* ed. Odie B. Faulk, appendix III, 165.

21. John Middagh, *Frontier Newspaper,* 7, 15–24; *Lone Star,* November 22, 1884; *El Paso Lone Star,* January 24, 1885.

22. Record of Appointment of Postmasters 1832–September 30, 1871, NAMS M841, NNEB-20; Metz, *Dallas Stoudenmire,* 2–3; "Rio Grande and El Paso Railroad," HTO; Deed Books F, H, and I, EPC Archives; *El Paso Herald,* May 6, 1883; *The Lone Star,* June 11, 1884 and September 27, 1884; "John B. Tays," *American Biography & Genealogy*, California Edition, ed. Robert J. Burdette, 293.

23. "John B. Tays," *American Biography & Genealogy,* 293–94; Bernice Bedford Conley, "Dreamers and Dwellers: Ontario and Neighbors," 100, 173–74; *An Illustrated History of Southern California* (Chicago: Lewis Publishing Co., 1890), 628.

24. "John B. Tays," *American Biography & Genealogy,* 294; *Brooklyn Daily Eagle,* June 21, 1900; *El Paso Daily Herald,* January 24, 1890 and June 19, 1900.

25. *El Paso Daily Herald,* April 16, 1885, April 10, 1888, July 16, 1909.

26. See Alexander, *Dangerous Dan Tucker* (Silver City, NM: High Lonesome Books, 2001).

27. Robert N. Mullin, "Here Lies John Kinney," *Journal of Arizona History* (Autumn 1973): 223–42.

28. Lockhart, "Gregorio Nancianceno Garcia, 1st," *Password* 40 (Fall 1995): 123–24.

29. Ringgold, *Frontier Days,* 15–18; U.S. Court of Claims, Indian Depredations

Case File No. 1561 Price Cooper v United States; *El Paso Times*, November 15, 1900. According to his obituary in the *El Paso Times*, Cooper was "not under 94 years old." The federal census only five months earlier listed his age as eighty-eight. Forty years before, in 1860, he was recorded as being just thirty-two. Twenty-seven years after that, in the Salt War, he put his age as twenty-nine years older, age sixty-one.

30. Dona Ana County District Court Criminal Case Files Nos. 879, 880, 881; *El Paso Herald*, December 31, 1919.

31. Indian War Veteran Pension Applications for Tomas Garcia (No. SC 10368), Carlos Garcia (No. SC 10447), Jesus Carbujal de Garcia (No. WA 15132), and Ascensión E. Guerra (WC 15123), Indian War Pension Files, 1865–1890; VA Records, RG 15.

32. Lockhart, "Gregorio Nancianceno Garcia, 2nd," *Password* 40 (Fall 1995): 126–28; *El Paso Times*, December 29, 1938; *El Paso Herald*, December 29, 1938; Indian War Veteran Pension Applications for Gregorio N. Garcia, filed February 21, 1921 and June 9, 1927, No. SC 17706.

33. Another century would pass before the voters of El Paso County elected a Mexican American sheriff. Deed book J:244, EPC Archives; Criminal Case No. 282, EPC Archives.

34. Cool, "Texas Ranger Mark Ludwick," *NOLA Quarterly* 25:4, 33–45

35. *El Paso Herald Post*, August 16, 1931; *El Paso Times*, January 31, 1937, February 14, 1937, December 29, 1938; *El Paso Herald*, December 29, 1938.

36. *El Paso Times*, January 31, 1937, February 14, 1937; *El Paso Daily Herald*, March 16, 1938.

37. *El Paso Herald*, December 29, 1938; *El Paso Times*, December 29 and December 30, 1938; *El Paso Daily Herald*, March 16, 1938; correspondence with Irma Perez regarding San Elizario cemetery records for Jesus Montes.

38. Thomas A. Zickefoose, Indian War Veteran Pension Application No. 1712967, Indian War Pension Files, 1865–1890; VA Records, RG 15; NARA; Correspondence with Larry Zickefoose, Monday, February 02, 2004; Bartlett, *Personal Narrative* I, 122.

Epilogue

1. *El Paso Herald*, June 9, 1923.

2. The self-guided tour of the historic district is online at http://www.epcounty.com/history/saneli.htm.

3. Robert R. Archibald, "San Elizario: A Place to Remember," April 27, 2002. http://www.co.el-paso.tx.us/comm3/archibald.htm

4. Ron Grossman, "Will Mexican Wave Divide U.S.?" *Chicago Tribune*, March 22, 2004, republished online at www.newspapertree.com.

BIBLIOGRAPHY

Manuscripts and Archival Collections

Brown, Frank. Annals of Travis County and of the City of Austin (From the earliest times to the close of 1875). 1875.

CENTER FOR AMERICAN HISTORY, UNIVERSITY OF TEXAS AT AUSTIN: William Wallace Mills Collection; Walter Prescott Webb Collection Files for "Texas Rangers"; El Paso County Commissioners' Court Minutes.

Clark, Harry Wilson. "The History, Archeology, and Oral Traditions of the San Elizario Jail." Thesis. University of Texas at El Paso, 2002.

COLORADO STATE ARCHIVES: Charles Kerber Military Records.

EL PASO CITY LIBRARY: Vertical Files; General Directory of the City of El Paso 1886–87.

EL PASO COUNTY ARCHIVES, SONNICHSEN COLLECTION, UNIVERSITY OF TEXAS AT EL PASO: Property Deeds; Criminal Case Files; Civil Case Files; Probate Case Files; Catholic Church Records of Baptisms, Confirmations, Marriages and Interments, San Elizario, Texas.

EL PASO HISTORICAL SOCIETY: Frank E Hunter. "Reminiscences: El Paso Salt War"; Reminiscence of John D. Davis, Files of El Paso Pioneer Association; Biographical and Historical Sketch Book of the El Paso Pioneers' Association.

Halla, Frank Louis, Jr. "El Paso, Texas, and Juarez, Mexico: A Study of a Bi-ethnic Community, 1846–1881." Ph.D. diss. University of Texas at Austin, 1978.

Hammon, Anne, "West Texas and the State Constitutional Convention of 1875." Thesis. Texas Tech., 1935.

HISTORICAL SOCIETY OF PENNSYLVANIA: Military Papers of Louis H. Carpenter, MS Collection 115.

LIBRARY OF CONGRESS: Papers of William Tecumseh Sherman.

NATIONAL ARCHIVES AND RECORDS ADMINISTRATION: RG 15: Civil War Pension Files; Indian Wars Pension Files; RG 28: Records of Appointments of Postmasters; RG 29: United States Census Records; RG 58: Internal Revenue Assessment Lists for Texas 1865–1866; RG 59: State Department—Secretary of State Correspondence; Despatches from U.S. Ministers to Mexico 1823–1906; Despatches from US Consuls in Ciudad Juarez (Paso del Norte) 1850–1906; RG 094: Records of the Adjutant General's Office (AGO), United States Army—LR 1871–1880; Records of the Appointment, Commission and Personnel Branch; LR by the Commission Branch; Returns from Military Posts, 1800–1916; Compiled Military Service Records; Pension Records Civil War and Indian Wars; RG 123: Court of Claims—Indian Depredations Files; RG 153: Registers of the

Records of General Courts Martial 1809–1890; Records of General Courts Martial; RG 391: Records of US Regular Army Mobile Units 1821–1942; Returns for Regular Army Infantry Regiments, June 1821–December 1916; RG 393: LS by the Dept of New Mex./District of New Mex. 1849–1890; Registers of LR & LR by Headquarters, District of New Mex. 1865–1890; Special Files of Headquarters, Division of the Missouri 1863–1885; Records of the Department of Texas and the 5th Military District 1865–1870.

NEW MEXICO STATE ARCHIVES: Criminal Cases — Dona Ana, Lincoln, & Grant Counties.

Roberts, Gary L. "Sand Creek: Tragedy and Symbol." Ph.D. diss. University of Oklahoma, 1984.

RUTHERFORD B. HAYES PRESIDENTIAL CENTER LIBRARY: Rutherford B. Hayes Letters; Scrapbook Collection — December, 1877, Cabinet meeting newspaper clippings, Telegrams — Executive Mansion, RBH Collection; Guy M Bryan Papers.

Tays, George H., "Diary of . . . while in the Texas Rangers," transcribed by Shirley Joy, 1983, private collection.

TEXAS GENERAL LAND OFFICE — Land Grant Search Records, http://www.glo.state.tx.us/.

TEXAS STATE ARCHIVES AND LIBRARY COMMISSION: Reconstruction Records — Records of the State Police; Records of the Adjutant General — Records of the Frontier Force, Records of the Frontier Battalion; Records of the Governors (Coke, Hubbard); Records of the Sec. of State — EPC Election Returns and Contests 1870–1878; Journals of the Legislature of the State of Texas: 12th-15th; Biographies of Leading Texans (MS).

Trujillo de Hernandez, Olga, "Charles Ellis: Salt War Victim," unpublished family memoir, in the author's possession.

UNIVERSITY OF TEXAS AT EL PASO, SONNICHSEN COLLECTION: Deposition of E. W. Bates, George B. Zimpelman v. Frank Howard, et al., Texas District Court, McLennan County (1887), R. T. Copenbarger MS Collection; Depositions of A. J. Fountain, Zimpelman v. Howard et al., C. L. Sonnichsen MS Collection; Montes, Elizario M. And Mrs. S. R. Skaggs, "The Salt War," Blanca Gomez Hill Collection.

Ward, Charles Francis. "Salt War of San Elizario." Thesis. University of Texas, 1932.

Webster, Michael Gordon. "Texan Manifest Destiny and the Mexican Border Conflict 1865–1880." Ph.D. diss. University of Indiana, 1972.

Government Publications and Documents

Adjutant General, *Records of California Men in the War of the Rebellion 1861 to 1867* (Sacramento: J. D. Young, Supt. State Printing, 1890), http://www.calweb.com/~arline/CoFcalvol.html.

Counterinsurgency, United States Army Field Manual FM3–24, Washington, DC, 2006.
1867 Voters' Registration, republished in *Rio Grande Researcher*, 19:1–4 (March–December 2005), http://www.rootsweb.com/txepgs/rgr_mar06.html.
El Paso Troubles in Texas ("EPT"), House of Representatives Executive Document No. 45: 45th Congress. Washington, DC, 1878.
Mexican Border Troubles ("MBT"), House of Representatives Executive Document No. 13, 45th Congress. Washington, DC, 1878.
Relations of United States with Mexico (includes *Texas Frontier Troubles*)("TFT"), House of Representatives Executive Document No. 701: 45th Congress. Washington, DC, 1878.
Report of the Adjutant General of the State of Texas: 1877.
Report of the United States and Mexican Boundary Survey by William H. Emory, Senate Executive Document No. 108, 34th Congress, 1st Session. Washington, DC, 1855.
Reports of Exploration and Survey to Ascertain the Most Practicable and Economical Route for a Railroad from the Mississippi River to the Pacific Ocean, vol. 2: *Report of Brevet Captain John Pope . . . [1854]*. Washington, DC, 1855, http://www.hti.umich.edu/m/moagrp/.
Reports of the Secretary of War, with Reconnaissances of Routes from San Antonio to El Paso . . . ," Senate Executive Document No. 64, 31st Congress, 1st Sess. Washington, DC, 1849.
Report to H.V. Brown [Postmaster General] on the opening and present condition of the United States overland mail route between San Antonio, Texas and San Diego, California by J.C. Wood. Washington, DC, 1858, http://www.hti.umich.edu/m/moagrp/.
Texas Land Title Abstracts, www.Ancestry.com.
War of the Rebellion: Official Records of the Union and Confederate Armies, 128 vols. Washington, DC, 1880–1901. Cornell University Making of America Multivolume Collection, http://cdl.library.cornell.edu/moa/browse.monographs/waro.html.

Books and Articles

Alberts, Don E. *The Battle of Glorieta: Union Victory in the West*. College Station: Texas A&M University Press, 1998.
Alexander, Bob. *Dangerous Dan Tucker*. Silver City, NM: High-Lonesome Books, 2001.
Askins, Charles. *Texans, Guns & History*. New York: Bonanza Books, 1970.
Austerman, Wayne R. *Sharps Rifles and Spanish Mules: The San Antonio–El Paso Mail, 1851–1881*. College Station: Texas A&M University Press, 1985.
Bancroft, Hubert Howe. *History of Mexico*, vol. VI. [1861–1887], New York: Arno Press, 1967.

Bartlett, John Russell. *Personal Narrative of Exploration and Incidents in Texas, New Mexico, California, Sonora, and Chihuahua.* Chicago: Rio Grande Press, Inc., reprint 1965.

Baylor, George W. *Into the Far Wild Country,* ed. Jerry D. Thompson. El Paso: Texas Western Press, 1996.

Bell, James G. "A Log of the Texas–California Cattle Trail, 1854," ed. J. Evetts Haley. *Southwestern Historical Quarterly* 35 (January–April 1932): 208–37, 290–91.

Biographical Encyclopedia of Texas. New York: Southern Publishing Co., 1880.

Black, Esther Boulton. *Stories of Old Upland.* Ontario, California: Chaffey Communities Cultural Center, 1979.

Bowden, J. J. "Magoffin Salt War." *Password* 7 (Summer 1962): 95–124.

Brand, Peter. "Sherman W. McMasters: The El Paso Salt War, Texas Rangers, & Tombstone." *Western Outlaw-Lawman History Association Journal* 8 (Winter 1999): 2–19.

Broaddus, J. Morgan. *Legal Heritage of El Paso.* El Paso: Texas Western College Press, 1963.

Brown, Richard Maxwell. *No Duty to Retreat.* New York: Oxford University Press, 1991; repr., Norman: University of Oklahoma Press, 1991.

Bruce, Robert V. *1877: Year of Violence.* Chicago: Ivan R. Dee, Inc., 1989.

Burdette, Robert J., ed. *American Biography & Genealogy,* California Edition. Lewis Pub. Co., 1900.

Butler, Benjamin F. *Butler's Book.* Harrisburg, PA: Archive Society, [1892]1997.

Callahan, Manuel. "Mexican Border Troubles: Social War, Settler Colonialism and the Production of Frontier Discourses, 1848–1880." PhD. diss. Austin: University of Texas, 2003. http://www.lib.utexas.edu/etd/d/2003/callahanm036/callahanm036.pdf.

Campbell, Randolph B. *Gone to Texas.* New York: Oxford University Press, 2003.

Carpenter, Steven. "Search for the San Elizario Road." *Journal of Big Bend Studies* 14 (2002): 141–62.

Claiborne, John M. "Brave George Barnhart Zimpelman." *Confederate Veteran* 19:3, 114. Archive of Terry's Texas Rangers, www.terrystexasrangers.org.

Conley, Bernice Bedford. "Dreamers and Dwellers: Ontario and Neighbors." Privately published, 1982.

Cool, Paul. "El Paso Salt War: A Review of the Historical Literature." *Journal of Big Bend Studies* 17 (2005): 49–82

——. "The Many Lives and Suggested Death of Jim McDaniels." *Revenge!* ed. Sharon Cunningham and Mark Boardman. Lafayette, IN: ScarletMask, 2004, 21–40.

——. "Texas Ranger Mark Ludwick." *Quarterly of the National Association for Outlaw and Lawman History* 25 (July–September and October–December 2001): 43–55 and 33–45.

——. "The Puzzling J.A. Tays." *The Dispatch*, 7 (Summer 2001), www.texasrangers.org.

Cunningham, Eugene. "Salt." *Hunter's Frontier Times* 7 (1930): 153, 156–58.

Davis, W[illiam] W[atts] H[art]. *El Gringo: Or, New Mexico and Her People.* Chicago: Rio Grande Press, reprinted 1962.

Dawson III, Joseph C. *Doniphan's Epic March.* Lawrence: University Press of Kansas, 1999.

De Bow, J. D. B. "Progress of the Great West in Population, Agriculture, Arts, and Commerce." *Commercial Review of the South and West* 4 (July 1847–January 1848). Making of America, http://www.hti.umich.edu/m/moagrp/.

De León, Arnoldo. *They Called Them Greasers.* Austin: University of Texas Press, 1983.

——. *Tejano Epic: Essays in Honor of Felix D. Almaráz, Jr.* Austin: Texas State Historical Association, 2005.

Early Marriages of El Paso County. http://freepages.genealogy.rootsweb.com/~janarmstrong/elpmgs.html.

Edrington, Thomas S. and John Taylor. *The Battle of Glorieta Pass.* Albuquerque: University of New Mexico Press, 1998.

Edwards, Frank S. *A Campaign in New Mexico with Colonel Doniphan.* University of Kansas, 1847, http://kuhttp.cc.ukans.edu/carrie/kancoll/.

Eickhoff, Randy Lee. *Exiled: The Tigua Indians of Ysleta del Sur.* Plano, TX: Republic of Texas Press, 1996.

Ford, John S. "Report of Dr. John S. Ford upon the Practicability of a Route from Austin to El Paso del Norte." Contributed by C.L. Greenwood. *Southwestern Historical Quarterly* 48 (October 1944): 262–68.

Freeman, Joanne B. *Affairs of Honor: National Politics in the New Republic.* New Haven, CT: Yale University Press, 2001.

Gibson, A.M. *Life and Death of Albert Jennings Fountain.* Norman: University of Oklahoma Press, 1965.

Gillett, James. *Six Years with the Texas Rangers.* New Haven, CT: Yale University Press, 1925; reprint Lincoln: Bison Books, University of Nebraska Press, 1976.

Glasrud, Bruce A. and Arnoldo De León. *Bibliophiling Tejano Scholarship: Secondary Sources on Hispanic Texas.* Alpine, TX: Center for Big Bend Studies, 2003.

Gregg, Josiah. *El Gringo, or New Mexico and Her People. 1857.* Books of the Southwest. http://southwest.library.arizona.edu/elgr/.

Gregg, Robert D. *Influence of Border Troubles on Relations Between The United States and Mexico 1876–1910.* Baltimore: Johns Hopkins Press, 1937.

Griffen, William B. *Utmost Good Faith: Patterns of Apache-Mexican Hostilities in Northern Chihuahua Border Warfare, 1821–1848.* Albuquerque: University of New Mexico Press, 1988.

Handbook of Texas Online, http://www.tsha.utexas.edu/handbook/online/.

Harvard Law Review Association. "Law, Race, and the Border: The El Paso Salt War of 1877." *Harvard Law Review* 117 (January 2004): 941–63.

Hatley, Allen G. *Bringing the Law to Texas: Crime and Violence in Nineteenth Century Texas*. LaGrange, TX: Centex Press, 2002.

Heitman, Francis B. *Historical Register and Dictionary of the US Army*. Urbana: University of Illinois Press, [1903]1963.

Hendricks, Rick. "Antonio Severo Borrajo—Hispano-Mexican Patriot Priest." *Southern New Mexico Historical Review* 9 (January 2002): 1–5.

Hendricks, Rick and W.H. Timmons. *San Elizario: Spanish Presidio to Texas County Seat*. El Paso: Texas Western Press, 1998.

Higgins, Noelle. "The Application of International Law to Wars of National Liberation." *Journal of Humanitarian Assistance*, www.jha.ac/articles/a132.pdf.

Joy, Shirley P[atterson]. *John Tays and Jane Ellis of Nova Scotia and Their Descendants*. Madeira, OH: privately published, 1984.

Julyan, Robert. *Place Names of New Mexico*. Albuquerque: University of New Mexico Press, 1998.

Kenner, Charles L. *Buffalo Soldiers and Officers of the Ninth Cavalry 1867–1898*. Norman: University of Oklahoma Press, 1999.

Kohlberg, Ernst. *Letters of Ernst Kohlberg*. El Paso: Texas Western Press, 1973.

Laurie, Clayton D. and Cole, Richard H. *The Role of Federal Military Forces in Domestic Disorders 1877–1945*. Washington: Center of Military History, 1997.

Leiker, James N. *Racial Borders: Black Soldiers along the Rio Grande*. College Station: Texas A&M University Press, 2002.

Lewis, William Ray. "The Hayes Administration and Mexico." *Southwestern Historical Quarterly* 24 (1920): 140–53.

Lockhart, Bill. "Gregorio Nancianceno Garcia, 1st: Indian Fighter and Politician." *Password* 40 (Fall 1995): 119–25.

——. "Gregorio Nancianceno Garcia 2nd: Lawman and Politician." *Password* 40 (Fall 1995): 126–28.

——. "The Incorporations of San Elizario, Texas." *Password* 40 (Fall 1995): 169–77.

López, David Pérez. "La Expulsion del Padre Borrajo." Translated by Manuella Cárdenas-Oellermann, http://utminers.utep.edu/beperez/padreborrajo.htm (accessed November 30, 2002).

Martin, G. A. "Old San Elizario." *Voice of the Mexican Border* (January 1934): 190–97.

Martinez, Oscar J. *Border Boom Town: Ciudad Juarez since 1848*. Austin: University of Texas Press, 1978.

——. *Troublesome Border*. Tucson: University of Arizona Press, 1995.

Kearney, Milo and Manuel Medrano, *Medieval Culture and the Mexican American Borderlands*. College Station, TX: Texas A&M University Press, 2001.

Mautz, Carl. *Biographies of Western Photographers*. Nevada City, CA: Carl Mautz Publishing, 1997.

McKay, Seth Shepard. *Debates in the Texas Constitutional Convention of 1875*. Austin: University of Texas Press, 1930.

Metz, Leon C. *Dallas Stoudenmire*. Norman: University of Oklahoma Press, 1969; reprint 1993.

Middagh, John. *Frontier Newspaper: The El Paso Times*. El Paso: Texas Western Press, 1958.

Miller, Darlis A. *California Column in New Mexico*. Albuquerque: University of New Mexico Press, 1982.

Miller, Thomas Lloyd. *Public Lands of Texas 1519–1970*. Norman: University of Oklahoma Press, 1972.

Mills, W[illiam] W[allace]. *Forty Years in El Paso 1858–1898*. El Paso: Carl Hertzog, 1962.

Moneyhon, Carl H. *Republicanism in Reconstruction Texas*. College Station: Texas A&M University Press, reprint 2001.

Mullin, Robert N. "Here Lies John Kinney." *Journal of Arizona History* (Autumn 1973): 223–42.

Nelson, Scott L. "Trailing Jerry Barton." *National Association for Outlaw and Lawman History Quarterly* 26 (January–March 2002): 36–42.

Nolan, Frederick. "Boss Rustler: The Life and Times of John Kinney." *True West* (September and October 1996): 14–21, 12–19.

——. "Saga of the San Augustine Ranch." *True West* (June 1999): 43–52.

——. *The West of Billy the Kid*. Norman: University of Oklahoma Press, 1998.

Ormsby, Waterford L. *Overland Butterfield Mail*, 1858 reprint. San Marino, CA: Henry E. Huntington Library and Art Gallery, 1942, Tenth printing 1998.

Osorio, Ruben. "Villismo: Nationalism and Popular Mobilization in Northern Mexico," in *Rural Revolt in Mexico*, ed. Daniel Nugent. Durham, NC: Duke University Press, 1998.

Owen, Gordon R. *The Two Alberts: Fountain and Fall*. Las Cruces, NM: Yucca Tree Press, 1996.

Paddock, B. B. *Biographies of North and West Texas*. Chicago: Lewis Publishing Co., 1906, http://www/genealogymagazine.com.

Paust, Jordan J. "There Is No Need to Revise the Laws of War in Light of September 11th." *American Society of International Law* (November 2002), www.asil.org/taskforce/paust.pdf.

Peterson, John A., Timothy B. Graves, and David V. Hill. *The 1995 San Elizario Plaza Archeological Project*. El Paso: University of Texas at El Paso, 2002.

Phelan, Martha M. and Edith C. Lane. "A 'Salt War' Experience." *New Mexico Highway Journal* 8 (December 1930): 24–25, 35.

Rasch, Philip J. *Gunsmoke in Lincoln County*, ed. Robert K. DeArment. Stillwater, OK: National Association for Outlaw and Lawman History, Inc., 1997.

——. *Desperadoes of Arizona Territory*, ed. Robert K. DeArment. Stillwater, OK: National Association for Outlaw and Lawman History, Inc., 1999.

Ringgold, Jennie Parks. *Frontier Days in the Southwest.* San Antonio: Naylor Company, 1952.

Roberts, Gary L. *Death Comes for the Chief Justice.* Niwot: University Press of Colorado, 1990.

Robinson, Charles M., III. *The Men Who Wear the Star.* New York: Random House, 2000.

Rodenbough, Theo [philus] F. and William L. Haskin. *The Army of the United States: Historical Sketches of Staff and Line with Portraits of Generals-in-Chief.* New York: Maynard, Merrill, & Co. 1896.

Romero, Mary. "El Paso Salt War: Mob Action or Political Struggle." *Atzlan Journal* 16 (Spring and Fall 1985): 119–43, http://www.sscnet.ucla.edu/csrc/library/azr-16–1-119.html.

Rosenbaum, Robert J. *Mexicano Resistance in the Southwest.* Dallas: Southern Methodist University Press, 1983.

Salpointe, John Baptist. *Soldier of the Cross,* ed. Odie B. Faulk. Tuscon: Diocese of Tucson, 1966.

Sonnichsen, C[harles] L[eland]. *El Paso Salt War of 1877.* El Paso: Carl Hertzog and the Texas Western Press, 1961.

——. *The Mescalero Apaches.* Norman: University of Oklahoma Press, 1958; reprint 1973.

——. *Pass of the North.* El Paso: Texas Western Press, 1968.

Strickland, Rex. *El Sabio Sembrador: El Paso in 1854.* El Paso: Texas Western Press, 1960.

Timmons, W. H. *James Wiley Magoffin.* El Paso: Texas Western Press, 1999.

Utley, Robert M. *Frontier Regulars: The United States Army and the Indian, 1866–1890.* New York: MacMillan, 1973.

——. *Frontiersmen in Blue: The United States Army and the Indian, 1848–1865.* New York: MacMillan, 1967; reprinted Lincoln: University of Nebraska Press, 1981.

——. *High Noon in Lincoln: Violence on the Western Frontier.* Albuquerque: University of New Mexico Press, 1987.

——. *Lone Star Justice: The First Century of the Texas Rangers.* New York: Oxford University Press, 2002.

Wadsworth, Richard. *Forgotten Fortress: Fort Millard Fillmore and Antebellum New Mexico.* Las Cruces, NM: Yucca Tree Press, 2002.

Wasserman, Mark. *Capitalists, Caciques and Revolution.* Chapel Hill, NC: University of North Carolina Press, 1984.

Webb, Walter Prescott. *The Texas Rangers: A Century of Self-Defense,* 2nd ed. Austin: University of Texas Press, 1965, reprint 1996.

White, Owen P. *Out of the Desert.* El Paso: McMath Co., 1923.

Wilkins, Frederick. *The Law Comes to Texas.* Austin: State House Press, 1999.

Williams, Clayton W. *Texas' Last Frontier: Fort Stockton and the Trans-Pecos, 1861–1895,* ed. Ernest Wallace. College Station: Texas A&M University Press, 1982.

Wilson, John P. *When the Texans Came.* Albuquerque: University of New Mexico Press, 2001.

Winfrey, Dorman H. and James M. Day. *Indian Papers of Texas and the Southwest 1825–1916.* Austin: Texas State Historical Association, 1995.

Wyatt-Brown, Bertram. *Honor and Violence in the Old South.* New York: Oxford University Press, 1986.

——. "Honor's History Across the Academy." *Historically Speaking* (June 2002), http://www.clas.ufl.edu/users/bwyattb/honorhistory.html.

INDEX

Page numbers shown in *italics* refer to photo captions.

ISBN-13: 978-1-60344-016-5
ISBN-10: 1-60344-016-X